Instrument Test Guide

Questions, Answers & Explanations in Numerical Order for the Instrument, Instrument Ground and Flight Instructor Ratings for Airplane, Helicopter, Airship, and Foreign Pilots… **plus the official *Computerized Testing Supplement* (CT-8080-3B)**

Aviation Supplies & Academics, Inc.
Newcastle, Washington

Instrument Test Guide

Aviation Supplies & Academics, Inc.
7005 132nd Place SE
Newcastle, Washington 98059-3153
(206) 235-1500

FAA Questions herein are from United States government sources and contain current information as of August 16, 1995.

None of the material in this publication supersedes any documents, procedures or regulations issued by the Federal Aviation Administration.

Important: This Test Guide should be sold with and used in conjunction with *Computerized Testing Supplement for Instrument Rating* (FAA-CT-8080-3B).

ASA reprints the FAA test figures and legends contained within this government document, and it is also sold separately and available from aviation retailers nationwide. Order #ASA-CT-8080-3B.

ISBN 1-56027-234-1
ASA-I040G

Printed in the United States of America

97 96 95 5 4 3 2 1

About the Contributors

Charles L. Robertson
Assistant Professor, UND Aerospace
University of North Dakota

Charles Robertson as flight instructor, assistant professor and manager of training at UND Aerospace, contributes a vital and substantial combination of pilot and educator to ASA's reviewing team. After graduating with education degrees from Florida State University in 1967, and Ball State University in 1975, he began his twenty-year career in the United States Air Force as Chief of avionics branch, 58th Military Airlift Squadron, and went on to flight instruction, training for aircraft systems, and airport managing, while gaining many thousands of hours flying international passenger and cargo, aerial refueling and airlift missions. As Division Chief in 1988, Robertson directed the USAF Strategic Air Command's "Alpha Alert Force" and coordinated its daily flight training operations. He holds the CFI Airplane Land, Multi-Engine, Single-Engine and Instrument, the ATP Airplane Land and Multi-Engine, Commercial Pilot, Advanced and Instrument Ground Instructor licenses.

Dr. Gerry Fairbairn
Professor of Aviation
Daniel Webster College

Dr. Gerry Fairbairn is an aviation educator with 25 years of experience in the field, has 6,500 hours of flight instruction, lectures for the AOPA Air Safety Foundation's flight instructor revalidation clinics, and is a National Accident Prevention Counselor. For four years after graduation (B.S., Aeronautical Operations, San Jose State College, 1967), he conducted certification tests in fighter and bomber aircraft for the U.S. Air Force Systems Command; he then went on to complete his M.S. Business Management (C.S.U. San Diego), and Ph.D. in Business Administration (U. of Santa Clara), while teaching aviation courses for 14 years. For six years until 1994, Dr. Fairbairn was Vice President of Academic Affairs and Dean of the Faculty at Daniel Webster College. As well as leader and educator, Dr. Fairbairn is a technical advisor to legal counsel in aviation matters, and an aviation consultant in the business community. He is an ATP, and holds the CFI (airplane, instrument, and glider ratings), Commercial (airplane and glider), Ground Instructor (Advanced, Instrument) and A&P Mechanic certificates.

Contents

Instructions

Preface v

Free Test Update Coupon vii

Description of the Tests ix

Cheating or Other Unauthorized Conduct x

Taking a Knowledge Test by Computer x

Computer Testing Designees x

Testing and Scoring xi

Retesting Procedures xi

Eligibility Requirements for the Instrument Rating xii

Certificate and Logbook Endorsements xiii

Test-Taking Tips xiv

Suggested Materials for the Instrument Rating xv

ASA Test Guide Layout xvi

Instrument Rating Questions 1-72

Answers, References & Explanations 1-86

Cross-References:

Answer, Subject Matter Knowledge Code & Category 1-8

Subject Matter Knowledge Code & Question Number 1-10

Preface

Welcome to ASA's Test Guide Series. ASA's test books have been helping pilots prepare for the FAA Knowledge Tests since 1984 with great success. We are confident that with proper use of this book, you will score very well on any of the instrument rating tests.

All of the questions in the FAA Instrument Rating Test Question Bank are included here. We suggest you begin by reading the book cover-to-cover. Then go back to the beginning and place emphasis on those questions most likely to be included in your test (identified by the aircraft category above each question).

It is important to answer every question assigned on your FAA Knowledge Test. If in their ongoing review, the FAA authors decide a question has no correct answer, is no longer applicable, or is otherwise defective, your answer will be marked correct no matter which one you chose. However, you will not be given the automatic credit unless you have marked an answer. Unlike some other exams you may have taken, there is no penalty for "guessing" in this instance.

The FAA does not supply the correct answers to questions reproduced in this book, is not responsible for answers contained herein, and will not reveal what they consider the correct answers to be. Our answers are based on research from the reference sources provided with each question. If your study leads you to question an answer choice, we recommend you seek the assistance of a local ground or flight instructor. If you still believe the answer needs review, please forward your questions, recommendations, or concerns to:

Aviation Supplies & Academics, Inc.
7005 132nd Place SE
Newcastle, WA 98059-3153

Phone (206) 235-1500
Fax (206) 235-0128
CompuServe 102105,223

Technical research contributed by:
Charles Robertson, University of North Dakota
Gerry Fairbairn, Daniel Webster College

Editor:
Jennie Trerise, ASA

Technical Editor:
Jackie Spanitz, ASA

Free Test Update Coupon

Complete this coupon and mail to ASA to receive updates to the FAA computerized tests.
The update will include information from FedWorld on the changes made to each computerized test, and will be mailed to you as changes take place.

This will be your mailing label, so please print clearly.

Please send me ASA's Free Test Update:

Name ______________________________

Address ______________________________

City ______________ State ______ Zip ________

Mail this coupon to:

Aviation Supplies & Academics, Inc.
7005 132nd Place SE
Newcastle, WA 98059-3153

Which ASA Test Book do you currently use?
❑ Private ❑ Commercial ❑ Instrument ❑ CFI ❑ FE ❑ ATP

Does this book meet all your needs? Yes ❑ No ❑
If no, please explain:

____

____

General comments or suggestions:

____

____

____

I040G / 8-16-95

Description of the Tests

All test questions are the objective, multiple-choice type, with three answer choices. Each question can be answered by the selection of a single response. Each test question is independent of other questions; that is, a correct response to one does not depend upon, or influence the correct response to another.

A significant number of the questions are "category-specific" and appear *only* on the airplane test, the helicopter test, or the airship test. The 20-question "added rating" tests are composed mostly of these category-specific questions. A 20-question "added rating" test is administered to an applicant who already holds an instrument rating in one category (airplane or helicopter) and wishes to meet the knowledge requirements for the other category. The category-specific questions pertain to such knowledge areas as recency of experience and weather minimums.

Tests developed from the instrument rating knowledge bank of questions:

Instrument Rating — Airplane

Instrument Rating — Rotorcraft/Helicopter

Instrument Rating — Airplane (Added Rating)

Instrument Rating — Rotorcraft/Helicopter (Added Rating)

Instrument Rating — Foreign Pilot

Instrument Flight Instructor — Airplane

Instrument Flight Instructor — Rotorcraft/Helicopter

Instrument Flight Instructor — Airplane (Added Rating)

Instrument Flight Instructor — Rotorcraft/ Helicopter (Added Rating)

Ground Instructor — Instrument

Ground Instructor—Instrument applicants should be prepared to answer any question that appears in the instrument question bank as they are expected to teach all instrument ratings.

The Instrument Rating—Airplane and Helicopter have 60 questions each and 2.5 hours is allowed for taking each test.

The Instrument Flight Instructor—Airplane and Helicopter, the Ground Instructor—Instrument, and the Instrument Rating—Foreign Pilot tests have 50 questions each and 2.5 hours is allowed for taking each test.

All added-rating tests have 20 questions each and 1.0 hour is allowed for taking each test.

A score of 70 percent must be attained to successfully pass each test.

Cheating or Other Unauthorized Conduct

Computer testing centers follow rigid testing procedures established by the FAA. This includes test security. When entering the test area, you are permitted to take with you only scratch paper furnished by the test administrator, and an authorized aviation computer, plotter, etc., approved for use in accordance with FAA Order 8080.6, "Conduct of Airman Knowledge Testing via the Computer Medium, and AC 60-11, Aids Authorized for Use by Airman Written Test Applicants." The FAA has directed testing centers to stop a test any time a test administrator suspects a cheating incident has occurred. An FAA investigation will then follow. If the investigation determines that cheating or other unauthorized conduct has occurred, any airman certificate that you hold may be revoked, and you may not be allowed to take a test for 1 year.

Taking a Knowledge Test By Computer

Testing center personnel cannot begin the test until you provide them with the proper authorization. You should always check with your instructor or local Flight Standards District Office if you are not sure what kind of authorization you need to bring to the testing facility.

The next step is the actual registration process. Most computer testing centers require that all applicants contact a central 1-800 phone number. At this time you should select a testing site of your choice, schedule a test date, and make financial arrangements for test payment. You may register for tests several weeks in advance of the proposed testing date. You may also cancel your appointment up to 2 business days before test time, without financial penalty. After that time, you may be subject to a cancellation fee as determined by the testing center.

You are now ready to take the test. Remember, you always have an opportunity to take a sample test before the actual test begins. For the actual test you will be under a time limit, but if you know your material, there should be sufficient time to complete and review your test.

Computer Testing Designees

The following is a list of the computer testing designees authorized to give FAA knowledge tests. The services on this list should be helpful in choosing where to register for a test or for requesting additional information.

Aviation Business Services

1-800-947-4228

Outside U.S. (415) 259-8550

Drake Prometric

1-800-359-3278

Outside U.S. (612) 896-7702

Sylvan Learning Systems, Inc.

1-800-967-1100

Outside U.S. (410) 880-0880, Extension 8890

Testing and Scoring

Within moments of completing the test, you will receive an airman test report, which contains your score. It will list the subject matter knowledge areas in which you answered questions incorrectly. The total number of subject matter knowledge codes shown on the test report is not necessarily a reflection of the total number of incorrectly-answered questions. You can study these knowledge areas to improve your understanding of the subject matter. See the *Subject Matter Knowledge Code/Question Number Cross-Reference* in the back of this book, for a complete list of which questions apply to each subject matter knowledge code.

Your instructor is required to review each of the knowledge areas listed on your airman test report with you, and complete an endorsement that remedial study was conducted in these deficient areas. The examiner may also quiz you on these areas of deficiency during the practical test.

The airman test report, which must show the computer testing company's embossed seal, is an important document. DO NOT LOSE THE AIRMAN TEST REPORT as you will need to present it to the examiner prior to taking the practical test. Loss of this report means that you will have to request a duplicate copy from the FAA in Oklahoma City, which is costly and time-consuming.

Retesting Procedures

If the score on the airman test report is 70 percent or above, it is valid for 24 calendar months. The ground instructor instrument and instrument foreign pilot tests do not have an expiration date. You may elect to retake any test, in anticipation of a better score, after 30 days from the date your last test was taken. Prior to retesting, you must give your current airman test report to the computer testing administrator. Remember, the score of the latest test you take will become the official test score. The FAA will not consider allowing anyone with a passing score to retake a test before the 30-day remedial study period.

A person who fails a knowledge test may apply for retesting before 30 days of the last test providing that person presents the failed test report and an endorsement from an authorized instructor certifying that additional instruction has been given, and the instructor finds the person competent to pass the test. A person may retake a failed test after 30 days without an endorsement from an authorized instructor.

Eligibility Requirements for the Instrument Rating

To be eligible for an instrument rating, a pilot must:

1. Hold at least a current Private Pilot Certificate with a current medical certificate.
2. Be able to read, speak, and understand English.
3. Score at least 70 percent on the FAA Knowledge Test on the appropriate subjects.
4. Pass an oral and flight check on the subjects and maneuvers outlined in the Instrument Practical Test Standards (#ASA-8081-4B).
5. Have a total of 125 hours pilot time, of which 50 hours are as pilot-in-command in cross-country flight in a powered aircraft with other than a student pilot certificate (each flight must have a landing at least 50 NM from the departure point).
6. Have 40 hours of simulated or actual instrument time (not more than 20 hours may be instruction in a ground trainer).
7. Have 15 hours instrument flight instruction by an authorized flight instructor (at least 5 hours in an airplane or helicopter as appropriate).
8. Must have received ground instruction or logged home-study pertaining to IFR regulations, procedures, various methods of navigation, weather report procurement and use, and the *Aeronautical Information Manual*, among others. A complete list of requirements can be found in FAR §61.65.

Certifcate and Logbook Endorsements

When you go to take your FAA Knowledge Test, you will be required to show proper identification and have certification of your preparation for the examination, signed by an appropriately Certified Flight or Ground Instructor. Ground Schools will have issued the endorsements as you complete the course. If you choose a home-study for your Knowledge Test, you can either get an endorsement from your instructor or submit your home-study materials to an FAA Office for review and approval prior to taking the test.

Instrument Rating Endorsement

Endorsement for aeronautical knowledge: FAR §61.65(b)

I certify that Mr./Ms. ______________________________ has received the ground instruction required for the instrument (airplane/helicopter) rating by FAR §61.65(b)(1) through (4).

Signed__________________________________ Date ____________

CFI Number______________________________ Expires __________

Flight Instructor Instrument Rating Endorsement

Endorsement for aeronautical knowledge: FAR §61.185(a) and (b)

I certify that Mr./Ms. ______________________________ has completed the course of instruction required by FAR §61.185(a)(1) through (6), and has logged the ground instruction required by FAR §61.65(b) in preparation for the flight instructor instrument rating.

Signed__________________________________ Date ____________

CFI Number______________________________ Expires __________

Ground Instructor Instrument Rating Endorsement

Endorsement for aeronautical knowledge: FAR §143.11

I certify that Mr./Ms. ______________________________ is competent to pass the written exam based on the material necessary in FAR §143.11.

Signed__________________________________ Date ____________

CFI Number______________________________ Expires __________

Test-Taking Tips

The following test-taking hints have proven helpful in creating a smooth process for taking the test:

1. Take a sign-off from an instructor, photo I.D., the testing fee, calculator, flight computer (ASA's E6-B or CX-1a Pathfinder), plotter, magnifying glass, and a sharp pointer like a safety pin to the testing center.
2. Your first action upon sitting down should be to write on the scratch paper the weight and balance and any other formulas and information you can remember from your study. Remember, some of the formulas may be on your E6-B.
3. Answer each question in accordance with the latest regulations and procedures.
4. Read each question carefully before looking at the possible answers. You should clearly understand the problem before attempting to solve it.
5. After formulating an answer, determine which answer choice corresponds the closest with your answer. The answer chosen should completely resolve the problem.
6. From the answer choices given, it may appear that there is more than one possible answer. However, there is only one answer that is correct and complete. The other answers are either incomplete or are derived from popular misconceptions.
7. If a certain question is difficult for you, it is best to mark it for RECALL and proceed to the other questions. After you answer the less difficult questions, return to those which you marked for recall and answer them. The recall marking procedure will be explained to you prior to starting the test. Although the computer should alert you to unanswered questions, make sure every question has an answer recorded. This procedure will enable you to use the available time to the maximum advantage.
8. Perform each math calculation twice to confirm your answer. If adding or subtracting a column of numbers, reverse your direction the second time to reduce errors.
9. When solving a calculation problem, select the answer nearest to your solution. The problem has been checked with various types of calculators; therefore, if you have solved it correctly, your answer will be closer to the correct answer than any of the other choices.
10. Remember that information is provided in the FAA Legends and FAA Figures.
11. Be cautious about changing an answer, as your first impression is right more often than not. If in doubt, answer "C", but only if that is not the totally incorrect answer.
12. Remember to answer every question, even the ones with no completely correct answer, to ensure the FAA gives you credit for a bad question.
13. Take your time and be thorough but relaxed. Take a minute off every half-hour or so to relax the brain and the body. Get a drink of water halfway through the test.
14. The same questions are used for the Instrument Rating (airplane/helicopter) as well as the CFII (Instrument Flight Instructor) and IGI (Instrument Ground Instructor) Knowledge Tests. For those persons planning to acquire all three of these ratings within a two-year span, we recommend preparing for all three exams at the same time — this will save time in studying. However, the FAA does require all three tests be taken separately; one test session cannot be applied to all three tests.

Suggested Materials for the Instrument Rating

The following are some of the publications and products recommended for the Instrument Rating. All are reprinted by ASA and available from authorized ASA dealers and distributors.

ASA-APT	*The Complete Advanced Pilot*, by Bob Gardner
ASA-AC00-6A	*Aviation Weather*
ASA-AC00-45D	*Aviation Weather Services*
ASA-AC61-13B	*Basic Helicopter Handbook*
ASA-FR-AM-BK	*1995 Federal Aviation Regulations and Aeronautical Information Manual* (combined)
ASA-AC61-21A	*Flight Training Handbook*
ASA-FTS-I	*Flight Training Syllabus — Instrument*
ASA-PM-3	*Instrument Flying*, by Trevor Thom
ASA-PM-S-I	*Instrument Rating Syllabus*
ASA-AC61-27C	*Instrument Flying Handbook*
ASA-8081-4B	*Instrument Practical Test Standards* (Airplane & Helicopter)
ASA-8081-9A	*Instrument Flight Instructor Practical Test Standards* (Airplane & Helicopter)
ASA-OEG-I	*Instrument Oral Exam Guide*
ASA-CP-IFR	Instrument Plotter
ASA-H2G	Jiffyhood (Instrument Training Hood)
ASA-AC67-2	*Medical Handbook for Pilots*
ASA-AC61-23B	*Pilot's Handbook of Aeronautical Knowledge*
ASA-AC91-23A	*Pilot's Weight and Balance Handbook*

ASA Test Guide Layout

The FAA Instrument test questions are taken directly from FedWorld (the government's Internet site) and are reprinted in numerical order in the first section of this Test Guide. In the next section, the answers, codes, references and explanations to these FAA questions are listed in numerical order. *See* the EXAMPLES below.

Some FAA Questions refer to Figures or Legends immediately following the question number, i.e., "4201. (Refer to Figure 14.)." These are FAA Figures and Legends which can be found in the separate booklet: *Computerized Testing Supplement* (CT-8080-XX). This supplement is bundled with the Test Guide and is the exact material you will have access to when you take your computerized test. We provide it separately, so you will become accustomed to referring to the FAA Figures and Legends as you would during the test.

Figures referenced by the Explanation and pertinent to the understanding of that particular question are labeled by their corresponding Question number. For example: the caption "Questions 4245 and 4248" means the figure accompanies the Explanations for both Question 4245 and 4248.

Answers to each question are found within brackets [X] above each Explanation, and in the Cross-Reference at the back of this book.

FAA QUESTION EXAMPLE:

4201. (Refer to Figure 14.) The four forces acting on an airplane in flight are

A— lift, weight, thrust, and drag.
B— lift, weight, gravity, and thrust.
C— lift, gravity, power, and friction.

See separate book: *Computerized Testing Supplement* (CT-8080-XX)

FAA answer choices

ASA EXPLANATION EXAMPLE:

Rating. This question may be found on tests for these ratings.*

ALL, AIR, RTC, GLI, LTA
4201 [A] — (H01) — AC 61-23B, Chapter 1
Lift, weight, thrust, and drag are the four basic aerodynamic forces acting on an aircraft in flight.

Answer/Code line: Correct answer [in brackets], followed by the FAA's Subject Matter Knowledge Code (in parentheses), followed by further references where applicable.

Explanation

* **Note:** The FAA does *not* identify which questions are on the different ratings' tests. Unless the wording of a question is pertinent to only one rating category, it may be found on *any* of the tests.

Instrument Rating
Knowledge Test Questions

Note: Figures referred to in these FAA questions can be found in the separate booklet, *Computerized Testing Supplement for Instrument Rating* (#ASA-CT-8080-3B).

4001. No pilot may act as pilot in command of an airship under IFR nor in weather conditions less than the minimums prescribed for VFR unless that pilot has, within the past 6 months, completed at least

A—three instrument approaches and logged 3 hours.
B—six instrument flights under actual IFR conditions.
C—six instrument approaches and logged 6 hours instrument time, or passed an instrument competency check in an airship.

4002. What limitation is imposed on a newly certificated commercial airplane pilot if that person does not hold an instrument pilot rating?

A—The carrying of passengers or property for hire on cross-country flights at night is limited to a radius of 50 nautical miles (NM).
B—The carrying of passengers for hire on cross-country flights is limited to 50 NM for night flights, but not limited for day flights.
C—The carrying of passengers for hire on cross-country flights is limited to 50 NM and the carrying of passengers for hire at night is prohibited.

4003. Before beginning any flight under IFR, the pilot in command must become familiar with all available information concerning that flight. In addition, the pilot must

A—be familiar with all instrument approaches at the destination airport.
B—list an alternate airport on the flight plan and confirm adequate takeoff and landing performance at the destination airport.
C—be familiar with the runway lengths at airports of intended use, and the alternatives available if the flight cannot be completed.

4004. The use of certain portable electronic devices is prohibited on aircraft that are being operated under

A—IFR.
B—VFR.
C—DVFR.

4005. If weather conditions are such that it is required to designate an alternate airport on your IFR flight plan, you should plan to carry enough fuel to arrive at the first airport of intended landing, fly from that airport to the alternate airport, and fly thereafter for

A—30 minutes at slow cruising speed.
B—45 minutes at normal cruising speed.
C—1 hour at normal cruising speed.

4006. Except when necessary for takeoff or landing or unless otherwise authorized by the Administrator, the minimum altitude for IFR flight is

A—3,000 feet over all terrain.
B—3,000 feet over designated mountainous terrain; 2,000 feet over terrain elsewhere.
C—2,000 feet above the highest obstacle over designated mountainous terrain; 1,000 feet above the highest obstacle over terrain elsewhere.

4007. If the aircraft's transponder fails during flight within Class B airspace,

A—the pilot should immediately request clearance to depart the Class B airspace.
B—ATC may authorize deviation from the transponder requirement to allow aircraft to continue to the airport of ultimate destination.
C—aircraft must immediately descend below 1,200 feet AGL and proceed to destination.

4008. If a pilot enters the condition of flight in the pilot logbook as simulated instrument conditions, what qualifying information must also be entered?

A—Place and type of each instrument approach completed and name of safety pilot.
B—Number and type of instrument approaches completed and route of flight.
C—Name and pilot certificate number of safety pilot and type of approaches completed.

4009. What portion of dual instruction time may a certificated instrument flight instructor log as instrument flight time?

A—All time during which the instructor acts as instrument instructor, regardless of weather conditions.
B—All time during which the instructor acts as instrument instructor in actual instrument weather conditions.
C—Only the time during which the instructor flies the aircraft by reference to instruments.

4010. Which flight time may be logged as instrument time when on an instrument flight plan?

A—All of the time the aircraft was not controlled by ground references.
B—Only the time you controlled the aircraft solely by reference to flight instruments.
C—Only the time you were flying in IFR weather conditions.

4011. What are the minimum qualifications for a person who occupies the other control seat as safety pilot during simulated instrument flight?

A—Appropriately rated in the aircraft.
B—Private pilot.
C—Private pilot with instrument rating.

4012. The minimum instrument time required, within the last 6 months, to be current for IFR is

A—6 hours; at least 3 of the 6 hours in the category of aircraft to be flown.
B—6 hours in the same category aircraft.
C—6 hours in the same category aircraft, and at least 3 of the 6 hours in actual conditions.

4013. After your recent IFR experience lapses, how much time do you have before you must pass an instrument competency check to act as pilot in command under IFR?

A—6 months.
B—90 days.
C—12 months.

4014. An instrument rated pilot, who has not logged any instrument time in 1 year or more, cannot serve as pilot in command under IFR, unless the pilot

A—completes the required 6 hours and six approaches, followed by an instrument competency check given by an FAA-designated examiner.
B—passes an instrument competency check in the category of aircraft involved, given by an approved FAA examiner, instrument instructor, or FAA inspector.
C—passes an instrument competency check in the category of aircraft involved, followed by 6 hours and six instrument approaches, 3 of those hours in the category of aircraft involved.

4015. A pilot's recent IFR experience expires on July 1 of this year. What is the latest date the pilot can meet the IFR experience requirement without having to take an instrument competency check?

A—December 31, this year.
B—June 30, next year.
C—July 31, this year.

4016. What additional approaches, if any, must you perform to maintain IFR currency in a helicopter?

Within the past 6 months, you have accomplished:
One approach in a helicopter.
Two approaches in an airplane.
Two approaches in an approved simulator.

A—None.
B—One approach in an airplane, helicopter, or approved simulator.
C—Two approaches in a helicopter and one approach in an approved simulator.

4017. What minimum conditions are necessary for the instrument approaches required for IFR currency?

A—The approaches may be made in an aircraft, approved instrument ground trainer, or any combination of these.
B—At least three approaches must be made in the same category of aircraft to be flown.
C—At least three approaches must be made in the same category and class of aircraft to be flown.

4018. Under which condition may you act as pilot in command of a helicopter under IFR?

Your certificates and ratings: Private Pilot Certificate with AMEL and Airplane instrument, rotorcraft category rating, and helicopter class rating.

A—If a certificated helicopter instrument flight instructor is on board.
B—If you meet the recent helicopter IFR experience requirements.
C—If you acquire a helicopter instrument rating and meet IFR currency requirements.

4019. What additional flight hours within the past 6 months are required to maintain IFR currency in a helicopter, if you already have 3 hours in an instrument simulator?

A—3 hours of actual or simulated instrument time in the same type helicopter.
B—3 hours of actual or simulated instrument time in a helicopter.
C—3 hours of simulated instrument time in any aircraft.

4020. How may a pilot satisfy the recent instrument experience requirement necessary to act as pilot in command in IMC?

A—Log 6 hours of instrument time under actual or simulated IFR conditions within the last 6 months, including six instrument approaches of any kind. Three of the 6 hours must be in flight in the category of aircraft involved.
B—Log 6 hours of instrument time under actual or simulated IFR conditions within the last 6 months, including at least three instrument approaches of any kind. Three of the 6 hours must be in flight in the category of aircraft involved.
C—Log 6 hours of instrument time under actual or simulated IFR conditions within the last 3 months, including at least six instrument approaches of any kind. Three of the 6 hours must be in flight in any category aircraft.

4021. How long does a pilot remain current for IFR flight after successfully completing an instrument competency check if no further IFR flights are made?

A—90 days.
B—6 months.
C—12 months.

4022. Which additional IFR experience allows you to meet the recent IFR experience requirements to act as pilot in command of a helicopter under IFR?

Your instrument experience within the past 6 months is:

2 hours and one instrument approach in an approved simulator

4 hours simulated IFR and one instrument approach in an airplane

A—3 hours of simulated instrument flight time in a helicopter and four instrument approaches in a simulator.
B—1 hour of simulated instrument flight time and four instrument approaches in a helicopter.
C—Four instrument approaches in an airplane or helicopter.

4023. What minimum conditions are necessary for the instrument approaches required for IFR currency in an airplane?

A—A minimum of six in any approved instrument ground trainer or aircraft within the past 6 months.
B—A minimum of six, at least three of which must be in an aircraft within the past 6 months.
C—A minimum of six in an aircraft, at least three of which must be in the same category within the past 6 months.

4024. Under which condition are you required to have an instrument rating for flight in VMC?

A—Flight through an MOA.
B—Flight into an ADIZ.
C—Flight in a positive control area.

4025. The pilot in command of a civil aircraft must have an instrument rating only when operating

A—under IFR in controlled airspace and in a positive control area or positive control route segment.
B—under IFR, in weather conditions less than the minimum for VFR flight, and in a positive control area or route segment.
C—in weather conditions less than the minimum prescribed for VFR flight.

4026. Which additional IFR experience allows you to meet the recent IFR experience requirements to act as pilot in command of an airplane under IFR?

Your present instrument experience within the past 6 months is:

3 hours and one instrument approach in a simulator

3 hours and one instrument approach in an airplane

A—Four instrument approaches in an approved simulator.
B—Three hours of simulated or actual instrument flight time in a helicopter and two instrument approaches in an airplane or helicopter.
C—Three instrument approaches in a helicopter.

4027. To meet the minimum required instrument experience to remain current for IFR operations, you must accomplish during the past 6 months at least six instrument approaches

A—and 6 hours of instrument time; 3 of the 6 hours in flight in the category of aircraft to be flown.
B—and 6 hours of instrument time in any aircraft.
C—three of which must be in the same category and class of aircraft to be flown, and 6 hours of instrument time in any aircraft.

4028. A certificated commercial pilot who carries passengers for hire in an airplane at night is required to have at least

A—an associated type rating if the airplane is of the multiengine class.
B—a First-Class Medical Certificate.
C—an airplane instrument pilot rating.

4029. You intend to carry passengers for hire on a night VFR flight in a single-engine airplane within a 25-mile radius of the departure airport. You are required to possess at least which rating(s)?

A—A Commercial Pilot Certificate with a single-engine land rating.
B—A Commercial Pilot Certificate with a single-engine and instrument (airplane) rating.
C—A Private Pilot Certificate with a single-engine land and instrument airplane rating.

4030. Do regulations permit you to act as pilot in command of a helicopter in IMC if you hold a Private Pilot Certificate with ASEL, airplane instrument rating, rotorcraft category, and helicopter class rating?

A—Yes, if you comply with the recent IFR experience requirements for a helicopter.
B—No, you must hold either an unrestricted Airline Transport Pilot-Helicopter Certificate or a helicopter instrument rating.
C—No, however, you may do so if you hold an Airline Transport Pilot-Helicopter Certificate, limited to VFR.

4031. Under which condition must the pilot in command of a civil aircraft have at least an instrument rating?

A—When operating in the Continental Control Area.
B—For a flight in VFR conditions while on an IFR flight plan.
C—For any flight above an altitude of 1,200 feet AGL, when the visibility is less than 3 miles.

4032. What are the minimum fuel requirements for airplanes in IFR conditions, if the first airport of intended landing is forecast to have a 1,500-foot ceiling and 3 miles visibility at flight-planned ETA? Fuel to fly to the first airport of intended landing,

A—and fly thereafter for 45 minutes at normal cruising speed.
B—fly to the alternate, and fly thereafter for 45 minutes at normal cruising speed.
C—fly to the alternate, and fly thereafter for 30 minutes at normal cruising speed.

4033. Before beginning any flight under IFR, the pilot in command must become familiar with all available information concerning that flight. In addition, the pilot must

A—list an alternate airport on the flight plan and become familiar with the instrument approaches to that airport.
B—list an alternate airport on the flight plan and confirm adequate takeoff and landing performance at the destination airport.
C—be familiar with the runway lengths at airports of intended use, and the alternatives available if the flight cannot be completed.

4034. Which limitation is imposed on the holder of a Commercial Pilot Certificate if that person does not hold an instrument rating?

A—That person is limited to private pilot privileges at night.
B—The carrying of passengers or property for hire on cross-country flights at night is limited to a radius of 50 NM.
C—The carrying of passengers for hire on cross-country flights is limited to 50 NM and the carrying of passengers for hire at night is prohibited.

4035. To carry passengers for hire in an airplane on cross-country flights of more than 50 NM from the departure airport, the pilot in command is required to hold at least

A—a Category II pilot authorization.
B—a First-Class Medical certificate.
C—a Commercial Pilot Certificate with an instrument rating.

4036. When must an operational check on the aircraft VOR equipment be accomplished to operate under IFR?

A—Within the preceding 10 days or 10 hours of flight time.
B—Within the preceding 30 days or 30 hours of flight time.
C—Within the preceding 30 days.

4037. In the 48 contiguous states, excluding the airspace at or below 2,500 feet AGL, an operable coded transponder equipped with Mode C capability is required in all controlled airspace at and above

A—12,500 feet MSL.
B—10,000 feet MSL.
C—Flight level (FL) 180.

4038. A coded transponder equipped with altitude reporting capability is required in all controlled airspace

A—at and above 10,000 feet MSL, excluding at and below 2,500 feet AGL.
B—at and above 2,500 feet above the surface.
C—below 10,000 feet MSL, excluding at and below 2,500 feet AGL.

4039. Who is responsible for determining that the altimeter system has been checked and found to meet FAR requirements for a particular instrument flight?

A—Owner.
B—Operator.
C—Pilot in command.

4040. What are the fuel requirements for a night IFR flight in a helicopter when an alternate airport is not required?

A—Enough fuel to complete the flight to the first airport of intended landing and fly after that for 30 minutes.
B—Enough fuel to complete the flight to the first airport of intended landing, and fly thereafter for 45 minutes at cruising speed.
C—Enough fuel to complete the flight to the first airport of intended landing, make an approach, and fly thereafter for 45 minutes at cruising speed.

4041. When an alternate airport is required for helicopters on the flight plan, you must have sufficient fuel to complete the flight to the first airport of intended landing, fly to the alternate, and thereafter fly for at least

A—30 minutes at normal cruising speed.
B—45 minutes at holding speed.
C—45 minutes at normal cruising speed.

4042. If an unpressurized aircraft is operated above 12,500 feet MSL, but not more than 14,000 feet MSL, for a period of 2 hours 20 minutes, how long during that time is the minimum flightcrew required to use supplemental oxygen?

A—2 hours 20 minutes.
B—1 hour 20 minutes.
C—1 hour 50 minutes.

4043. A helicopter being operated under IFR is required to have, in addition to the equipment required for VFR and night, at least

A—distance measuring equipment.
B—dual VOR receivers.
C—a slip-skid indicator.

4044. Which data must be recorded in the aircraft log or other appropriate log by a pilot making a VOR operational check for IFR operations?

A—VOR name or identification, date of check, amount of bearing error, and signature.
B—Place of operational check, amount of bearing error, date of check, and signature.
C—Date of check, VOR name or identification, place of operational check, and amount of bearing error.

4045. What is the maximum cabin pressure altitude at which a pilot can fly for longer than 30 minutes without using supplemental oxygen?

A—10,500 feet.
B—12,000 feet.
C—12,500 feet.

4046. What record shall be made in the aircraft log or other permanent record by the pilot making the VOR operational check?

A—The date, place, bearing error, and signature.
B—The date, frequency of VOR or VOT, number of flight hours since last check, and signature.
C—The date, place, satisfactory or unsatisfactory, and signature.

4047. Your aircraft had the static pressure system and altimeter tested and inspected on January 5, of this year, and was found to comply with FAA standards. These systems must be reinspected and approved for use in controlled airspace under IFR by

A—January 5, next year.
B—January 5, 2 years hence.
C—January 31, 2 years hence.

4048. Which checks and inspections of flight instruments or instrument systems must be accomplished before an aircraft can be flown under IFR?

A—VOR within 30 days, altimeter systems within 24 calendar months, and transponder within 24 calendar months.
B—ELT test within 30 days, altimeter systems within 12 calendar months, and transponder within 24 calendar months.
C—VOR within 24 calendar months, transponder within 24 calendar months, and altimeter system within 12 calendar months.

4049. An aircraft altimeter system test and inspection must be accomplished within

A—12 calendar months.
B—18 calendar months.
C—24 calendar months.

4050. Where is DME required under IFR?

A—At or above 24,000 feet MSL if VOR navigational equipment is required.
B—In positive control areas.
C—Above 18,000 feet MSL.

4051. An aircraft operated during IFR under FAR Part 91 is required to have which of the following?

A—Radar altimeter.
B—Dual VOR system.
C—Gyroscopic direction indicator.

4052. What is the maximum IFR altitude you may fly in an unpressurized aircraft without providing passengers with supplemental oxygen?

A—12,500 feet.
B—14,000 feet.
C—15,000 feet.

4053. What is the oxygen requirement for an unpressurized aircraft at 15,000 feet?

A—All occupants must use oxygen for the entire time at this altitude.
B—Crew must start using oxygen at 12,000 feet and passengers at 15,000 feet.
C—Crew must use oxygen for the entire time above 14,000 feet and passengers must be provided supplemental oxygen only above 15,000 feet.

4054. When making an airborne VOR check, what is the maximum allowable tolerance between the two indicators of a dual VOR system (units independent of each other except the antenna)?

A—4° between the two indicated bearings of a VOR.
B—Plus or minus 4° when set to identical radials of a VOR.
C—6° between the two indicated radials of a VOR.

4055. What minimum navigation equipment is required for IFR flight?

A—VOR/LOC receiver, transponder, and DME.
B—VOR receiver and, if in ARTS III environment, a coded transponder equipped for altitude reporting.
C—Navigation equipment appropriate to the ground facilities to be used.

4056. You check the flight instruments while taxiing and find that the vertical speed indicator (VSI) indicates a descent of 100 feet per minute. In this case, you

A—must return to the parking area and have the instrument corrected by an authorized instrument repairman.
B—may take off and use 100 feet descent as the zero indication.
C—may not take off until the instrument is corrected by either the pilot or a mechanic.

4057. To meet the requirements for flight under IFR, a helicopter must be equipped with certain operable instruments and equipment. One of those required is

A—a clock with sweep-second pointer or digital presentation.
B—a radar altimeter.
C—a transponder with altitude reporting capability.

4058. How is your flight plan closed when your destination airport has IFR conditions and there is no control tower or flight service station (FSS) on the field?

A—The ARTCC controller will close your flight plan when you report the runway in sight.
B—You may close your flight plan any time after starting the approach by contacting any FSS or ATC facility.
C—Upon landing, you must close your flight plan by radio or by telephone to any FSS or ATC facility.

4059. When may a pilot file a composite flight plan?

A—When requested or advised by ATC.
B—Any time a portion of the flight will be VFR.
C—Any time a landing is planned at an intermediate airport.

4060. When filing a composite flight plan where the first portion of the flight is IFR, which fix(es) should be indicated on the flight plan form?

A—All points of transition from one airway to another, fixes defining direct route segments, and the clearance limit fix.
B—Only the fix where you plan to terminate the IFR portion of the flight.
C—Only those compulsory reporting points on the IFR route segment.

4061. What is the recommended procedure for transitioning from VFR to IFR on a composite flight plan?

A—Prior to transitioning to IFR, contact the nearest FSS, close the VFR portion, and request ATC clearance.
B—Upon reaching the proposed point for change to IFR, contact the nearest FSS and cancel your VFR flight plan, then contact ARTCC and request an IFR clearance.
C—Prior to reaching the proposed point for change to IFR, contact ARTCC, request your IFR clearance, and instruct them to cancel the VFR flight plan.

4062. When is an IFR flight plan required?

A—When less than VFR conditions exist in either Class E or Class G airspace and in Class A airspace.
B—In all Class E airspace when conditions are below VFR, in Class A airspace, and in defense zone airspace.
C—In Class E airspace when IMC exists or in Class A airspace.

4063. Prior to which operation must an IFR flight plan be filed and an appropriate ATC clearance received?

A—Flying by reference to instruments in controlled airspace.
B—Entering controlled airspace when IMC exists.
C—Takeoff when IFR weather conditions exist.

4064. To operate under IFR below 18,000 feet, a pilot must file an IFR flight plan and receive an appropriate ATC clearance prior to

A—entering controlled airspace.
B—entering weather conditions below VFR minimums.
C—takeoff.

4065. To operate an aircraft under IFR, a flight plan must have been filed and an ATC clearance received prior to

A—controlling the aircraft solely by use of instruments.
B—entering weather conditions in any airspace.
C—entering controlled airspace.

4066. When is an IFR clearance required during VFR weather conditions?

A—When operating in the Class E airspace.
B—When operating in a Class A airspace.
C—When operating in airspace above 14,500 feet.

4067. Operation in which airspace requires filing an IFR flight plan?

A—Any airspace when the visibility is less than 1 mile.
B—Class E airspace with IMC and class A airspace.
C—Positive control area, Continental Control Area, and all other airspace, if the visibility is less than 1 mile.

4068. When departing from an airport located outside controlled airspace during IMC, you must file an IFR flight plan and receive a clearance before

A—takeoff.
B—entering IFR conditions.
C—entering Class E airspace.

4069. What is a waypoint when used for an IFR flight?

A—A predetermined geographical position used for an RNAV route or an RNAV instrument approach.
B—A reporting point defined by the intersection of two VOR radials.
C—A location on a victor airway which can only be identified by VOR and DME signals.

4070. Preferred IFR routes beginning with a fix indicate that departing aircraft will normally be routed to the fix via

A—the established airway(s) between the departure airport and the fix.
B—a standard instrument departure (SID), or radar vectors.
C—direct route only.

4071. For which speed variation should you notify ATC?

A—When the groundspeed changes more than 5 knots.
B—When the average true airspeed changes 5 percent or 10 knots, whichever is greater.
C—Any time the groundspeed changes 10 MPH.

4072. (Refer to Figure 1.) Which item(s) should be checked in block 1 for a composite flight plan?

A—VFR with an explanation in block 11.
B—IFR with an explanation in block 11.
C—VFR and IFR.

4073. (Refer to Figure 1.) The time entered in block 12 for an IFR flight should be based on which fuel quantity?

A—Total fuel required for the flight.
B—Total useable fuel on board.
C—The amount of fuel required to fly to the destination airport, then to the alternate, plus a 45-minute reserve.

4074. (Refer to Figure 1.) What information should be entered in block 7 of an IFR flight plan if the flight has three legs, each at a different altitude?

A—Altitude for first leg.
B—Altitude for first leg and highest altitude.
C—Highest altitude.

4075. (Refer to Figure 1.) Which equipment determines the code to be entered in block 3 as a suffix to aircraft type on the flight plan form?

A—DME, ADF, and airborne radar.
B—DME, transponder, and ADF.
C—DME, transponder, and RNAV.

4076. When may a pilot cancel the IFR flight plan prior to completing the flight?

A—Any time.
B—Only if an emergency occurs.
C—Only in VFR conditions outside positive controlled airspace.

4077. Which airspaces are depicted on the En Route Low Altitude Chart?

A—Class D, Class C, Class B, Class E and special use airspace.
B—Class A, special use airspace, Class D and Class E.
C—Special use airspace, Class E, Class D, Class A, Class B and Class C.

4078. Where are the compulsory reporting points, if any, on a direct flight not flown on radials or courses of established airways or routes?

A—Fixes selected to define the route.
B—There are no compulsory reporting points unless advised by ATC.
C—At the changeover points.

4079. Which sources of aeronautical information, when used collectively, provide the latest status of airport conditions (e.g., runway closures, runway lighting, snow conditions)?

A—Airman's Information Manual, aeronautical charts, and Distant (D) Notice to Airmans (NOTAM's).
B—Airport Facility Directory, FDC NOTAM's, and Local (L) NOTAM's.
C—Airport Facility Directory, Distant (D) NOTAM's, and Local (L) NOTAM's.

4080. What is the purpose of FDC NOTAMs?

A—To provide the latest information on the status of navigation facilities to all FSS facilities for scheduled broadcasts.
B—To issue notices for all airports and navigation facilities in the shortest possible time.
C—To advise of changes in flight data which affect instrument approach procedure (IAP), aeronautical charts, and flight restrictions prior to normal publication.

4081. What minimum weather conditions must be forecast for your ETA at an airport that has only a VOR approach with standard alternate minimums, for the airport to be listed as an alternate on the IFR flight plan?

A—800-foot ceiling and 1 statute mile (SM) visibility.
B—800-foot ceiling and 2 SM visibility.
C—1,000-foot ceiling and visibility to allow descent from minimum en route altitude (MEA), approach, and landing under basic VFR.

4082. Is an alternate airport required for an IFR flight to ATL (Atlanta Hartsfield) if the proposed ETA is 1930Z?

ATL FT AMD 1 161615 1630Z C20 BKN 6RW VRBL 20 SCT C40 BKN OCNL 4RW 3315. 18Z 40 SCT C100 BKN 3110 OCNL C20 BKN 6RW CHC C10 OVC 3TRW 3310G20 AFT 21Z. 09Z MVFR CIG R.

A—An alternate is required because the ceiling could fall below 2,000 feet within 2 hours before to 2 hours after the ETA.
B—An alternate is not required because the ceiling and visibility are forecast to remain at or above 1,000 feet and 3 miles, respectively.
C—An alternate is not required because the ceiling and visibility are forecast to be at or above 2,000 feet and 3 miles within 1 hour before to 1 hour after the ETA.

4083. What minimum conditions must exist at the destination airport to avoid listing an alternate airport on an IFR flight plan when a standard IAP is available?

A—From 2 hours before to 2 hours after ETA, forecast ceiling 2,000, and visibility 2 and 1/2 miles.
B—From 2 hours before to 2 hours after ETA, forecast ceiling 3,000, and visibility 3 miles.
C—From 1 hour before to 1 hour after ETA, forecast ceiling 2,000, and visibility 3 miles.

4084. Under what condition are you not required to list an alternate airport on an IFR flight plan for an IFR flight in a helicopter?

A—When the ceiling is forecast to be at least 1,000 feet above the lowest of the MEA, MOCA, or initial approach altitude within 2 hours of your ETA at the destination airport.
B—When the weather reports or forecasts indicate the ceiling and visibility will be at least 2,000 feet and 3 miles for 1 hour before to 1 hour after your ETA at the destination airport.
C—When the ceiling is forecast to be at least 1,000 feet above the lowest of the MEA, MOCA, or initial approach altitude and the visibility is 2 miles more than the minimum landing visibility within 2 hours of your ETA at the destination airport.

4085. What standard minimums are required to list an airport as an alternate on an IFR flight plan if the airport has VOR approach only?

A—Ceiling and visibility at ETA, 800 feet and 2 miles, respectively.
B—Ceiling and visibility from 2 hours before until 2 hours after ETA, 800 feet and 2 miles, respectively.
C—Ceiling and visibility at ETA, 600 feet and 2 miles, respectively.

4086. What are the minimum weather conditions that must be forecast to list an airport as an alternate when the airport has no approved IAP?

A—The ceiling and visibility at ETA, 2,000 feet and 3 miles, respectively.
B—The ceiling and visibility from 2 hours before until 2 hours after ETA, 2,000 feet and 3 miles, respectively.
C—The ceiling and visibility at ETA must allow descent from MEA, approach, and landing, under basic VFR.

4087. What minimum weather conditions must be forecast for your ETA at an airport that has a precision approach procedure, with standard alternate minimums, in order to list it as an alternate for the IFR flight?

A—600-foot ceiling and 2 SM visibility at your ETA.
B—600-foot ceiling and 2 SM visibility from 2 hours before to 2 hours after your ETA.
C—800-foot ceiling and 2 SM visibility at your ETA.

4088. Which publication covers the procedures required for aircraft accident and incident reporting responsibilities for pilots?

A—FAR Part 61.
B—FAR Part 91.
C—NTSB Part 830.

4089. Under what condition is pressure altitude and density altitude the same value?

A—At standard temperature.
B—When the altimeter setting is 29.92" Hg.
C—When indicated, and pressure altitudes are the same value on the altimeter.

4090. Under which condition will pressure altitude be equal to true altitude?

A—When the atmospheric pressure is 29.92" Hg.
B—When standard atmospheric conditions exist.
C—When indicated altitude is equal to the pressure altitude.

4091. Which condition would cause the altimeter to indicate a lower altitude than actually flown (true altitude)?

A—Air temperature lower than standard.
B—Atmospheric pressure lower than standard.
C—Air temperature warmer than standard.

4092. Which is true regarding the use of airborne weather-avoidance radar for the recognition of certain weather conditions?

A—The radarscope provides no assurance of avoiding instrument weather conditions.
B—The avoidance of hail is assured when flying between and just clear of the most intense echoes.
C—The clear area between intense echoes indicates that visual sighting of storms can be maintained when flying between the echoes.

4093. When an altimeter is changed from 30.11" Hg to 29.96" Hg, in which direction will the indicated altitude change and by what value?

A—Altimeter will indicate 15 feet lower.
B—Altimeter will indicate 150 feet lower.
C—Altimeter will indicate 150 feet higher.

4094. A common type of ground or surface based temperature inversion is that which is produced by

A—warm air being lifted rapidly aloft in the vicinity of mountainous terrain.
B—the movement of colder air over warm air, or the movement of warm air under cold air.
C—ground radiation on clear, cool nights when the wind is light.

4095. How much colder than standard temperature is the actual temperature at 9,000 feet, as indicated in the following excerpt from the Winds and Temperature Aloft Forecast?

FT	6000	9000
	0737-04	1043-10

A—3° C.
B—10° C.
C—7° C.

4096. The primary cause of all changes in the Earth's weather is

A—variation of solar energy received by the Earth's regions.
B—changes in air pressure over the Earth's surface.
C—movement of the air masses.

4097. A characteristic of the stratosphere is

A—an overall decrease of temperature with an increase in altitude.
B—a relatively even base altitude of approximately 35,000 feet.
C—relatively small changes in temperature with an increase in altitude.

4098. Steady precipitation, in contrast to showers, preceding a front is an indication of

A—stratiform clouds with moderate turbulence.
B—cummuliform clouds with little or no turbulence.
C—stratiform clouds with little or no turbulence.

4099. The presence of ice pellets at the surface is evidence that

A—there are thunderstorms in the area.
B—a cold front has passed.
C—there is freezing rain at a higher altitude.

4100. Which conditions result in the formation of frost?

A—The temperature of the collecting surface is at or below freezing and small droplets of moisture are falling.
B—When dew forms and the temperature is below freezing.
C—Temperature of the collecting surface is below the dewpoint of surrounding air and the dewpoint is colder than freezing.

4101. To which meteorological condition does the term "dewpoint" refer?

A—The temperature to which air must be cooled to become saturated.
B—The temperature at which condensation and evaporation are equal.
C—The temperature at which dew will always form.

4102. What temperature condition is indicated if wet snow is encountered at your flight altitude?

A—The temperature is above freezing at your altitude.
B—The temperature is below freezing at your altitude.
C—You are flying from a warm air mass into a cold air mass.

4103. The amount of water vapor which air can hold largely depends on

A—relative humidity.
B—air temperature.
C—stability of air.

4104. Clouds, fog, or dew will always form when

A—water vapor condenses.
B—water vapor is present.
C—the temperature and dewpoint are equal.

4105. What causes surface winds to flow across the isobars at an angle rather than parallel to the isobars?

A—Coriolis force.
B—Surface friction.
C—The greater density of the air at the surface.

4106. Winds at 5,000 feet AGL on a particular flight are southwesterly while most of the surface winds are southerly. This difference in direction is primarily due to

A—a stronger pressure gradient at higher altitudes.
B—friction between the wind and the surface.
C—stronger Coriolis force at the surface.

4107. What relationship exists between the winds at 2,000 feet above the surface and the surface winds?

A—The winds at 2,000 feet and the surface winds flow in the same direction, but the surface winds are weaker due to friction.
B—The winds at 2,000 feet tend to parallel the isobars while the surface winds cross the isobars at an angle toward lower pressure and are weaker.
C—The surface winds tend to veer to the right of the winds at 2,000 feet and are usually weaker.

4108. Which force, in the Northern Hemisphere, acts at a right angle to the wind and deflects it to the right until parallel to the isobars?

A—Centrifugal.
B—Pressure gradient.
C—Coriolis.

4109. Under what condition will true altitude be lower than indicated altitude with an altimeter setting of 29.92" Hg?

A—In warmer than standard air temperature.
B—In colder than standard air temperature.
C—When density altitude is higher than indicated altitude.

4110. Which of the following defines the type of altitude used when maintaining FL 210?

A—Indicated.
B—Pressure.
C—Calibrated.

4111. Altimeter setting is the value to which the scale of the pressure altimeter is set so the altimeter indicates

A—true altitude at field elevation.
B—pressure altitude at field elevation.
C—pressure altitude at sea level.

4112. The most frequent type of ground- or surface-based temperature inversion is that produced by

A—radiation on a clear, relatively still night.
B—warm air being lifted rapidly aloft in the vicinity of mountainous terrain.
C—the movement of colder air under warm air, or the movement of warm air over cold air.

4113. If the air temperature is +8° C at an elevation of 1,350 feet and a standard (average) temperature lapse rate exists, what will be the approximate freezing level?

A—3,350 feet MSL.
B—5,350 feet MSL.
C—9,350 feet MSL.

4114. What feature is associated with a temperature inversion?

A—A stable layer of air.
B—An unstable layer of air.
C—Air mass thunderstorms.

4115. What type of clouds will be formed if very stable moist air is forced upslope?

A—First stratified clouds and then vertical clouds.
B—Vertical clouds with increasing height.
C—Stratified clouds with little vertical development.

4116. The general characteristics of unstable air are

A—good visibility, showery precipitation, and cumuliform-type clouds.
B—good visibility, steady precipitation, and stratiform-type clouds.
C—poor visibility, intermittent precipitation, and cumuliform-type clouds.

4117. Which is a characteristic of stable air?

A—Fair weather cumulus clouds.
B—Stratiform clouds.
C—Unlimited visibility.

4118. What type clouds can be expected when an unstable air mass is forced to ascend a mountain slope?

A—Layered clouds with little vertical development.
B—Stratified clouds with considerable associated turbulence.
C—Clouds with extensive vertical development.

4119. What are the characteristics of stable air?

A—Good visibility, steady precipitation, and stratus-type clouds.
B—Poor visibility, intermittent precipitation, and cumulus-type clouds.
C—Poor visibility, steady precipitation, and stratus-type clouds.

4120. What are some characteristics of unstable air?

A—Nimbostratus clouds and good surface visibility.
B—Turbulence and poor surface visibility.
C—Turbulence and good surface visibility.

4121. Stability can be determined from which measurement of the atmosphere?

A—Low-level winds.
B—Ambient lapse rate.
C—Atmospheric pressure.

4122. What determines the structure or type of clouds which form as a result of air being forced to ascend?

A—The method by which the air is lifted.
B—The stability of the air before lifting occurs.
C—The amount of condensation nuclei present after lifting occurs.

4123. Which of the following combinations of weather producing variables would likely result in cumuliform-type clouds, good visibility, rain showers, and possible clear-type icing in clouds?

A—Unstable, moist air, and no lifting mechanism.
B—Stable, dry air, and orographic lifting.
C—Unstable, moist air, and orographic lifting.

4124. Unsaturated air flowing upslope will cool at the rate of approximately (dry adiabatic lapse rate)

A—3° C per 1,000 feet.
B—2° C per 1,000 feet.
C—2.5° C per 1,000 feet.

4125. A temperature inversion will normally form only

A—in stable air.
B—in unstable air.
C—when a stratiform layer merges with a cumuliform mass.

4126. Which weather phenomenon signals the beginning of the mature stage of a thunderstorm?

A—The start of rain at the surface.
B—Growth rate of cloud is maximum.
C—Strong turbulence in the cloud.

4127. Frontal waves normally form on

A—slow moving cold fronts or stationary fronts.
B—slow moving warm fronts and strong occluded fronts.
C—rapidly moving cold fronts or warm fronts.

4128. Which are characteristics of an unstable cold air mass moving over a warm surface?

A—Cumuliform clouds, turbulence, and poor visibility.
B—Cumuliform clouds, turbulence, and good visibility.
C—Stratiform clouds, smooth air, and poor visibility.

4129. Which clouds have the greatest turbulence?

A—Towering cumulus.
B—Cumulonimbus.
C—Altocumulus castellanus.

4130. Standing lenticular clouds, in mountainous areas, indicate

A—an inversion.
B—unstable air.
C—turbulence.

4131. The suffix "nimbus", used in naming clouds, means a

A—cloud with extensive vertical development.
B—raincloud.
C—dark massive, towering cloud.

4132. The presence of standing lenticular altocumulus clouds is a good indication of

A—a jetstream.
B—very strong turbulence.
C—heavy icing conditions.

4133. Which family of clouds is least likely to contribute to structural icing on an aircraft?

A—Low clouds.
B—High clouds.
C—Clouds with extensive vertical development.

4134. What are the four families of clouds?

A—Stratus, cumulus, nimbus, and cirrus.
B—Clouds formed by updrafts, fronts, cooling layers of air, and precipitation into warm air.
C—High, middle, low, and those with extensive vertical development.

4135. Where can wind shear associated with a thunderstorm be found? Choose the most complete answer.

A—In front of the thunderstorm cell (anvil side) and on the right side of the cell.
B—In front of the thunderstorm cell and directly under the cell.
C—On all sides of the thunderstorm cell and directly under the cell.

4136. Which weather phenomenon is always associated with the passage of a frontal system?

A—A wind change.
B—An abrupt decrease in pressure.
C—Clouds, either ahead or behind the front.

4137. Where do squall lines most often develop?

A—In an occluded front.
B—In a cold air mass.
C—Ahead of a cold front.

4138. Where does wind shear occur?

A—Exclusively in thunderstorms.
B—Wherever there is an abrupt decrease in pressure and/or temperature.
C—With either a wind shift or a windspeed gradient at any level in the atmosphere.

4139. What is an important characteristic of wind shear?

A—It is primarily associated with the lateral vortices generated by thunderstorms.
B—It usually exists only in the vicinity of thunderstorms, but may be found near a strong temperature inversion.
C—It may be associated with either a wind shift or a windspeed gradient at any level in the atmosphere.

4140. Which is a characteristic of low-level wind shear as it relates to frontal activity?

A—With a warm front, the most critical period is before the front passes the airport.
B—With a cold front, the most critical period is just before the front passes the airport.
C—Turbulence will always exist in wind-shear conditions.

4141. What is indicated by the term "embedded thunderstorms"?

A—Severe thunderstorms are embedded within a squall line.
B—Thunderstorms are predicted to develop in a stable air mass.
C—Thunderstorms are obscured by massive cloud layers and cannot be seen.

4142. If squalls are reported at your destination, what wind conditions should you anticipate?

A—Sudden increases in windspeed of at least 15 knots to a peak of 20 knots or more, lasting for at least 1 minute.
B—Peak gusts of at least 35 knots for a sustained period of 1 minute or longer.
C—Rapid variation in wind direction of at least 20° and changes in speed of at least 10 knots between peaks and lulls.

4143. During the life cycle of a thunderstorm, which stage is characterized predominately by downdrafts?

A—Cumulus.
B—Dissipating.
C—Mature.

4144. Which weather phenomenon is always associated with a thunderstorm?

A—Lightning.
B—Heavy rain showers.
C—Supercooled raindrops.

4145. Which thunderstorms generally produce the most severe conditions, such as heavy hail and destructive winds?

A—Warm front.
B—Squall line.
C—Air mass.

4146. Which procedure is recommended if a pilot should unintentionally penetrate embedded thunderstorm activity?

A—Reverse aircraft heading or proceed toward an area of known VFR conditions.
B—Reduce airspeed to maneuvering speed and maintain a constant altitude.
C—Set power for recommended turbulence penetration airspeed and attempt to maintain a level flight attitude.

4147. What is an indication that downdrafts have developed and the thunderstorm cell has entered the mature stage?

A—The anvil top has completed its development.
B—Precipitation begins to fall from the cloud base.
C—A gust front forms.

4148. What are the requirements for the formation of a thunderstorm?

A—A cumulus cloud with sufficient moisture.
B—A cumulus cloud with sufficient moisture and an inverted lapse rate.
C—Sufficient moisture, an unstable lapse rate, and a lifting action.

4149. Fair weather cumulus clouds often indicate

A—turbulence at and below the cloud level.
B—poor visibility.
C—smooth flying conditions.

4150. What is an important characteristic of wind shear?

A—It is an atmospheric condition that is associated exclusively with zones of convergence.
B—The Coriolis phenomenon in both high- and low-level air masses is the principal generating force.
C—It is an atmospheric condition that may be associated with a low-level temperature inversion, a jet stream, or a frontal zone.

4151. Why is frost considered hazardous to flight operation?

A—Frost changes the basic aerodynamic shape of the airfoil.
B—Frost decreases control effectiveness.
C—Frost causes early airflow separation resulting in a loss of lift.

4152. In which meteorological environment is aircraft structural icing most likely to have the highest rate of accumulation?

A—Cumulonimbus clouds.
B—High humidity and freezing temperature.
C—Freezing rain.

4153. What is an operational consideration if you fly into rain which freezes on impact?

A—You have flown into an area of thunderstorms.
B—Temperatures are above freezing at some higher altitude.
C—You have flown through a cold front.

4154. The average height of the troposphere in the middle latitudes is

A—20,000 feet.
B—25,000 feet.
C—37,000 feet.

4155. A jetstream is defined as wind of

A—30 knots or greater.
B—40 knots or greater.
C—50 knots or greater.

4156. Under which condition does advection fog usually form?

A—Moist air moving over colder ground or water.
B—Warm, moist air settling over a cool surface under no-wind conditions.
C—A land breeze blowing a cold air mass over a warm water current.

4157. A high cloud is composed mostly of

A—ozone.
B—condensation nuclei.
C—ice crystals.

4158. An air mass is a body of air that

A—has similar cloud formations associated with it.
B—creates a wind shift as it moves across the Earth's surface.
C—covers an extensive area and has fairly uniform properties of temperature and moisture.

4159. What enhances the growth rate of precipitation?

A—Advective action.
B—Upward currents.
C—Cyclonic movement.

4160. If you fly into severe turbulence, which flight condition should you attempt to maintain?

A—Constant airspeed (V_A).
B—Level flight attitude.
C—Constant altitude and constant airspeed.

4161. Which precipitation type normally indicates freezing rain at higher altitudes?

A—Snow.
B—Hail.
C—Ice pellets.

4162. Which weather condition can be expected when moist air flows from a relatively warm surface to a colder surface?

A—Increased visibility.
B—Convective turbulence due to surface heating.
C—Fog.

4163. Fog is usually prevalent in industrial areas because of

A—atmospheric stabilization around cities.
B—an abundance of condensation nuclei from combustion products.
C—increased temperatures due to industrial heating.

4164. In which situation is advection fog most likely to form?

A—An air mass moving inland from the coast in winter.
B—A light breeze blowing colder air out to sea.
C—Warm, moist air settling over a warmer surface under no-wind conditions.

4165. In what localities is advection fog most likely to occur?

A—Coastal areas.
B—Mountain slopes.
C—Level inland areas.

4166. What types of fog depend upon a wind in order to exist?

A—Steam fog and downslope fog.
B—Precipitation-induced fog and ground fog.
C—Advection fog and upslope fog.

4167. What situation is most conducive to the formation of radiation fog?

A—Warm, moist air over low, flatland areas on clear, calm nights.
B—Moist, tropical air moving over cold, offshore water.
C—The movement of cold air over much warmer water.

4168. The strength and location of the jetstream is normally

A—stronger and farther north in the winter.
B—weaker and farther north in the summer.
C—stronger and farther north in the summer.

4169. Which conditions are favorable for the formation of radiation fog?

A—Moist air moving over colder ground or water.
B—Cloudy sky and a light wind moving saturated warm air over a cool surface.
C—Clear sky, little or no wind, small temperature/dewpoint spread, and over a land surface.

4170. The body of a Terminal Forecast covers a geographical area within

A—a 5-mile radius of the center of a runway complex.
B—25 miles of the center of an airport.
C—10 miles of the station originating the FT.

4171. Test data indicate that ice, snow, or frost having a thickness and roughness similar to medium or coarse sandpaper on the leading edge and upper surface of a wing can

A—reduce lift by as much as 50 percent and increase drag by as much as 50 percent.
B—increase drag and reduce lift by as much as 25 percent.
C—reduce lift by as much as 30 percent and increase drag by 40 percent.

4172. What wind direction and speed is represented by the entry 9900+00 for 9,000 feet, on an Winds and Temperatures Aloft Forecast (FD)?

A—Light and variable; less than 5 knots.
B—Vortex winds exceeding 200 knots.
C—Light and variable; less than 10 knots.

4173. What conclusion(s) can be drawn from a 500-millibar Constant Pressure Chart for a planned flight at FL 180?

A—Winds aloft at FL 180 generally flow across the height contours.
B—Observed temperature, wind, and temperature/dew-point spread along the proposed route can be approximated.
C—Upper highs, lows, troughs, and ridges will be depicted by the use of lines of equal pressure.

4174. What important information is provided by the Radar Summary Chart that is not shown on other weather charts?

A—Lines and cells of hazardous thunderstorms.
B—Types of precipitation between reporting stations.
C—Areas of cloud cover and icing levels within the clouds.

4175. The section of the Area Forecast entitled "significant clouds and weather" contains a

A—summary of cloudiness and weather significant to flight operations broken down by states or other geographical areas.
B—summary of forecast sky cover, cloud tops, visibility, and obstructions to vision along specific routes.
C—statement of Airmen's Meteorological Information (AIRMET's) and Significant Meteorological Information (SIGMET) still in effect at the time of issue.

4176. Which primary source should be used to obtain forecast weather information at your destination for the planned ETA?

A—Area Forecast.
B—Radar Summary and Weather Depiction Charts.
C—Terminal Forecast.

4177. Omission of a wind entry in a Terminal Forecast specifically implies that the wind is expected to be less than

A—5 knots.
B—6 knots.
C—8 knots.

4178. The absence of a visibility entry in a Terminal Forecast specifically implies that the surface visibility

A—exceeds 10 miles.
B—exceeds 6 miles.
C—is at least 15 miles in all directions from the center of the runway complex.

4179. The word "wind" in the categorical outlook in the Terminal Forecast means that the wind during that period is forecast to be

A—10 knots or stronger.
B—less than 25 knots.
C—25 knots or stronger.

4180. What expected windspeed is specifically implied at 2200Z by this Terminal Forecast for Memphis?

MEM 251010 C5 X 1/2F 1710 OCNL C0 X 1/2F.
16Z C25 BKN 11/2F 1720. 22Z 20 SCT. 00Z CLR.

A—Less than 6 knots.
B—Less than 10 knots.
C—Calm and variable.

4181. SIGMET's are issued as a warning of weather conditions potentially hazardous

A—particularly to light aircraft.
B—to all aircraft.
C—only to light aircraft operations.

4182. What significant cloud coverage is reported by a pilot in this SA?

MOB...M9...OVC 2LF 131/44/43/3212/991/UA/OV 15NW MOB 1355/SK OVC 025/045 OVC 090

A—Three separate overcast layers exist with bases at 2,500, 7,500, and 13,500 feet.
B—The top of lower overcast is 2,500 feet; base and top of second overcast layer is 4,500 and 9,000 feet, respectively.
C—The base of second overcast layer is 2,500 feet; top of second overcast layer is 7,500 feet; base of third layer is 13,500 feet.

4183. Which meteorological condition is issued in the form of a SIGMET (WS)?

A—Widespread sand or duststorms lowering visibility to less than 3 miles.
B—Moderate icing.
C—Sustained winds of 30 knots or greater at the surface.

4184. A pilot planning to depart at 1100Z on an IFR flight is particularly concerned about the hazard of icing. What sources reflect the most accurate information on icing conditions (current and forecast) at the time of departure?

A—Low-Level Significant Weather Prognostic Chart, and the Area Forecast.
B—The Area Forecast, and the Freezing Level Chart.
C—Pilot weather reports (PIREP's), AIRMET's, and SIGMET's.

4185. Which forecast provides specific information concerning expected sky cover, cloud tops, visibility, weather, and obstructions to vision in a route format?

A—DFW FA 131240.
B—MEM FT 132222.
C—249 TWEB 252317.

4186. The section of the Area Forecast entitled "Hazards/Flight Precautions" contains

A—a 12-hour forecast that identifies and locates aviation weather hazards.
B—a statement listing those AIRMET's still in effect.
C—a summary of general weather conditions over several states.

4187. What is the maximum forecast period for AIRMET's?

A—Two hours.
B—Four hours.
C—Six hours.

4188. When is the temperature at one of the forecast altitudes omitted at a specific location or station in the Winds and Temperatures Aloft Forecast (FD)?

A—When the temperature is standard for that altitude.
B—For the 3,000-foot altitude (level) or when the level is within 2,500 feet of station elevation.
C—Only when the winds are omitted for that altitude (level).

4189. When is the wind-group at one of the forecast altitudes omitted at a specific location or station in the Winds and Temperatures Aloft Forecast (FD)? When the wind

A—is less than 5 knots.
B—is less than 10 knots.
C—at the altitude is within 1,500 feet of the station elevation.

4190. Decode the excerpt from the Winds and Temperature Aloft Forecast (FD) for OKC at 39,000 feet.

FT	3000	6000	39000
OKC			830558

A—Wind 130° at 50 knots, temperature -58° C.
B—Wind 330° at 105 knots, temperature -58° C.
C—Wind 330° at 205 knots, temperature -58° C.

4191. Which values are used for winds aloft forecasts?

A—Magnetic direction and knots.
B—Magnetic direction and MPH.
C—True direction and knots.

4192. (Refer to Figure 2.) What approximate wind direction, speed, and temperature (relative to ISA) should a pilot expect when planning for a flight over PSB at FL 270?

A—260° magnetic at 93 knots; ISA +7° C.
B—280° true at 113 knots; ISA +3° C.
C—255° true at 93 knots; ISA +6° C.

4193. (Refer to Figure 2.) What approximate wind direction, speed, and temperature (relative to ISA) should a pilot expect when planning for a flight over ALB at FL 270?

A—270° magnetic at 97 knots; ISA -4° C.
B—260° true at 110 knots; ISA +5° C.
C—275° true at 97 knots; ISA +4° C.

4194. (Refer to Figure 2.) What approximate wind direction, speed, and temperature (relative to ISA) should a pilot expect when planning for a flight over EMI at FL 270?

A—265° true; 100 knots; ISA +3° C.
B—270° true; 110 knots; ISA +5° C.
C—260° magnetic; 100 knots; ISA -5° C.

4195. What flight planning information can a pilot derive from constant pressure charts?

A—Clear air turbulence and icing conditions.
B—Levels of widespread cloud coverage.
C—Winds and temperatures aloft.

4196. The station originating the following weather report has a field elevation of 1,800 feet MSL. If the sky cover is one continuous layer, what is its thickness?

W8X1FK 174/74/73/0000/004/OVC/35

A—900 feet.
B—1,700 feet.
C—2,700 feet.

4197. (Refer to Figure 9.) The Severe Weather Outlook Chart depicts

A—areas of probable severe thunderstorms by the use of single hatched areas on the chart.
B—areas of forecast, severe or extreme turbulence, and areas of severe icing for the next 24 hours.
C—areas of general thunderstorm activity (excluding severe) by the use of hatching on the chart.

4198. Interpret the PIREP.

UA/OV 20S ATL 1620 FL050/TP BE 18/IC MDT RIME ICE

A—20 NM south of Atlanta at 1620Z, a pilot flying at 5,000 feet in a Beech 18 reported moderate rime ice.
B—20 minutes after the hour snow began at Atlanta, wind 160° at 20 knots; a Beech 18 reported moderate rime ice at 5,000 feet.
C—Snow encountered at 2,000 feet over Atlanta at 1620Z; a Beech 18 encountered rime ice at 5,000 feet.

4199. A station is forecasting wind and temperature aloft at FL 390 to be 300° at 200 knots; temperature -54° C. How would this data be encoded in the FD?

A—300054.
B—809954.
C—309954.

4200. Which weather conditions should be expected beneath a low-level temperature inversion layer when the relative humidity is high?

A—Smooth air and poor visibility due to fog, haze, or low clouds.
B—Light wind shear and poor visibility due to haze and light rain.
C—Turbulent air and poor visibility due to fog, low stratus-type clouds, and showery precipitation.

4201. What is the single source reference that contains information regarding frontal movement, turbulence, and icing conditions for a specific area?

A—Terminal Forecast.
B—Weather Depiction Chart.
C—Area Forecast.

4202. A ceiling is defined as the height of the

A—highest layer of clouds or obscuring phenomena aloft that covers over 6/10 of the sky.
B—lowest layer of clouds that contributed to the overall overcast.
C—lowest layer of clouds or obscuring phenomena aloft that is reported as broken or overcast.

4203. The reporting station originating this Surface Aviation Weather Report has a field elevation of 1,000 feet. If the reported sky cover is one continuous layer, what is its thickness?

MDW RS 1856 M7 OVC11/2R+F 990/63/61/3205/980/... UA.../SK OVC 65

A—4,800 feet.
B—5,000 feet.
C—5,800 feet.

4204. What is the significance of the "F2" in the remarks portion of this Surface Aviation Weather Report for CLE?

CLE SP 1350-X E80 BKN 150 OVC 1GF 169/67/67/2105/003/R23LVV11/2 F2

A—The restriction to visibility is caused by fog and the prevailing visibility is 2 SM.
B—The partial obscuration is caused by fog and the visibility value is variable, 1 and 1/2 to 2 SM.
C—Fog is obscuring 2/10 of the sky.

4205. What is meant by the entry in the remarks section of this Surface Aviation Weather Report for BOI?

BOI SA 1854 -X M7 OVC1 1/2R+F 990/63/61/3205/980/RF2 RB12

A—Runway fog, visibility 2 miles; base of the rainclouds 1,200 feet.
B—Rain and fog obscuring 2/10 of the sky; rain began 12 minutes before the hour.
C—Rain and fog obscuring 2/10 of the sky; rain began at 1812.

4206. (Refer to Figure 4.) The Weather Depiction Chart indicates that northern Illinois and southern Wisconsin are reporting

A—marginal VFR conditions due to reduced visibility in drizzle and fog.
B—low IFR conditions due to ceilings below 500 feet with drizzle.
C—IFR conditions due to overcast ceilings less than 1,000 feet with reduced visibility in rain and rain showers.

4207. (Refer to Figure 4.) The Weather Depiction Chart indicates the heaviest precipitation along the front is occurring in

A—Missouri.
B—Illinois.
C—Kansas.

4208. (Refer to Figure 4.) The Weather Depiction Chart indicates that the coastal sections of Texas and Louisiana are reporting

A—all ceilings at or above 20,000 feet with visibilities of 20 miles or more.
B—marginal VFR conditions due to broken ceilings of 2,000 feet.
C—VFR conditions with scattered clouds at 2,000 feet and higher cirroform.

4209. The Surface Analysis Chart depicts

A—actual pressure systems, frontal locations, cloud tops, and precipitation at the time shown on the chart.
B—frontal locations and expected movement, pressure centers, cloud coverage, and obstructions to vision at the time of chart transmission.
C—actual frontal positions, pressure patterns, temperature, dewpoint, wind, weather, and obstructions to vision at the valid time of the chart.

4210. A pilot reporting turbulence that momentarily causes slight, erratic changes in altitude and/or attitude should report it as

A—light turbulence.
B—moderate turbulence.
C—light chop.

4211. The Low-Level Significant Weather Prognostic Chart depicts weather conditions

A—that are forecast to exist at a valid time shown on the chart.
B—as they existed at the time the chart was prepared.
C—that existed at the time shown on the chart which is about 3 hours before the chart is received.

4212. Which meteorological conditions are depicted by a prognostic chart?

A—Conditions existing at the time of the observation.
B—Interpretation of weather conditions for geographical areas between reporting stations.
C—Conditions forecast to exist at a specific time shown on the chart.

4213. (Refer to Figure 5.) What is the meaning of the symbol depicted as used on the U.S. Low-Level Significant Weather Prog Chart?

A—Showery precipitation (e.g. rain showers) embedded in an area of continuous rain covering half or more of the area.
B—Continuous precipitation (e.g. rain) covering half or more of the area.
C—Showery precipitation (e.g. thunderstorms/rain showers) covering half or more of the area.

4214. A prognostic chart depicts the conditions

A—existing at the surface during the past 6 hours.
B—which presently exist from the 1,000-millibar through the 700-millibar level.
C—forecast to exist at a specific time in the future.

4215. What information is provided by a Convective Outlook (AC)?

A—It describes areas of probable severe icing and severe or extreme turbulence during the next 24 hours.
B—It provides prospects of both general and severe thunderstorm activity during the following 24 hours.
C—It indicates areas of probable convective turbulence and the extent of instability in the upper atmosphere (above 500 MB).

4216. (Refer to Figure 18.) A planned low altitude flight from central Oklahoma to western Tennessee at 1800Z is likely to encounter

A—continuous or intermittent rain or rain showers, moderate turbulence, and freezing temperatures below 8,000 feet.
B—a stationary front with intermittent rain showers in central Oklahoma and VFR weather in western Tennessee.
C—showery precipitation covering less than half the area, no turbulence below 18,000 feet, and freezing temperatures above 12,000 feet.

4217. (Refer to Figure 18. SFC-400MB) The 12-Hour Significant Weather Surface Prognosis Chart at 1800Z indicates that northeastern West Virginia will likely experience

A—ceilings 1,000 to 3,000 feet, visibility 3 to 5 miles.
B—thunderstorms and rain showers covering half or more of the area.
C—continuous rain covering less than half of the area.

4218. (Refer to Figure 18.) The 24-Hour Significant Weather Surface Prognosis Chart at 0600Z indicates that southeastern Colorado and western Kansas can expect ceilings

A—less than 1,000 feet and/or visibility less than 3 miles in rain showers and thunderstorms.
B—less than 1,000 feet and/or visibility less than 3 miles, and moderate turbulence below 10,000 feet MSL.
C—less than 1,000 feet and/or visibility less than 3 miles, and moderate turbulence above 10,000 feet MSL.

4219. (Refer to Figure 18.) The chart symbols over the middle of California coastline on the 12-Hour Significant Weather Prognosis 400MB Chart at 1800Z indicate

A—expected ceilings 1,000 to 3,000 feet with visibilities 3 to 5 miles.
B—expected base of moderate turbulent layer to be 10,000 feet MSL.
C—light turbulence expected above 12,000 feet MSL.

4220. Interpret this PIREP.

UA/OVR MRB FL060/SK INTMTLY BL/TB MDT/RM R TURBC INCRS WWD.

A—Ceiling 6,000 feet intermittently below moderate thundershowers; turbulence increasing westward.
B—FL 60,000, intermittently below clouds; moderate rain, turbulence increasing with the wind.
C—At 6,000 feet; intermittently between layers; moderate turbulence; moderate rain; turbulence increasing westward.

4221. (Refer to Figure 7.) What weather conditions are depicted within the area indicated by arrow E?

A—Frequent embedded thunderstorms, less than one-eighth coverage, tops at FL 370.
B—Frequent lightning in thunderstorms at FL 370.
C—Frequent cumulonimbus, five-eighths to eight-eighths coverage, bases below 24,000 feet MSL and tops at 37,000 feet MSL.

4222. (Refer to Figure 7.) What weather conditions are depicted within the area indicated by arrow D?

A—Existing isolated cumulonimbus, tops above 41,000 feet and less than one-fifth coverage.
B—Forecast isolated embedded cumulonimbus, bases below 24,000 feet MSL, tops at 41,000 feet MSL and less than one-eighth coverage.
C—Forecast isolated thunderstorms, tops at FL 410, less than two-fifths coverage.

4223. (Refer to Figure 7.) What weather conditions are depicted within the area indicated by arrow C?

A—Severe CAT forecast within the area outlined by dashes from 32,000 feet MSL to below the lower limit of the chart.
B—Moderate turbulence at FL 320 within the area outlined by dashes.
C—Moderate to severe CAT has been reported at FL 320.

4224. (Refer to Figure 7.) What weather conditions are depicted within the area indicated by arrow B?

A—Light to moderate turbulence at and above 33,000 feet MSL.
B—Moderate to severe turbulence from below 24,000 feet MSL to 33,000 feet MSL.
C—Moderate to severe CAT is forecast to exist at FL 330.

4225. (Refer to Figure 7.) What information is indicated by arrow A?

A—The height of the tropopause in meters above sea level.
B—The height of the existing layer of CAT.
C—The height of the tropopause in hundreds of feet above MSL.

4226. Which weather forecast describes prospects for an area coverage of both severe and general thunderstorms during the following 24 hours?

A—Terminal Forecast.
B—Convective outlook.
C—Severe Weather Watch Bulletin.

4227. Which feature is associated with the tropopause?

A—Absence of wind and turbulent conditions.
B—Absolute upper limit of cloud formation.
C—Abrupt change in temperature lapse rate.

4228. From which primary source should you obtain information regarding the weather expected to exist at your destination at your estimated time of arrival?

A—Weather Depiction Chart.
B—Radar Summary and Weather Depiction Chart.
C—Terminal Forecast.

4229. (Refer to Figure 7.) What weather conditions are depicted within the area indicated by arrow F?

A—Two-eighths to six-eighths coverage, occasional embedded thunderstorms, tops at FL 510.
B—One-eighth to four-eighths coverage, occasional embedded thunderstorms, maximum tops at 51,000 feet MSL.
C—Occasionally embedded cumulonimbus, bases from 18,000 feet to 51,000 feet.

4230. (Refer to Figure 8.) What weather conditions are depicted in the area indicated by arrow A on the Radar Summary Chart?

A—Moderate to strong echoes; echo tops 30,000 feet MSL; line movement toward the northwest.
B—Weak to moderate echoes; average echo bases 30,000 feet MSL; cell movement toward the southeast; rain showers with thunder.
C—Strong to very strong echoes; echo tops 30,000 feet MSL; thunderstorms and rain showers.

4231. (Refer to Figure 8.) What weather conditions are depicted in the area indicated by arrow D on the Radar Summary Chart?

A—Echo tops 4,100 feet MSL, strong to very strong echoes within the smallest contour, and area movement toward the northeast at 50 knots.
B—Intense to extreme echoes within the smallest contour, echo tops 29,000 feet MSL, and cell movement toward the northeast at 50 knots.
C—Strong to very strong echoes within the smallest contour, echo bases 29,000 feet MSL, and cell in northeast Nebraska moving northeast at 50 knots.

4232. (Refer to Figure 8.) What weather conditions are depicted in the area indicated by arrow C on the Radar Summary Chart?

A—Average echo bases 2,800 feet MSL, thundershowers, and intense to extreme echo intensity.
B—Cell movement toward the northwest at 20 knots, intense echoes, and echo bases 28,000 feet MSL.
C—Area movement toward the northeast at 20 knots, strong to very strong echoes, and echo tops 28,000 feet MSL.

4233. (Refer to Figure 8.) What weather conditions are depicted in the area indicated by arrow B on the Radar Summary Chart?

A—Weak echoes, heavy rain showers, area movement toward the southeast.
B—Weak to moderate echoes, rain showers increasing in intensity.
C—Strong echoes, moderate rain showers, no cell movement.

4234. (Refer to Figure 8.) What weather conditions are depicted in the area indicated by arrow E on the Radar Summary Chart?

A—Highest echo tops 30,000 feet MSL, weak to moderate echoes, thunderstorms and rain showers, and cell movement toward northwest at 15 knots.
B—Echo bases 29,000 to 30,000 feet MSL, strong echoes, rain showers increasing in intensity, and area movement toward northwest at 15 knots.
C—Thundershowers decreasing in intensity; area movement toward northwest at 15 knots; echo bases 30,000 feet MSL.

4235. For most effective use of the Radar Summary Chart during preflight planning, a pilot should

A—consult the chart to determine more accurate measurements of freezing levels, cloud cover, and wind conditions between reporting stations.
B—compare it with the charts, reports, and forecasts of a three-dimensional picture of clouds and precipitation.
C—utilize the chart as the only source of information regarding storms and hazardous conditions existing between reporting stations.

4236. (Refer to Figure 8.) What weather conditions are depicted in the area indicated by arrow G on the Radar Summary Chart?

A—Echo bases 10,000 feet MSL; cell movement toward northeast at 15 knots; weak to moderate echoes; rain.
B—Area movement toward northeast at 15 knots; rain decreasing in intensity; echo bases 1,000 feet MSL; strong echoes.
C—Strong to very strong echoes; area movement toward northeast at 15 knots; echo tops 10,000 feet MSL; light rain.

4237. (Refer to Figure 8.) What weather conditions are depicted in the area indicated by arrow F on the Radar Summary Chart?

A—Line of echoes; thunderstorms; highest echo tops 46,000 feet MSL; no line movement indicated.
B—Echo bases vary from 15,000 feet to 46,000 feet MSL; thunderstorms increasing in intensity; line of echoes moving rapidly toward the north.
C—Line of severe thunderstorms moving from south to north; echo bases vary from 4,400 feet to 4,600 feet MSL; extreme echoes.

4238. Hazardous wind shear is commonly encountered near the ground

A—during periods when the wind velocity is stronger than 35 knots.
B—during periods when the wind velocity is stronger than 35 knots and near mountain valleys.
C—during periods of strong temperature inversion and near thunderstorms.

4239. (Refer to Figure 9.) The Severe Weather Outlook Chart which is used primarily for advance planning, provides what information?

A—An 18-hour categorical outlook with a 48-hour valid time for severe weather watch, thunderstorm lines, and areas of expected tornado activity.
B—A preliminary 12-hour outlook for severe thunderstorm activity and probable convective turbulence.
C—A preliminary 24-hour severe weather outlook for general and severe thunderstorm activity, tornadoes, and watch areas.

4240. (Refer to Figure 9.) What is the significance of the annotations "MDT" and "SLGT" on the 00Z-12Z panel of the AC?

A—Moderate risk area, surrounded by a slight risk area, of possible severe turbulence.
B—Slight to moderate chance of low-level wind shear.
C—Moderate risk of severe thunderstorms surrounded by a slight risk area.

4241. If you hear a SIGMET alert, how can you obtain the information in the SIGMET?

A—ATC will announce the hazard and advise you when to listen to an FSS broadcast.
B—Contact a weather watch station.
C—Contact the nearest FSS and ascertain whether the advisory is pertinent to your flight.

4242. (Refer to Figure 10.) The symbol on the TROP WIND SHEAR PROG indicated by arrow A represents the

A—wind direction at the tropopause (300°).
B—FL of the tropopause.
C—height of maximum wind shear (30,000 feet).

4243. (Refer to Figure 11.) What is the pressure in millibars on the tropopause pressure, temperature, and winds chart for ABQ?

A—150.
B—172.
C—200.

4244. (Refer to Figure 11.) What is the approximate wind direction and velocity at ABQ?

A—220°/200 knots.
B—060°/61 knots.
C—240°/115 knots.

4245. (Refer to Figure 10.) The reference to 6K (see arrow B) indicates

A—vertical wind shear per 1,000 feet.
B—wind velocity at 34,000 feet.
C—vertical wind velocity at 34,000 feet.

4246. (Refer to Figure 12.) What is the approximate wind direction and velocity at 34,000 feet (see arrow C)?

A—290°/50 knots.
B—330°/50 knots.
C—090°/48 knots.

4247. (Refer to Figure 12.) The wind direction and velocity on the Observed Winds Aloft Chart (see arrow A) is indicated from the

A—northeast at 35 knots.
B—northwest at 47 knots.
C—southwest at 35 knots.

4248. (Refer to Figure 9.) The crosshatched area on the Severe Weather Outlook Chart indicates

A—moderate risk of thunderstorms.
B—forecast risk of tornados.
C—forecast risk of heavy thunderstorms.

4249. (Refer to Figure 12.) What is the approximate wind direction and velocity at CVG at 34,000 feet (see arrow A)?

A—040°/35 knots.
B—097°/40 knots.
C—230°/35 knots.

4250. (Refer to Figure 12, arrow B.) What is the approximate wind direction and velocity at BOI (see arrow B)?

A—270°/55 knots.
B—250°/95 knots.
C—080°/95 knots.

4251. What is the expected duration of an individual microburst?

A—Two minutes with maximum winds lasting approximately 1 minute.
B—One microburst may continue for as long as 2 to 4 hours.
C—Seldom longer than 15 minutes from the time the burst strikes the ground until dissipation.

4252. Maximum downdrafts in a microburst encounter may be as strong as

A—8,000 feet per minute.
B—7,000 feet per minute.
C—6,000 feet per minute.

4253. An aircraft that encounters a headwind of 45 knots, within a microburst, may expect a total shear across the microburst of

A—40 knots.
B—80 knots.
C—90 knots.

4254. (Refer to Figure 13.) If involved in a microburst encounter, in which aircraft positions will the most severe downdraft occur?

A—4 and 5.
B—2 and 3.
C—3 and 4.

4255. (Refer to Figure 13.) When penetrating a microburst, which aircraft will experience an increase in performance without a change in pitch or power?

A—3.
B—2.
C—1.

4256. (Refer to Figure 13.) The aircraft in position 3 will experience which effect in a microburst encounter?

A—Decreasing headwind.
B—Increasing tailwind.
C—Strong downdraft.

4257. (Refer to Figure 13.) What effect will a microburst encounter have upon the aircraft in position 4?

A—Strong tailwind.
B—Strong updraft.
C—Significant performance increase.

4258. (Refer to Figure 13.) How will the aircraft in position 4 be affected by a microburst encounter?

A—Performance increasing with a tailwind and updraft.
B—Performance decreasing with a tailwind and downdraft.
C—Performance decreasing with a headwind and downdraft.

4259. (Refer to Figures 21 and 21A, 22 and 22A, 23 and 23A, 24, 25 and 25A, and 26 and 26A.) After departing GJT and arriving at Durango Co., La Plata Co. Airport, you are unable to land because of weather. How long can you hold over DRO before departing for return flight to the alternate, Grand Junction Co., Walker Field Airport?

Total useable fuel on board, 68 gallons.
Average fuel consumption 15 GPH.
Wind and velocity at 16,000, 2308-16°.

A—1 hour 33 minutes.
B—1 hour 37 minutes.
C—1 hour 42 minutes.

4260. (Refer to FD excerpt below, and use the wind entry closest to the flight planned altitude.) Determine the time to be entered in block 10 of the flight from GJT to DRO.

Route of flight .... Figure 21
Flight log & MAG VAR .... Figure 22
En route chart .... Figure 24

FT	12,000	18,000
FNM	2408-05	2208-21

A—1 hour 08 minutes.
B—1 hour 03 minutes.
C—58 minutes.

4261. (Refer to Figure 24.) Proceeding southbound on V187, (vicinity of Cortez VOR) contact is lost with Denver Center. What frequency should be used to contact Denver Center?

A—133.425.
B—108.4.
C—122.2.

4262. (Refer to Figures 22 and 24.) For planning purposes, what would the highest MEA be on the flight planned between Grand Junction, Walker Airport, and Durango, La Plata Co. Airport?

A—12,000 feet.
B—15,000 feet.
C—16,000 feet.

4263. (Refer to Figure 24.) At what point should a VOR changeover be made from JNC VOR to MANCA intersection southbound on V187?

A—36 NM south of JNC.
B—52 NM south of JNC.
C—74 NM south of JNC.

4264. (Refer to Figure 24.) What is the MOCA between JNC and MANCA intersection on V187?

A—10,900 feet MSL.
B—12,000 feet MSL.
C—13,700 feet MSL.

4265. (Refer to Figures 21, 22, and 24.) What fuel would be consumed on the flight between Grand Junction Co. and Durango, Co. if the average fuel consumption is 15 GPH.

A—17 gallons.
B—20 gallons.
C—25 gallons.

4266. (Refer to Figure 27.) What aircraft equipment code should be entered in block 3 of the flight plan?

A—T.
B—U.
C—A.

4267. (Refer to Figures 27 and 28.) What CAS must be used to maintain the filed TAS at the flight planned altitude if the outside air temperature is -5° C?

A—134 KCAS.
B—139 KCAS.
C—142 KCAS.

4268. (Refer to the FD excerpt below, and use the wind entry closest to the flight planned altitude.) Determine the time to be entered in block 10 of the flight plan.

Route of flight .... Figures 27, 28, 29 and 29A, 30, and 31
Flight log & MAG VAR .... Figure 28
GNATS ONE DEPARTURE
and Excerpt from AFD .... Figure 30

FT	3000	6000	9000
OTH	0507	2006+03	2215-05

A—1 hour 10 minutes.
B—1 hour 15 minutes.
C—1 hour 20 minutes.

4269. (Refer to Figure 30.) During the arc portion of the SID (GNATS1.MOURN), a left crosswind is encountered. Where should the bearing pointer of an RMI be referenced relative to the wingtip to compensate for wind drift and maintain the 15 DME arc?

A—Behind the right wingtip reference point.
B—On the right wingtip reference point.
C—Behind the left wingtip reference point.

4270. (Refer to Figure 30.) Using an average groundspeed of 120 knots, what minimum rate of climb must be maintained to meet the required climb rate (feet per NM) to 4,100 feet as specified on the SID?

A—400 feet per minute.
B—500 feet per minute.
C—800 feet per minute.

4271. (Refer to Figure 30.) Which restriction to the use of the OED VORTAC would be applicable?

A—R-333 beyond 30 NM below 6,500 feet.
B—R-210 beyond 35 NM below 8,500 feet.
C—R-251 within 15 NM below 6,100 feet.

4272. (Refer to Figures 30 and 30A.) What is your position relative to GNATS intersection and the SID departure routing?

A—On departure course and past GNATS.
B—Right of departure course and past GNATS.
C—Left of departure course and have not passed GNATS.

4273. (Refer to Figures 27 and 30.) To which maximum service volume distance from the OED VORTAC should you expect to receive adequate signal coverage for navigation at the flight planned altitude?

A—100 NM.
B—80 NM.
C—40 NM.

4274. (Refer to Figures 29 and 29A.) What is the TDZ elevation for RWY 16 on Eugene/Mahlon Sweet Field?

A—363 feet MSL.
B—365 feet MSL.
C—396 feet MSL.

4275. (Refer to Figure 29.) What are the hours of operation (local standard time) of the control tower at Eugene/Mahlon Sweet Field?

A—0800 - 2300.
B—0600 - 0000.
C—0700 - 0100.

4276. (Refer to Figure 29.) Using a groundspeed of 90 knots on the ILS final approach course, what rate of descent should be used as a reference to maintain the ILS glide slope?

A—415 feet per minute.
B—480 feet per minute.
C—555 feet per minute.

4277. (Refer to Figure 32.) What aircraft equipment code should be entered in block 3 of the flight plan?

A—A.
B—C.
C—R.

4278. (Refer to Figure 32.) What CAS must be used to maintain the filed TAS at the flight planned altitude if the outside air temperature is +8° C?

A—154 KCAS.
B—157 KCAS.
C—163 KCAS.

4279. (Refer to the FD excerpt below, and use the wind entry closest to the flight planned altitude.) Determine the time to be entered in block 10 of the flight plan.

Route of flight .... Figures 32,33,34,35,36
Flight log & MAG VAR .... Figure 33
RNAV RWY 33 & Excerpt from AFD .... Figure 36

FT	3000	6000	9000	12000
DAL	2027	2239+13	2240+08	2248+05

A—1 hour 35 minutes.
B—1 hour 41 minutes.
C—1 hour 46 minutes.

4280. (Refer to Figure 34.) At which altitude and location on V573 would you expect the navigational signal of the HOT VOR/DME to be unreliable?

A—3,000 feet at APINE intersection.
B—2,600 feet at MARKI intersection.
C—4,000 feet at ELMMO intersection.

4281. (Refer to Figures 35 and 37.) What is your position relative to the CONNY intersection on the BUJ.BUJ3 transition?

A—Left of the TXK R-272 and approaching the BUJ R-059°.
B—Left of the TXK R-266 and past the BUJ R-065.
C—Right of the TXK R-270 and approaching the BUJ R-245.

4282. (Refer to Figures 36A and 36B.) Under which condition should the missed approach procedure for the RNAV RWY 33 approach be initiated?

A—Immediately upon reaching the 5.0 DME from the FAF.
B—When passage of the MAP waypoint is shown on the ambiguity indicator.
C—After the MDA is reached and 1.8 DME fix from the MAP waypoint.

4283. (Refer to Figures 32, 36, 36A, and 36B.) What is the MDA and visibility criteria respectively for the S 33 approach procedure?

A—1,240 feet MSL; 1 SM.
B—1,280 feet MSL; 1 and 1/4 SM.
C—1,300 feet MSL; 1 SM.

4284. How can a pilot determine if a Global Positioning System (GPS) installed in an aircraft is approved for IFR enroute and IFR approaches?

A—Flight manual supplement.
B—GPS operator's manual.
C—Aircraft owner's handbook.

4285. (Refer to Figures 36A and 36B.) What is the minimum number of waypoints required for the complete RNAV RWY 33 approach procedure including the IAF's and missed approach procedure?

A—One waypoint.
B—Two waypoints.
C—Three waypoints.

4286. (Refer to Figures 35 and 35A.) At which point does the BUJ.BUJ3 arrival begin?

A—At the TXK VORTAC.
B—At BOGAR intersection.
C—At the BUJ VORTAC.

4287. (Refer to Figure 34.) For planning purposes, what is the highest useable altitude for an IFR flight on V573 from the HOT VORTAC to the TXK VORTAC?

A—16,000 feet MSL.
B—14,500 feet MSL.
C—13,999 feet MSL.

4288. (Refer to Figure 38.) What aircraft equipment code should be entered in block 3 of the flight plan?

A—C.
B—R.
C—A.

4289. (Refer to Figure 38.) What CAS must be used to maintain the filed TAS at the flight planned altitude if the outside air temperature is +05° C?

A—129 KCAS.
B—133 KCAS.
C—139 KCAS.

4290. (Refer to the FD excerpt below, and use the wind entry closest to the flight planned altitude.) Determine the time to be entered in block 10 of the flight plan.

Route of flight .... Figures 38, 39, and 40
Flight log & MAG VAR .... Figure 39
ACTON TWO ARRIVAL .... Figure 41

FT	3000	6000	9000	12000
ABI		2033+13	2141+09	2142+05

A—1 hour 24 minutes.
B—1 hour 26 minutes.
C—1 hour 31 minutes.

4291. (Refer to Figure 40.) For planning purposes, what is the highest useable altitude for an IFR flight on V16 between the BGS and ABI VORTACs?

A—17,000 feet MSL.
B—18,000 feet MSL.
C—6,500 feet MSL.

4292. (Refer to Figures 41 and 41A.) At which point does the AQN.AQN2 arrival begin?

A—ABI VORTAC.
B—ACTON VORTAC.
C—CREEK intersection.

4293. (Refer to Figures 41 and 41A.) Which frequency would you anticipate using to contact Regional Approach Control? (ACTON TWO ARRIVAL).

A—119.05.
B—124.15.
C—125.8.

4294. (Refer to Figures 41 and 41A.) On which heading should you plan to depart CREEK intersection?

A—010°.
B—040°.
C—350°.

4295. (Refer to Figures 41, 42, 42A and 42B.) Which frequency should you expect to use for Regional Approach Control, control tower, and ground control respectively at DFW?

A—119.05; 126.55; 121.65.
B—119.05; 124.15; 121.8.
C—125.8; 124.15; 121.8.

4296. (Refer to Figures 42A and 42B.) Which navigational information and services would be available to the pilot when using the localizer frequency?

A—Localizer and glide slope, DME, TACAN with no voice capability.
B—Localizer information only, ATIS and DME are available.
C—Localizer and glide slope, DME, and no voice capability.

4297. (Refer to Figures 42, 42A and 42B.) What is the difference in elevation (in feet MSL) between the airport elevation and the TDZE for RWY 36L?

A—15 feet.
B—18 feet.
C—22 feet.

4298. (Refer to Figures 42A and 42B.) What rate of descent should you plan to use initially to establish the glidepath for the ILS RWY 36L approach? (Use 120 knots groundspeed.)

A—425 feet per minute.
B—530 feet per minute.
C—635 feet per minute.

4299. (Refer to Figures 42A, 42B and 43.) What is your position relative to CHARR intersection? The aircraft is level at 3,000 feet MSL.

A—Right of the localizer course approaching CHARR intersection and approaching the glide slope.
B—Left of the localizer course approaching CHARR intersection and below the glide slope.
C—Right of the localizer course, past CHARR intersection and above the glide slope.

4300. (Refer to Figure 44.) What aircraft equipment code should be entered in block 3 of the flight plan?

A—A.
B—C.
C—R.

4301. (Refer to Figure 44.) What CAS must be used to maintain the filed TAS at the flight planned altitude if the outside air temperature is +5° C?

A—147 KCAS.
B—150 KCAS.
C—154 KCAS.

4302. Determine the time to be entered in block 10 of the flight plan. (Refer to the FD excerpt below, and use the wind entry closest to the flight planned altitude.)

Route of flight .... Figures 44, 45, 46, and 47
Flight log & MAG VAR .... Figure 45
GROMO TWO DEPARTURE
and Excerpt from AFD .... Figure 46

FT	3000	6000	9000	12000
YKM	1615	1926+12	2032+08	2035+05

A—54 minutes.
B—1 hour 02 minutes.
C—1 hour 07 minutes.

4303. (Refer to Figure 46.) Using an average groundspeed of 140 knots, what minimum indicated rate of climb must be maintained to meet the required climb rate (feet per NM) to 6,300 feet as specified on the SID?

A—350 feet per minute.
B—583 feet per minute.
C—816 feet per minute.

4304. (Refer to Figures 46 and 48.) What is your position relative to the 9 DME ARC and the 206° radial of the SID?

A—On the 9 DME arc and approaching R-206.
B—Outside the 9 DME arc and past R-206.
C—Inside the 9 DME arc and approaching R-206.

4305. (Refer to Figure 46.) What are the hours of operation (local time) of the ATIS for the Yakima Air Terminal when daylight savings time is in effect?

A—0500 to 2100 local.
B—0600 to 2200 local.
C—0700 to 2300 local.

4306. (Refer to Figures 49 and 49A.) What should be used to establish the MAP on the LORAN RNAV RWY 10 approach at Portland International Airport?

A—6 NM from FAF.
B—3.5 NM from PONCO waypoint.
C—2.5 NM from FAF.

4307. (Refer to Figures 44, 49 and 49A.) What is the MDA and visibility criteria during the LORAN RNAV RWY 10 approach at Portland International?

A—900 MSL; visibility RVR/24.
B—900 MSL; visibility RVR/40.
C—900 MSL; visibility 1 and 1/4 SM.

4308. (Refer to Figures 49 and 49A.) How many waypoints are required to fly the complete LORAN RNAV RWY 10 (Portland International) approach procedure, including the missed approach procedure?

A—2.
B—3.
C—4.

4309. (Refer to Figures 49 and 49A.) What is the TDZE for RWY 10R at Portland International Airport?

A—900 feet MSL.
B—26 feet MSL.
C—20 feet MSL.

4310. (Refer to Figures 49 and 49A.) The aircraft used on this flight has vertical guidance capability with the RNAV equipment. What is the angle of the glidepath from the FAF to the 2.5 DME from the MAP waypoint?

A—2.88°.
B—3.00°.
C—3.20°.

4311. (Refer to Figures 49 and 49A.) At which point or fix may descent be initiated to the MDA on the LORAN RNAV RWY 10R approach (Portland International)?

A—WETTR waypoint.
B—3.5 NM from PONCO waypoint.
C—6 NM from the MAP waypoint.

4312. (Refer to Figure 50.) What aircraft equipment code should be entered in block 3 of the flight plan?

A—R.
B—T.
C—U.

4313. (Refer to Figure 50.) What CAS must be used to maintain the filed TAS at the flight planned altitude? (Temperature 0° C.)

A—136 KCAS.
B—140 KCAS.
C—147 KCAS.

4314. Determine the time to be entered in block 10 of the flight plan. (Refer to the FD excerpt below, and use the wind entry closest to the flight planned altitude.)

Route of flight .... Figures 50, 51, 52, and 53
Flight log and MAG VAR .... Figure 51
HABUT ONE DEPARTURE
and Excerpt from AFD .... Figure 52

FT	3000	6000	9000
SBA	0610	2115+05	2525+00

A—43 minutes.
B—46 minutes.
C—51 minutes.

4315. (Refer to Figures 52 and 54.) What is the aircraft's position relative to the HABUT intersection? (The VOR-2 is tuned to 116.5.)

A—South of the localizer and past the GVO R-163.
B—North of the localizer and approaching the GVO R-163.
C—South of the localizer and approaching the GVO R-163.

4316. (Refer to Figure 52.) Using an average groundspeed of 100 knots, what minimum rate of climb would meet the required minimum climb rate per NM as specified by the SID?

A—425 feet per minute.
B—580 feet per minute.
C—642 feet per minute.

4317. (Refer to Figure 53.) Where is the VOR COP on V27 between the GVO and MQO VORTACs?

A—20 DME from GVO VORTAC.
B—20 DME from MQO VORTAC.
C—30 DME from SBA VORTAC.

4318. (Refer to Figure 53.) What service is indicated by the solid square in the radio aids to navigation box for PRB VORTAC?

A—VOR with TACAN compatible DME.
B—Availability of HIWAS.
C—En Route Flight Advisory Service available.

4319. (Refer to Figures 55 and 55A.) Using an average groundspeed of 90 knots, what constant rate of descent from 2,400 feet MSL at the 6 DME fix would enable the aircraft to arrive at 2,000 feet MSL at the FAF?

A—200 feet per minute.
B—400 feet per minute.
C—600 feet per minute.

4320. (Refer to Figures 55 and 55A.) As a guide in making range corrections, how many degrees of relative bearing change should be used for each one-half mile deviation from the desired arc?

A—2° to 3°.
B—5° maximum.
C—10° to 20°.

4321. (Refer to Figures 55 and 55A.) Under which condition should a missed approach procedure be initiated if the runway environment (Paso Robles Municipal Airport) is not in sight?

A—After descending to 1,440 feet MSL.
B—After descent to 1,440 feet or reaching the 1 NM DME, whichever occurs first.
C—When you reach the established missed approach point and determine the visibility is less than 1 mile.

4322. (Refer to Figure 56.) What aircraft equipment code should be entered in block 3 of the flight plan?

A—U.
B—A.
C—R.

4323. (Refer to Figure 56.) What CAS should be used to obtain the filed TAS at the flight planned altitude if the outside air temperature is +5° ?

A—97 KCAS.
B—101 KCAS.
C—103 KCAS.

4324. Determine the time to be entered in block 10 of the flight plan. (Refer to the FD excerpt below, and use the wind entry closest to the flight planned altitude.)

Route of flight .... Figures 56, 57, 58, and 59
Flight log & MAG VAR .... Figure 57
Excerpt from AFD (CLL) .... Figure 58

FT	3000	6000	9000
HOU	0507	1015+05	2225+00

A—1 hour 06 minutes.
B—1 hour 10 minutes.
C—1 hour 14 minutes.

4325. (Refer to Figure 58.) On which frequencies could you receive the College Station FSS?

A—122.65, 122.2, 122.1, 113.3.
B—122.65, 122.2.
C—118.5, 122.65, 122.2.

4326. (Refer to Figure 58.) Which indications on the VOR receivers and DME at the Easterwood Field VOR receiver checkpoint would meet the regulatory requirement for this flight?

	VOR No. 1	TO/FROM	VOR No. 2	TO/FROM	DME
A—	097°	FROM	101°	FROM	3.3
B—	097°	TO	096°	TO	3.2
C—	277°	FROM	280°	FROM	3.3

4327. (Refer to Figures 59 and 60.) What are the operating hours (local standard time) of the Houston EFAS?

A—0600 to 2200.
B—0700 to 2300.
C—1800 to 1000.

4328. (Refer to Figures 56, 60A and 60B.) To which value may the visibility criteria be reduced, if any, for the S-ILS 4 approach?

A—RVR 20.
B—RVR 16.
C—RVR 12.

4329. (Refer to Figure 59.) Unless otherwise authorized by ATC, what is the minimum equipment for navigation of helicopters on an IFR cross-country flight when in the immediate vicinity of the HUMBLE VORTAC?

A—VOR receiver, transponder with 3/A 4096 code capability and two-way communications.
B—Transponder with 3/A 4096 code capability and two-way communications.
C—VOR (or TACAN) and two-way communications.

4330. (Refer to Figures 60A and 60B.) What is the DH and visibility criteria for a helicopter for the ILS RWY 4 approach if the MM is inoperative?

A—246/24.
B—246/12.
C—460/12.

4331. (Refer to Figures 60A, 60B, and 61.) What is your position relative to the PLATS intersection, glide slope, and the localizer course?

A—Past PLATS, below the glide slope, and right of the localizer course.
B—Approaching PLATS, above the glide slope, and left of the localizer course.
C—Past PLATS, above the glide slope, and right of the localizer course.

4332. (Refer to Figures 60A and 60B.) What is the elevation of the TDZE for RWY 4?

A—70 feet MSL.
B—54 feet MSL.
C—46 feet MSL.

4333. (Refer to Figure 62.) What aircraft equipment code should be entered in block 3 of the flight plan?

A—U.
B—A.
C—R.

4334. (Refer to Figure 62.) What CAS should be used to maintain the filed TAS at the flight planned altitude if the outside air temperature is +15° C?

A—91 KCAS.
B—96 KCAS.
C—101 KCAS.

4335. Determine the time to be entered in block 10 of the flight plan. (Refer to the FD excerpt below, and use the wind entry closest to the flight planned altitude.)

Route of flight .... Figures 62, 63, 64, and 65
Flight log & MAG VAR .... Figure 63
Excerpt from AFD (LFT) .... Figure 64

FT	3000	6000	9000
MSY	1422	1626+15	1630+10

A—56 minutes.
B—1 hour 02 minutes.
C—1 hour 07 minutes.

4336. (Refer to Figure 65.) Which point would be the appropriate VOR COP on V552 from the LFT to the TBD VORTACs?

A—CLYNT intersection.
B—HATCH intersection.
C—34 DME from the LFT VORTAC.

4337. (Refer to Figure 64.) The course deviation indicator (CDI) are centered. Which indications on the No. 1 and No. 2 VOR receivers over the Lafayette Regional Airport would meet the requirements for the VOR receiver check?

	VOR No. 1	TO/FROM	VOR No. 2	TO/FROM
A—	162°	TO	346°	FROM
B—	160°	FROM	162°	FROM
C—	341°	FROM	330°	FROM

4338. (Refer to Figures 65 and 66.) What is your position relative to GRICE intersection?

A—Right of V552 and approaching GRICE intersection.
B—Right of V552 and past GRICE intersection.
C—Left of V552 and approaching GRICE intersection.

4339. (Refer to Figures 65 and 67.) What is the significance of the symbol at GRICE intersection?

A—It signifies a localizer-only approach is available at Harry P. Williams Memorial.
B—The localizer has an ATC function in addition to course guidance.
C—GRICE intersection also serves as the FAF for the ILS approach procedure to Harry P. Williams Memorial.

4340. (Refer to Figure 68 or 68A.) Upon which maximum airspeed is the COPTER VOR/DME 117° approach category based?

A—80 knots.
B—90 knots.
C—100 knots.

4341. (Refer to Figure 68 or 68A.) What minimum visibility must exist to execute the COPTER VOR/DME 117° approach procedure?

A—3/4 mile.
B—1/2 mile.
C—1/4 mile.

4342. (Refer to Figure 68 or 68A.) What is the VASI approach slope angle for RWY 12 at Houma-Terrebonne?

A—3.0°.
B—2.8°.
C—2.5°.

4343. (Refer to Figure 68 or 68A.) What would be the approach minimums if you must use the Moisant Field altimeter settings?

A—440-1.
B—480 and 1/2.
C—580 and 1/2.

4344. (Refer to Figure 69.) What aircraft equipment code should be entered in block 3 of the flight plan?

A—A.
B—B.
C—U.

4345. (Refer to Figure 69.) What CAS should be used to maintain the filed TAS if the outside air temperature is +05° C?

A—119 KCAS.
B—124 KCAS.
C—126 KCAS.

4346. Determine the time to be entered in block 10 of the flight plan. (Refer to the FD excerpt below, and use the wind entry closest to the flight planned altitude.)

Route of flight .... Figures 69, 70, and 71
Flight log and MAG VAR .... Figure 70
JUDDS TWO ARRIVAL
and Excerpt from AFD .... Figures 72 and 72A

FT	3000	6000	9000
BDL	3320	3425+05	3430+00

A—1 hour 14 minutes.
B—58 minutes.
C—50 minutes.

4347. (Refer to Figures 71 and 71A.) What is your position relative to the Flosi intersection Northbound on V213?

A—West of V213 and approaching the Flosi intersection.
B—East of V213 and approaching the Flosi intersection.
C—West of V213 and past the Flosi intersection.

4348. (Refer to Figures 70 and 71.) Which VORTAC along the proposed route of flight could provide HIWAS information?

A—SPARTA VORTAC.
B—HUGUENOT VORTAC.
C—KINGSTON VORTAC.

4349. (Refer to Figure 72 or 72A.) At which location or condition does the IGN.JUDDS2 arrival begin?

A—JUDDS intersection.
B—IGN VORTAC.
C—BRISS intersection.

4350. (Refer to Figure 72.) How many precision approach procedures are published for Bradley International Airport?

A—One.
B—Three.
C—Four.

4351. (Refer to Figure 73 or 73A.) What is the minimum altitude at which you should intercept the glide slope on the ILS RWY 6 approach procedure?

A—3,000 feet MSL.
B—1,800 feet MSL.
C—1,690 feet MSL.

4352. (Refer to Figure 73 or 73A.) At which indication or occurrence should you initiate the published missed approach procedure for the ILS RWY 6 approach provided the runway environment is not in sight?

A—When reaching 374 feet MSL indicated altitude.
B—When 3 minutes (at 90 knots groundspeed) have expired or reaching 374 feet MSL, whichever occurs first.
C—Upon reaching 374 feet AGL.

4353. (Refer to Figure 73 or 73A.) Which sequence of marker beacon indicator lights, and their respective codes, will you receive on the ILS RWY 6 approach procedure to the MAP?

A—Blue—alternate dots and dashes; amber—dashes.
B—Amber—alternate dots and dashes; blue—dashes.
C—Blue—dashes; amber—alternate dots and dashes.

4354. (Refer to Figure 73 or 73A.) Using an average groundspeed of 90 knots on the final approach segment, what rate of descent should be used initially to establish the glidepath for the ILS RWY 6 approach procedure?

A—395 feet per minute.
B—480 feet per minute.
C—555 feet per minute.

4355. (Refer to Figure 73 or 73A.) What is the touchdown zone elevation for RWY 6?

A—174 feet MSL.
B—200 feet AGL.
C—270 feet MSL.

4356. (Refer to Figure 73 or 73A.) After passing the OM, Bradley Approach Control advises you that the MM on the ILS RWY 6 approach is inoperative. Under these circumstances, what adjustments, if any, are required to be made to the DH and visibility?

A—DH 424/24.
B—No adjustments are required.
C—DH 374/24.

4357. (Refer to Figure 73 or 73A.) Which runway and landing environment lighting is available for approach and landing on RWY 6 at Bradley International?

A—HIRL, REIL, and VASI.
B—HIRL and VASI.
C—ALSF2 and HIRL.

4358. (Refer to Figure 74.) What aircraft equipment code should be entered in block 3 of the flight plan?

A—T.
B—U.
C—A.

4359. (Refer to Figure 74.) What CAS should be used to maintain the filed TAS at the flight planned altitude if the outside air temperature is +5° C?

A—129 KCAS.
B—133 KCAS.
C—139 KCAS.

4360. Determine the time to be entered in block 10 of the flight plan. (Refer to the FD excerpt below, and use the wind entry closest to the flight planned altitude.)

Route of flight .... Figures 74, 75, 76, 77, and 78
Flight log & MAG VAR .... Figure 75
VOR indications
and Excerpts from AFD .... Figure 76

FT	6000	9000	12000	18000
BIL	2414	2422+11	2324+05	2126-11

A—1 hour 15 minutes.
B—1 hour 20 minutes.
C—1 hour 25 minutes.

4361. (Refer to Figure 77.) At which point does the basic SID terminate?

A—When Helena Departure Control establishes radar contact.
B—At STAKK intersection.
C—Over the BOZEMAN VOR.

4362. (Refer to Figure 76.) Which indication would be an acceptable accuracy check of both VOR receivers when the airplane is located on the VOR receiver checkpoint at the Helena Regional Airport?

A—A.
B—B.
C—C.

4363. (Refer to Figure 77.) At which minimum altitude should you cross the STAKK intersection?

A—6,500 feet MSL.
B—1,400 feet MSL.
C—10,200 feet MSL.

4364. (Refer to Figure 77.) Using an average groundspeed of 140 knots, what minimum rate of climb would meet the required minimum climb rate per NM as specified on the SID?

A—350 feet per minute.
B—475 feet per minute.
C—700 feet per minute.

4365. (Refer to Figures 76 and 78.) Which en route low altitude navigation chart would cover the proposed routing at the BOZEMAN VORTAC?

A—L-2.
B—L-7.
C—L-9.

4366. (Refer to Figure 78.) What is the maximum altitude that you may flight plan an IFR flight on V-86 EASTBOUND between BOZEMAN and BILLINGS VORTACs?

A—14,500 feet MSL.
B—17,000 feet MSL.
C—18,000 feet MSL.

4367. (Refer to Figures 78 and 79.) What is your position relative to the VOR COP southeast bound on V86 between the BOZEMAN and LIVINGSTON VORTACs? The No. 1 VOR is tuned to 116.1 and the No. 2 VOR is tuned to 112.2.

A—Past the LVM R-246 and west of the BZN R-110.
B—Approaching the LVM R-246 and west of the BZN R-110.
C—Past the LVM R-246 and east of the BZN R-110.

4368. (Refer to Figures 74 and 80.) Which aircraft approach category should be used for a circling approach for a landing on RWY 25?

A—A.
B—B.
C—C.

4369. (Refer to Figure 80.) How many initial approach fixes serve the VOR/DME RWY 27R (Billings Logan) approach procedure?

A—Three.
B—Four.
C—Five.

4370. (Refer to Figure 78.) What is the minimum crossing altitude over the BOZEMAN VORTAC for a flight southeast bound on V86?

A—8,500 feet MSL.
B—9,300 feet MSL.
C—9,700 feet MSL.

4371. (Refer to Figure 80.) What is the TDZE for landing on RWY 27R?

A—3,649 feet MSL.
B—3,514 feet MSL.
C—3,450 feet MSL.

4372. What is the maximum tolerance allowed for an operational VOR equipment check when using a VOT?

A—Plus or minus 4°.
B—Plus or minus 6°.
C—Plus or minus 8°.

4373. When is a pilot on an IFR flight plan responsible for avoiding other aircraft?

A—At all times when not in radar contact with ATC.
B—When weather conditions permit, regardless of whether operating under IFR or VFR.
C—Only when advised by ATC.

4374. You are flying an airship under an IFR flight plan and experience two-way communications radio failure while in VFR conditions. In this situation, you should continue your flight under

A—VFR and land as soon as practicable.
B—VFR and proceed to your flight plan destination.
C—IFR and maintain the last assigned route and altitude to your flight plan destination.

4375. The aircraft's transponder fails during flight within Class D airspace.

A—The pilot should immediately request clearance to depart the Class D airspace.
B—No deviation is required because a transponder is not required in Class D airspace.
C—Pilot must immediately request priority handling to proceed to destination.

4376. When using VOT to make a VOR receiver check, the CDI should be centered and the OBS should indicate that the aircraft is on the

A—090 radial.
B—180 radial.
C—360 radial.

4377. How should the pilot make a VOR receiver check when the aircraft is located on the designated checkpoint on the airport surface?

A—Set the OBS on 180° plus or minus 4°; the CDI should center with a FROM indication.
B—Set the OBS on the designated radial. The CDI must center within plus or minus 4° of that radial with a FROM indication.
C—With the aircraft headed directly toward the VOR and the OBS set to 000°, the CDI should center within plus or minus 4° of that radial with a TO indication.

4378. When the CDI needle is centered during an airborne VOR check, the omnibearing selector and the TO/FROM indicator should read

A—within 4° of the selected radial.
B—within 6° of the selected radial.
C—0° TO, only if you are due south of the VOR.

4379. What does declaring "minimum fuel" to ATC imply?

A—Traffic priority is needed to the destination airport.
B—Emergency handling is required to the nearest useable airport.
C—Merely an advisory that indicates an emergency situation is possible should any undue delay occur.

4380. When ATC has not imposed any climb or descent restrictions and aircraft are within 1,000 feet of assigned altitude, pilots should attempt to both climb and descend at a rate of between

A—500 feet per minute and 1,000 feet per minute.
B—500 feet per minute and 1,500 feet per minute.
C—1,000 feet per minute and 2,000 feet per minute.

4381. During an IFR flight in IMC, a distress condition is encountered, (fire, mechanical, or structural failure). The pilot should

A—not hesitate to declare an emergency and obtain an amended clearance.
B—wait until the situation is immediately perilous before declaring an emergency.
C—contact ATC and advise that an urgency condition exists and request priority consideration.

4382. (Refer to Figure 81.) When checking a dual VOR system by use of a VOT, which illustration indicates the VOR's are satisfactory?

A—1.
B—2.
C—4.

4383. While airborne, what is the maximum permissible variation between the two indicated bearings when checking one VOR system against the other?

A—Plus or minus 4° when set to identical radials of a VOR.
B—4° between the two indicated bearings to a VOR.
C—Plus or minus 6° when set to identical radials of a VOR.

4384. How should the pilot make a VOR receiver check when the airplane is located on the designated checkpoint on the airport surface?

A—With the aircraft headed directly toward the VOR and the OBS set to 000°, the CDI should center within plus or minus 4° of that radial with a TO indication.
B—Set the OBS on the designated radial. The CDI must center within plus or minus 4° of that radial with a FROM indication.
C—Set the OBS on 180° plus or minus 4°; the CDI should center with a FROM indication.

4385. (Refer to Figure 82.) Which is an acceptable range of accuracy when performing an operational check of dual VOR's using one system against the other?

A—1.
B—2.
C—4.

4386. Where can the VOT frequency for a particular airport be found?

A—On the IAP Chart and in the Airport/Facility Directory.
B—Only in the Airport/Facility Directory.
C—In the Airport/Facility Directory and on the A/G Voice Communication Panel of the En Route Low Altitude Chart.

4387. Which indications are acceptable tolerances when checking both VOR receivers by use of the VOT?

A—360° TO and 003° TO, respectively.
B—001° FROM and 005° FROM, respectively.
C—176° TO and 003° FROM, respectively.

4388. In which publication can the VOR receiver ground checkpoint(s) for a particular airport be found?

A—Airman's Information Manual.
B—En Route Low Altitude Chart.
C—Airport/Facility Directory.

4389. Which is the maximum tolerance for the VOR indication when the CDI is centered and the airplane is directly over the airborne checkpoint?

A—Plus or minus 6° of the designated radial.
B—Plus or minus 7° of the designated radial.
C—Plus or minus 8° of the designated radial.

4390. When should your transponder be on Mode C while on an IFR flight?

A—Only when ATC requests Mode C.
B—At all times if the equipment has been calibrated, unless requested otherwise by ATC.
C—When passing 12,500 feet MSL.

4391. When making an airborne VOR check, what is the maximum allowable tolerance between the two indicators of a dual VOR system (units independent of each other except the antenna)?

A—4° between the two indicated radials of a VOR.
B—Plus or minus 4° when set to identical radials of a VOR.
C—6° between the two indicated radials of a VOR.

4392. What is the significance of an ATC clearance which reads "... CRUISE SIX THOUSAND ..."?

A—The pilot must maintain 6,000 feet until reaching the IAF serving the destination airport, then execute the published approach procedure.
B—Climbs may be made to, or descents made from, 6,000 feet at the pilot's discretion.
C—The pilot may utilize any altitude from the MEA/MOCA to 6,000 feet, but each change in altitude must be reported to ATC.

4393. What is the recommended climb procedure when a nonradar departure control instructs a pilot to climb to the assigned altitude?

A—Maintain a continuous optimum climb until reaching assigned altitude and report passing each 1,000 foot level.
B—Climb at a maximum angle of climb to within 1,000 feet of the assigned altitude, then 500 feet per minute the last 1,000 feet.
C—Maintain an optimum climb on the centerline of the airway without intermediate level offs until 1,000 feet below assigned altitude, then 500 to 1500 feet per minute.

4394. When departing from an airport not served by a control tower, the issuance of a clearance containing a void time indicates that

A—ATC will assume the pilot has not departed if no transmission is received before the void time.
B—the pilot must advise ATC as soon as possible, but no later than 30 minutes, of their intentions if not off by the void time.
C—ATC will protect the airspace only to the void time.

4395. What response is expected when ATC issues an IFR clearance to pilots of airborne aircraft?

A—Read back the entire clearance as required by regulation.
B—Read back those parts containing altitude assignments or vectors and any part requiring verification.
C—Read-back should be unsolicited and spontaneous to confirm that the pilot understands all instructions.

4396. Which clearance items are always given in an abbreviated IFR departure clearance? (Assume radar environment.)

A—Altitude, destination airport, and one or more fixes which identify the initial route of flight.
B—Destination airport, altitude, and SID Name-Number-Transition, if appropriate.
C—Clearance limit, and SID Name, Number, and/or Transition, if appropriate.

4397. Which distance is displayed by the DME indicator?

A—Slant range distance in NM.
B—Slant range distance in SM.
C—Line-of-sight direct distance from aircraft to VORTAC in SM.

4398. On the runup pad, you receive the following clearance from ground control:

CLEARED TO DALLAS-LOVE AIRPORT AS FILED — MAINTAIN SIX THOUSAND — SQUAWK ZERO SEVEN ZERO FOUR JUST BEFORE DEPARTURE — DEPARTURE CONTROL WILL BE ONE TWO FOUR POINT NINER.

An abbreviated clearance, such as this, will always contain the

A—departure control frequency.
B—transponder code.
C—assigned altitude.

4399. Where does the DME indicator have the greatest error between ground distance to the VORTAC and displayed distance?

A—High altitudes far from the VORTAC.
B—High altitudes close to the VORTAC.
C—Low altitudes far from the VORTAC.

4400. For operations off established airways at 17,000 feet MSL in the contiguous U.S., (H) Class VORTAC facilities used to define a direct route of flight should be no farther apart than

A—75 NM.
B—100 NM.
C—200 NM.

4401. What does the Runway Visual Range (RVR) value, depicted on certain straight-in IAP Charts, represent?

A—The slant range distance the pilot can see down the runway while crossing the threshold on glide slope.
B—The horizontal distance a pilot should see down the runway from the approach end of the runway.
C—The slant visual range a pilot should see down the final approach and during landing.

4402. How should you preflight check the altimeter prior to an IFR flight?

A—Set the altimeter to 29.92" Hg. With current temperature and the altimeter indication, determine the true altitude to compare with the field elevation.
B—Set the altimeter first with 29.92" Hg and then the current altimeter setting. The change in altitude should correspond to the change in setting.
C—Set the altimeter to the current altimeter setting. The indication should be within 75 feet of the actual elevation for acceptable accuracy.

4403. When are ATIS broadcasts updated?

A—Every 30 minutes if weather conditions are below basic VFR; otherwise, hourly.
B—Upon receipt of any official weather, regardless of content change or reported values.
C—Only when the ceiling and/or visibility changes by a reportable value.

4404. Absence of the sky condition and visibility on an ATIS broadcast specifically implies that

A—the ceiling is more than 5,000 feet and visibility is 5 miles or more.
B—the sky condition is clear and visibility is unrestricted.
C—the ceiling is at least 3,000 feet and visibility is 5 miles or more.

4405. The most current en route and destination flight information for planning an instrument flight should be obtained from

A—the ATIS broadcast.
B—the FSS.
C—Notices to Airmen (Class II).

4406. From what source can you obtain the latest FDC NOTAM's?

A—In Notices to Airmen (Class II NOTAM's).
B—At an FAA FSS.
C—In the Airport/Facility Directory.

4407. When may ATC request a detailed report of an emergency even though a rule has not been violated?

A—When priority has been given.
B—Any time an emergency occurs.
C—When the emergency occurs in controlled airspace.

4408. The operation of an airport rotating beacon during daylight hours may indicate that

A—the in-flight visibility is less than 3 miles and the ceiling is less than 1,500 feet within Class E airspace.
B—the ground visibility is less than 3 miles and/or the ceiling is less than 1,000 feet in Class B, C, or D airspace.
C—an IFR clearance is required to operate within the airport traffic area.

4409. What service is provided by departure control to an IFR flight when operating from an airport within Class C airspace?

A—Separation from all aircraft operating in Class C airspace.
B—Position and altitude of all traffic within 2 miles of the IFR pilot's line of flight and altitude.
C—Separation from all IFR aircraft and participating VFR aircraft.

4410. What indication should a pilot receive when a VOR station is undergoing maintenance and may be considered unreliable?

A—No coded identification, but possible navigation indications.
B—Coded identification, but no navigation indications.
C—A voice recording on the VOR frequency announcing that the VOR is out of service for maintenance.

4411. A particular VOR station is undergoing routine maintenance. This is evidenced by

A—removal of the navigational feature.
B—broadcasting a maintenance alert signal on the voice channel.
C—removal of the identification feature.

4412. What is the meaning of a single coded identification received only once approximately every 30 seconds from a VORTAC?

A—The VOR and DME components are operative.
B—VOR and DME components are both operative, but voice identification is out of service.
C—The DME component is operative and the VOR component is inoperative.

4413. Which DME indication should you receive when you are directly over a VORTAC site at approximately 6,000 feet AGL?

A—0.
B—1.
C—1.3.

4414. Which information is always given in an abbreviated clearance?

A—SID or transition name and altitude to maintain.
B—Name of destination airport or specific fix and altitude.
C—Altitude to maintain and code to squawk.

4415. If a control tower and an FSS are located on the same airport, which tower function is assumed by the FSS during those periods when the tower is closed?

A—Automatic closing of the IFR flight plan.
B—Approach control clearance.
C—Airport Advisory Service.

4416. Which service is provided for IFR arrivals by a FSS located on an airport without a control tower?

A—Automatic closing of the IFR flight plan.
B—Airport advisories.
C—All functions of approach control.

4417. What action is recommended if a pilot does not wish to use a SID?

A—Advise clearance delivery or ground control before departure.
B—Advise departure control upon initial contact.
C—Enter "No SID" in the REMARKS section of the IFR flight plan.

4418. A particular SID requires a minimum climb rate of 210 feet per NM to 8,000 feet. If you climb with a groundspeed of 140 knots, what is the rate of climb required in feet per minute?

A—210.
B—450.
C—490.

4419. Which procedure applies to SID's?

A—SID clearances will not be issued unless requested by the pilot.
B—The pilot in command must accept a SID when issued by ATC.
C—If a SID is accepted, the pilot must possess at least a textual description.

4420. During a takeoff into IFR conditions with low ceilings, when should the pilot contact departure control?

A—Before penetrating the clouds.
B—When advised by the tower.
C—Upon completing the first turn after takeoff or upon establishing cruise climb on a straight-out departure.

4421. During a flight, the controller advises "traffic 2 o'clock 5 miles southbound." The pilot is holding 20 correction for a crosswind from the right. Where should the pilot look for the traffic?

A—40° to the right of the airplane's nose.
B—20° to the right of the airplane's nose.
C—Straight ahead.

4422. What is meant when departure control instructs you to "resume own navigation" after you have been vectored to a Victor airway?

A—You should maintain the airway by use of your navigation equipment.
B—Radar service is terminated.
C—You are still in radar contact, but must make position reports.

4423. What does the ATC term "Radar Contact" signify?

A—Your aircraft has been identified and you will receive separation from all aircraft while in contact with this radar facility.
B—Your aircraft has been identified on the radar display and radar flight-following will be provided until radar identification is terminated.
C—You will be given traffic advisories until advised the service has been terminated or that radar contact has been lost.

4424. Upon intercepting the assigned radial, the controller advises you that you are on the airway and to "RESUME OWN NAVIGATION." This phrase means that

A—you are still in radar contact, but must make position reports.
B—radar services are terminated and you will be responsible for position reports.
C—you are to assume responsibility for your own navigation.

4425. Unless otherwise prescribed, what is the rule regarding altitude and course to be maintained by a helicopter during an off-airways IFR flight over nonmountainous terrain?

A—1,000 feet above the highest obstacle within 5 SM of course.
B—2,000 feet above the highest obstacle within 5 SM of course.
C—1,000 feet above the highest obstacle within 3 NM of course.

4426. In addition to a VOR receiver and two-way communications capability, which additional equipment is required for IFR operation in Class B airspace?

A—Another VOR and communications receiver and a coded transponder.
B—Standby communications receiver, DME, and coded transponder.
C—An operable coded transponder having Mode C capability.

4427. No person may operate an aircraft in controlled airspace under IFR unless he/she files a flight plan

A—and receives a clearance by telephone prior to takeoff.
B—prior to takeoff and requests the clearance upon arrival on an airway.
C—and receives a clearance prior to entering controlled airspace.

4428. Unless otherwise prescribed, what is the rule regarding altitude and course to be maintained by a helicopter during an IFR off-airways flight over mountainous terrain?

A—1,000 feet above the highest obstacle within a horizontal distance of 5 NM of course.
B—7,500 feet above the highest obstacle within a horizontal distance of 3 NM of course.
C—2,000 feet above the highest obstacle within 4 NM of course.

4429. What is the definition of MEA?

A—The lowest published altitude which meets obstacle clearance requirements and assures acceptable navigational signal coverage.
B—The lowest published altitude which meets obstacle requirements, assures acceptable navigational signal coverage, two-way radio communications, and provides adequate radar coverage.
C—An altitude which meets obstacle clearance requirements, assures acceptable navigation signal coverage, two-way radio communications, adequate radar coverage, and accurate DME mileage.

4430. What altitude may a pilot select upon receiving a VFR-on-Top clearance?

A—Any altitude at least 1,000 feet above the meteorological condition.
B—Any appropriate VFR altitude at or above the MEA in VFR weather conditions.
C—Any VFR altitude appropriate for the direction of flight at least 1,000 feet above the meteorological condition.

4431. When must a pilot fly at a cardinal altitude plus 500 feet on an IFR flight plan?

A—When flying above 18,000 feet in VFR conditions.
B—When flying in VFR conditions above clouds.
C—When assigned a VFR-on-Top clearance.

4432. The MEA assures acceptable navigational signal coverage and

A—DME response.
B—radar coverage.
C—meets obstacle clearance requirements.

4433. You have filed an IFR flight plan with a VFR-on-Top clearance in lieu of an assigned altitude. If you receive this clearance and fly a course of 180°, at what altitude should you fly? (Assume VFR conditions.)

A—Any IFR altitude which will enable you to remain in VFR conditions.
B—An odd thousand-foot MSL altitude plus 500 feet.
C—An even thousand-foot MSL altitude plus 500 feet.

4434. MOAs are established to

A—prohibit all civil aircraft because of hazardous or secret activities.
B—separate certain military activities from IFR traffic.
C—restrict civil aircraft during periods of high-density training activities.

4435. Reception of signals from an off-airway radio facility may be inadequate to identify the fix at the designated MEA. In this case, which altitude is designated for the fix?

A—MRA.
B—MCA.
C—MOCA.

4436. Which condition is guaranteed for all of the following altitude limits: MAA, MCA, MRA, MOCA, and MEA? (Non-mountainous area.)

A—Adequate navigation signals.
B—Adequate communications.
C—1,000-foot obstacle clearance.

4437. If no MCA is specified, what is the lowest altitude for crossing a radio fix, beyond which a higher minimum applies?

A—The MEA at which the fix is approached.
B—The MRA at which the fix is approached.
C—The MOCA for the route segment beyond the fix.

4438. What requirement must be met before ATC will authorize a deviation from the FAR 91 requirement for a transponder in class B airspace?

A—A request for the proposed flight must be made to ATC at least 1 hour before the flight.
B—The proposed flight must be conducted when operating under instrument flight rules.
C—The proposed flight must be conducted in visual meteorological conditions (VMC).

4439. A request for a deviation from the 4096 code transponder equipment requirement, when operating in Class B airspace, must be submitted to the

A—FAA Administrator at least 24 hours before the proposed operation.
B—nearest FAA General Aviation District Office 24 hours before the proposed operation.
C—controlling ATC facility at least 1 hour before the proposed flight.

4440. Which of the following is required equipment for operating an airplane within Class B airspace?

A—A 4096 code transponder with automatic pressure altitude reporting equipment.
B—A VOR receiver with DME.
C—A 4096 code transponder.

4441. Which procedure is recommended while climbing to an assigned altitude on the airway?

A—Climb on the centerline of the airway except when maneuvering to avoid other air traffic in VFR conditions.
B—Climb slightly on the right side of the airway when in VFR conditions.
C—Climb far enough to the right side of the airway to avoid climbing or descending traffic coming from the opposite direction if in VFR conditions.

4442. Which clearance procedures may be issued by ATC without prior pilot request?

A—SID's, standard terminal arrival route (STAR), and contact approaches.
B—Contact and visual approaches.
C—SID's, STAR's, and visual approaches.

4443. What is the significance of an ATC clearance which reads "...CRUISE SIX THOUSAND..."?

A—The pilot must maintain 6,000 until reaching the IAF serving the destination airport, then execute the published approach procedure.
B—It authorizes a pilot to conduct flight at any altitude from minimum IFR altitude up to and including 6,000.
C—The pilot is authorized to conduct flight at any altitude from minimum IFR altitude up to and including 6,000, but each change in altitude must be reported to ATC.

4444. What is the procedure for setting the altimeter when assigned an IFR altitude of 18,000 feet or higher on a direct flight off airways?

A—Set the altimeter to 29.92" Hg before takeoff.
B—Set the altimeter to the current altimeter setting until reaching the assigned altitude, then set to 29.92" Hg.
C—Set the altimeter to the current reported setting for climbout and 29.92" Hg upon reaching 18,000 feet.

4445. En route at FL 290, the altimeter is set correctly, but not reset to the local altimeter setting of 30.57" Hg during descent. If the field elevation is 650 feet and the altimeter is functioning properly, what is the approximate indication upon landing?

A—715 feet.
B—1,300 feet.
C—Sea level.

4446. While you are flying at FL 250, you hear ATC give an altimeter setting of 28.92" Hg in your area. At what pressure altitude are you flying?

A—24,000 feet.
B—25,000 feet.
C—26,000 feet.

4447. Where are VFR-on-Top operations prohibited?

A—In Class A airspace.
B—During off-airways direct flights.
C—When flying through Class B airspace.

4448. What action should you take if your DME fails at FL 240?

A—Advise ATC of the failure and land at the nearest available airport where repairs can be made.
B—Notify ATC that it will be necessary for you to go to a lower altitude, since your DME has failed.
C—Notify ATC of the failure and continue to the next airport of intended landing where repairs can be made.

4449. Which rules apply to the pilot in command when operating on a VFR-on-Top clearance?

A—VFR only.
B—VFR and IFR.
C—VFR when "in the clear" and IFR when "in the clouds."

4450. When can a VFR-on-Top clearance be assigned by ATC?

A—Only upon request of the pilot when conditions are indicated to be suitable.
B—Any time suitable conditions exist and ATC wishes to expedite traffic flow.
C—When VFR conditions exist, but there is a layer of clouds below the MEA.

4451. Which ATC clearance should instrument-rated pilots request in order to climb through a cloud layer or an area of reduced visibility and then continue the flight VFR?

A—To VFR on Top.
B—Special VFR to VFR Over-the-Top.
C—VFR Over-the-Top.

4452. When on a VFR-on-Top clearance, the cruising altitude is based on

A—true course.
B—magnetic course.
C—magnetic heading.

4453. In which airspace is VFR-on-Top operation prohibited?

A— Class B airspace.
B— Class E airspace.
C— Class A airspace.

4454. What cruising altitude is appropriate for VFR on Top on a westbound flight below 18,000 feet?

A—Even thousand-foot levels.
B—Even thousand-foot levels plus 500 feet, but not below MEA.
C—Odd thousand-foot levels plus 500 feet, but not below MEA.

4455. What reports are required of a flight operating on an IFR clearance specifying VFR on Top in a nonradar environment?

A—The same reports that are required for any IFR flight.
B—All normal IFR reports except vacating altitudes.
C—Only the reporting of any unforecast weather.

4456. Which report should be made to ATC without a specific request when not in radar contact?

A—Entering instrument meteorological conditions.
B—When leaving final approach fix inbound on final approach.
C—Correcting an E.T.A. any time a previous E.T.A. is in error in excess of 2 minutes.

4457. What minimums must be considered in selecting an altitude when operating with a VFR-on-Top clearance?

A—At least 500 feet above the lowest MEA, or appropriate MOCA, and at least 1,000 feet above the existing meteorological condition.
B—At least 1,000 feet above the lowest MEA, appropriate MOCA, or existing meteorological condition.
C—Minimum IFR altitude, minimum distance from clouds, and visibility appropriate to altitude selected.

4458. A "CRUISE FOUR THOUSAND FEET" clearance would mean that the pilot is authorized to

A—vacate 4,000 feet without notifying ATC.
B—climb to, but not descend from 4,000 feet, without further ATC clearance.
C—use any altitude from minimum IFR to 4,000 feet, but must report leaving each altitude.

4459. What is the procedure when the DME malfunctions at or above 24,000 feet MSL?

A—Notify ATC immediately and request an altitude below 24,000 feet.
B—Continue to your destination in VFR conditions and report the malfunction.
C—After immediately notifying ATC, you may continue to the next airport of intended landing where repairs can be made.

4460. What action should you take if your No. 1 VOR receiver malfunctions while operating in controlled airspace under IFR? Your aircraft is equipped with two VOR receivers. The No. 1 receiver has Omni/Localizer/Glide Slope capability, and the No. 2 has only Omni.

A—Report the malfunction immediately to ATC.
B—Continue the flight as cleared; no report is required.
C—Continue the approach and request a VOR or NDB approach.

4461. While on an IFR flight, a pilot has an emergency which causes a deviation from an ATC clearance. What action must be taken?

A—Notify ATC of the deviation as soon as possible.
B—Squawk 7700 for the duration of the emergency.
C—Submit a detailed report to the chief of the ATC facility within 48 hours.

4462. You enter a holding pattern at a fix, not the same as the approach fix, and receive an EFC time of 1530. At 1520 you experience complete two-way communications failure. Which procedure should you follow to execute the approach to a landing?

A—Depart the holding fix to arrive at the approach fix as close as possible to the EFC time and complete the approach.
B—Depart the holding fix at the EFC time, and complete the approach.
C—Depart the holding fix at the EFC time or earlier if your flight planned ETA is before the EFC.

4463. Which procedure should you follow if you experience two-way communications failure while holding at a holding fix with an EFC time? (The holding fix is not the same as the approach fix.)

A—Depart the holding fix to arrive at the approach fix as close as possible to the EFC time.
B—Depart the holding fix at the EFC time.
C—Proceed immediately to the approach fix and hold until EFC.

4464. You are in IMC and have two-way radio communications failure. If you do not exercise emergency authority, what procedure are you expected to follow?

A—Set transponder to code 7600, continue flight on assigned route and fly at the last assigned altitude or the MEA, whichever is higher.
B—Set transponder to code 7700 for 1 minute, then to 7600, and fly to an area with VFR weather conditions.
C—Set transponder to 7700 and fly to an area where you can let down in VFR conditions.

4465. Which procedure should you follow if, during an IFR flight in VFR conditions, you have two-way radio communications failure?

A—Continue the flight under VFR and land as soon as practicable.
B—Continue the flight at assigned altitude and route, start approach at your ETA, or, if late, start approach upon arrival.
C—Land at the nearest airport that has VFR conditions.

4466. What altitude and route should be used if you are flying in IMC and have two-way radio communications failure?

A—Continue on the route specified in your clearance, fly at an altitude that is the highest of last assigned altitude, altitude ATC has informed you to expect, or the MEA.
B—Fly direct to an area that has been forecast to have VFR conditions, fly at an altitude that is at least 1,000 feet above the highest obstacles along the route.
C—Descend to MEA and, if clear of clouds, proceed to the nearest appropriate airport. If not clear of clouds, maintain the highest of the MEA's along the clearance route.

4467. At what time are current AIRMET's broadcast by the FSS?

A—15 minutes after the hour only.
B—Every 15 minutes until the AIRMET is canceled.
C—15 and 45 minutes after the hour during the first hour after issuance.

4468. Pilots on IFR flights seeking ATC in-flight weather avoidance assistance should keep in mind that

A—ATC radar limitations and, frequency congestion may limit the controllers capability to provide this service.
B—circumnavigating severe weather can only be accommodated in the en route areas away from terminals because of congestion.
C—ATC Narrow Band Radar does not provide the controller with weather intensity capability.

4469. When are you required to establish communications with the tower, (Class D airspace) if you cancel your IFR flight plan 10 miles from the destination?

A—Immediately after canceling the flight plan.
B—When advised by ARTCC.
C—Before entering Class D airspace.

4470. What does the symbol T within a black triangle in the minimums section of the IAP for a particular airport indicate?

A—Takeoff minimums are 1 mile for aircraft having two engines or less and 1/2 mile for those with more than two engines.
B—Instrument takeoffs are not authorized.
C—Takeoff minimums are not standard and/or departure procedures are published.

4471. What responsibility does the pilot in command of an IFR flight assume upon entering VFR conditions?

A—Report VFR conditions to ARTCC so that an amended clearance may be issued.
B—Use VFR operating procedures.
C—To see and avoid other traffic.

4472. As a rule of thumb, to minimize DME slant range error, how far from the facility should you be to consider the reading as accurate?

A—Two miles or more for each 1,000 feet of altitude above the facility.
B—One or more miles for each 1,000 feet of altitude above the facility.
C—No specific distance is specified since the reception is line-of-sight.

4473. Which airspace is defined as a transition area when designated in conjunction with an airport which has a prescribed IAP?

A—The Class E airspace extending upward from 700 feet or more above the surface and terminating at the base of the overlying controlled airspace.
B—That Class D airspace extending from the surface and terminating at the base of the continental control area.
C—The Class C airspace extending from the surface to 700 or 1,200 feet AGL, where designated.

4474. The vertical extent of the Class A airspace throughout the conterminous U.S. extends from

A—18,000 feet to and including FL 450.
B—18,000 feet to and including FL 600.
C—12,500 feet to and including FL 600.

4475. Class G airspace is that airspace where

A—ATC does not control air traffic.
B—ATC controls only IFR flights.
C—the minimum visibility for VFR flight is 3 miles.

4476. What are the vertical limits of a transition area that is designated in conjunction with an airport having a prescribed IAP?

A—Surface to 700 feet AGL.
B—1,200 feet AGL to the base of the overlying controlled airspace.
C—700 feet AGL or more to the base of the overlying controlled airspace.

4477. How can you obtain the pressure altitude on flights below 18,000 feet?

A—Set your altimeter to 29.92" Hg.
B—Use your computer to change the indicated altitude to pressure altitude.
C—Contact an FSS and ask for the pressure altitude.

4478. How can you determine the pressure altitude at an airport without a tower or FSS?

A—Set the altimeter to 29.92" Hg and read the altitude indicated.
B—Set the altimeter to the current altimeter setting of a station within 100 miles and correct this indicated altitude with local temperature.
C—Use your computer and correct the field elevation for temperature.

4479. Which altitude is indicated when the altimeter is set to 29.92" Hg ?

A—Density.
B—Pressure.
C—Standard.

4480. If you are departing from an airport where you cannot obtain an altimeter setting, you should set your altimeter

A—on 29.92" Hg.
B—on the current airport barometric pressure, if known.
C—to the airport elevation.

4481. En route at FL 290, your altimeter is set correctly, but you fail to reset it to the local altimeter setting of 30.26" Hg during descent. If the field elevation is 134 feet and your altimeter is functioning properly, what will it indicate after landing?

A—100 feet MSL.
B—474 feet MSL.
C—206 feet below MSL.

4482. How does a pilot normally obtain the current altimeter setting during an IFR flight in Class E airspace below 18,000 feet?

A—The pilot should contact ARTCC at least every 100 NM and request the altimeter setting.
B—FSS's along the route broadcast the weather information at 15 minutes past the hour.
C—ATC periodically advises the pilot of the proper altimeter setting.

4483. (Refer to Figure 83.) Which altimeter depicts 12,000 feet?

A—2.
B—3.
C—4.

4484. (Refer to Figure 84.) Which altimeter depicts 8,000 feet?

A—1.
B—2.
C—3.

4485. Unless otherwise specified on the chart, the minimum en route altitude along a jet route is

A—18,000 feet MSL.
B—24,000 feet MSL.
C—10,000 feet MSL.

4486. An abbreviated departure clearance "...CLEARED AS FILED..." will always contain the name

A—and number of the STAR to be flown when filed in the flight plan.
B—of the destination airport filed in the flight plan.
C—of the first compulsory reporting point if not in a radar environment.

4487. As a rule of thumb, to minimize DME slant range error, how far from the facility should you be to consider the reading as accurate?

A—Two miles or more for each 1,000 feet of altitude above the facility.
B—One or more miles for each 1,000 feet of altitude above the facility.
C—No specific distance is specified since the reception is line-of-sight.

4488. (Refer to Figures 85 and 86.) Which combination of indications confirm that you are approaching WAGGE intersection slightly to the right of the LOC centerline on departure?

A—1 and 3.
B—1 and 4.
C—2 and 3.

4489. (Refer to Figure 85.) What route should you take if cleared for the Washoe Two Departure and your assigned route is V6?

A—Climb on the LOC south course to WAGGE where you will be vectored to V6.
B—Climb on the LOC south course to cross WAGGE at 9,000, turn left and fly direct to FMG VORTAC and cross at or above 10,000, and proceed on FMG R-241.
C—Climb on the LOC south course to WAGGE, turn left and fly direct to FMG VORTAC. If at 10,000 turn left and proceed on FMG R-241; if not at 10,000 enter depicted holding pattern and climb to 10,000 before proceeding on FMG R-241.

4490. (Refer to Figure 85.) What procedure should be followed if communications are lost before reaching 9,000 feet?

A—At 9,000, turn left direct to FMG VORTAC, then via assigned route if at proper altitude; if not, climb in holding pattern until reaching the proper altitude.
B—Continue climb to WAGGE INT, turn left direct to FMG VORTAC, then if at or above MCA, proceed on assigned route; if not, continue climb in holding pattern until at the proper altitude.
C—Continue climb on LOC course to cross WAGGE INT at or above 9,000, turn left direct to FMG VORTAC to cross at 10,000 or above, and continue on assigned course.

4491. (Refer to Figure 85.) What is the minimum rate climb per NM to 9,000 feet required for the WASH2 WAGGE Departure?

A—400 feet.
B—750 feet.
C—875 feet.

4492. (Refer to Figure 85.) Of the following, which is the minimum acceptable rate of climb (feet per minute) to 9,000 feet required for the WASH2 WAGGE departure at a GS of 150 knots?

A—750 feet per minute.
B—825 feet per minute.
C—1,000 feet per minute.

4493. (Refer to Figure 87.) Where is the VOR COP when flying east on V306 from Daisetta to Lake Charles?

A—50 NM east of DAS.
B—40 NM east of DAS.
C—30 NM east of DAS.

4494. (Refer to Figure 87.) At STRUT intersection headed eastbound, ATC instructs you to hold on the 10 DME fix west of LCH on V306, standard turns, what entry procedure is recommended?

A—Direct.
B—Teardrop.
C—Parallel.

4495. (Refer to Figures 87 and 88.) What is your position with reference to FALSE intersection (V222) if your VOR receivers indicate as shown?

A—South of V222 and east of FALSE intersection.
B—North of V222 and east of FALSE intersection.
C—South of V222 and west of FALSE intersection.

4496. (Refer to Figure 87.) What is indicated by the localizer course symbol at Jefferson Co Field?

A—A published LDA localizer course is available.
B—A published SDF localizer course is avialable.
C—A published ILS localizer course with an ATC function is available.

4497. (Refer to Figure 87.) Which VHF frequencies, other than 121.5, can be used to receive Lake Charles FSS in the Lake Charles area?

A—122.1, 126.4.
B—123.6, 122.65.
C—122.2, 122.3.

4498. (Refer to Figure 87.) Why is the localizer back course at Jefferson County depicted?

A—The back course is not aligned with a runway.
B—The back course has a glide slope.
C—The back course has an ATC function.

4499. (Refer to Figure 87.) Where is the VOR changeover point on V20 between Beaumont and Hobby?

A—Halfway point.
B—MOCKS intersection.
C—Anahuac Beacon.

4500. (Refer to Figure 87.) While holding at the 10 DME fix east of LCH for an ILS approach to RWY 15 at Lake Charles Muni airport, ATC advises you to expect clearance for the approach at 1015. At 1000 you experience two-way radio communications failure. Which procedure should be followed?

A—Squawk 7600 and listen on the LOM frequency for instructions from ATC. If no instructions are received, start your approach at 1015.
B—Squawk 7700 for 1 minute, then 7600. After 1 minute, descend to the minimum final approach fix altitude. Start your approach at 1015.
C—Squawk 7600; plan to begin your approach at 1015.

4501. (Refer to Figure 89.) When flying from Milford Municipal to Bryce Canyon via V235 and V293, what minimum altitude should you be at when crossing Cedar City VOR?

A—11,400 feet.
B—12,000 feet.
C—13,000 feet.

4502. (Refer to Figure 89.) What VHF frequencies are available for communications with Cedar City FSS?

A—123.6, 121.5, 108.6, and 112.8.
B—122.2, 121.5, 122.6, and 112.1.
C—122.2, 121.5, 122.0, and 123.6.

4503. (Refer to Figure 89.) What are the oxygen requirements for an IFR flight northeast bound from Bryce Canyon on V382 at the lowest appropriate altitude in an unpressurized aircraft?

A—The required minimum crew must be provided and use supplemental oxygen for that part of the flight of more than 30 minutes.
B—The required minimum crew must be provided and use supplemental oxygen for that part of the flight of more than 30 minutes, and the passengers must be provided supplemental oxygen.
C—The required minimum crew must be provided and use supplemental oxygen, and all occupants must be provided supplemental oxygen for the entire flight above 15,000 feet.

4504. On what frequency should you obtain En Route Flight Advisory Service below FL 180?

A—122.1T/112.8R.
B—123.6.
C—122.0.

4505. In the event of two way radio communications failure while operating on an IFR clearance in VFR conditions the pilot should continue

A—by the route assigned in the last ATC clearance received.
B—the flight under VFR and land as soon as practical.
C—the flight by the most direct route to the fix specified in the last clearance.

4506. (Refer to Figure 89.) What is the ARTCC discrete frequency at the COP on V208 southwest bound from HVE to PGA VOR/DME?

A—122.1.
B—122.4.
C—133.6.

4507. (Refer to Figures 89 and 90.) What is your relationship to the airway while en route from BCE VORTAC to HVE VORTAC on V8?

A—Left of course on V8.
B—Left of course on V382.
C—Right of course on V8.

4508. (Refer to Figure 89.) What type airspace exists above Bryce Canyon Airport from the surface to 1200 feet AGL?

A—Class D airspace.
B—Class E airspace.
C—Class G airspace.

4509. (Refer to Figure 91.) What is the minimum crossing altitude at DBS VORTAC for a northbound IFR flight on V257?

A—7,500 feet.
B—8,600 feet.
C—11,100 feet.

4510. (Refer to Figure 91.) What are the two limiting cruising altitudes useable on V343 for a VFR-on-Top flight from DBS VORTAC to RANEY intersection?

A—14,500 and 16,500 feet.
B—15,000 and 17,000 feet.
C—15,500 and 17,500 feet.

4511. (Refer to Figure 91.) What should be the approximate elapsed time from BZN VOR to DBS VORTAC, if the wind is 24 knots from 260° and your intended TAS is 185 knots? (VAR 17° E.)

A—33 minutes.
B—37 minutes.
C—39 minutes.

4512. (Refer to Figure 91.) What lighting is indicated on the chart for Jackson Hole Airport?

A—Lights on prior request.
B—No lighting available.
C—Pilot controlled lighting.

4513. (Refer to Figure 91.) What are the oxygen requirements for an IFR flight eastbound on V520 from DBS VORTAC in an unpressurized aircraft at the MEA?

A—The required minimum crew must be provided and use supplemental oxygen for that part of the flight of more than 30 minutes.
B—The required minimum crew must be provided and use supplemental oxygen for that part of the flight of more than 30 minutes, and the passengers must be provided supplemental oxygen.
C—The required minimum crew must be provided and use supplemental oxygen.

4514. (Refer to Figure 91.) Southbound on V257, at what time should you arrive at DBS VORTAC if you crossed over CPN VORTAC at 0850 and over DIVID intersection at 0854?

A—0939.
B—0943.
C—0947.

4515. (Refer to Figure 91.) What is the function of the Great Falls RCO (Yellowstone vicinity)?

A—Long range communications outlet for Great Falls Center.
B—Remote communications outlet for Great Falls FSS.
C—Satellite remote controlled by Salt Lake Center with limited service.

4516. (Refer to Figure 91.) Where should you change VOR frequencies when en route from DBS VORTAC to JAC VOR/DME on V520?

A—35 NM from DBS VORTAC.
B—60 NM from DBS VORTAC.
C—60 NM from JAC VOR/DME.

4517. (Refer to Figure 91.) What is the minimum crossing altitude at SABAT intersection when eastbound from DBS VORTAC on V298?

A—8,300 feet.
B—11,100 feet.
C—13,000 feet.

4518. What is the minimum flight visibility and distance from clouds for flight at 10,500 feet with a VFR-on-Top clearance during daylight hours? (Class E airspace.)

A—3 SM, 1,000 feet above, 500 feet below, and 2,000 feet horizontal.
B—5 SM, 1,000 feet above, 1,000 feet below, and 1 mile horizontal.
C—5 SM, 1,000 feet above, 500 feet below, and 1 mile horizontal.

4519. What is the required flight visibility and distance from clouds if you are operating in Class E airspace at 9,500 feet MSL with a VFR-on-Top clearance during daylight hours?

A—3 SM, 1,000 feet above, 500 feet below, and 2,000 feet horizontal.
B—5 SM, 500 feet above, 1,000 feet below, and 2,000 feet horizontal.
C—3 SM, 500 feet above, 1,000 feet below, and 2,000 feet horizontal.

4520. (Refer to Figure 92.) What is the minimum in-flight visibility and distance from clouds required for a VFR-on-Top flight at 9,500 feet MSL (above 1,200 feet AGL) during daylight hours for area 3?

A—2,000 feet; (E) 1,000 feet; (F) 2,000 feet; (H) 500 feet.
B—5 miles; (E) 1,000 feet; (F) 2,000 feet; (H) 500 feet.
C—3 miles; (E) 1,000 feet; (F) 2,000 feet; (H) 500 feet.

4521. (Refer to Figure 92.) A flight is to be conducted in VFR-on-Top conditions at 12,500 feet MSL (above 1200 feet AGL). What is the in-flight visibility and distance from clouds required for operation in Class E airspace during daylight hours for area 1?

A—5 miles; (A) 1,000 feet; (B) 2,000 feet; (D) 500 feet.
B—5 miles; (A) 1,000 feet; (B) 1 mile; (D) 1,000 feet.
C—3 miles; (A) 1,000 feet; (B) 2,000 feet; (D) 1,000 feet.

4522. (Refer to Figure 92.) What is the minimum in-flight visibility and distance from clouds required in VFR conditions above clouds at 13,500 feet MSL (above 1,200 feet AGL) in Class G airspace during daylight hours for area 2?

A—5 miles; (A) 1,000 feet; (C) 2,000 feet; (D) 500 feet.
B—3 miles; (A) 1,000 feet; (C) 1 mile; (D) 1,000 feet.
C—5 miles; (A) 1,000 feet; (C) 1 mile; (D) 1,000 feet.

4523. (Refer to Figure 92.) What in-flight visibility and distance from clouds is required for a flight at 8,500 feet MSL (above 1,200 feet AGL) in Class G airspace in VFR conditions during daylight hours in area 4?

A—1 mile; (E) 1,000 feet; (G) 2,000 feet; (H) 500 feet.
B—3 miles; (E) 1,000 feet; (G) 2,000 feet; (H) 500 feet.
C—5 miles; (E) 1,000 feet; (G) 1 mile; (H) 1,000 feet.

4524. (Refer to Figure 92.) What is the minimum in-flight visibility and distance from clouds required for an airplane operating less than 1,200 feet AGL during daylight hours in area 6?

A—3 miles; (I) 1,000 feet; (K) 2,000 feet; (L) 500 feet.
B—1 mile; (I) clear of clouds; (K) clear of clouds; (L) clear of clouds.
C—1 mile; (I) 500 feet; (K) 1,000 feet; (L) 500 feet.

4525. (Refer to Figure 92.) What is the minimum in-flight visibility and distance from clouds required for an airplane operating less than 1,200 feet AGL under special VFR during daylight hours in area 5?

A—1 mile; (I) 2,000 feet; (J) 2,000 feet; (L) 500 feet.
B—3 miles; (I) clear of clouds; (J) clear of clouds; (L) 500 feet.
C—1 mile; (I) clear of clouds; (J) clear of clouds; (L) clear of clouds.

4526. (Refer to Figure 93.) What is the floor of Class E airspace when designated in conjunction with an airway?

A—700 feet AGL.
B—1,200 feet AGL.
C—1,500 feet AGL.

4527. (Refer to Figure 93.) Which altitude is the normal upper limit for Class D airspace?

A—1,000 feet AGL.
B—2,500 feet AGL.
C—4,000 feet AGL.

4528. (Refer to Figure 93.) What is the floor of Class E airspace when designated in conjunction with an airport which has an approved IAP?

A—500 feet AGL.
B—700 feet AGL.
C—1,200 feet AGL.

4529. (Refer to Figure 93.) Which altitude is the upper limit for Class A airspace?

A—14,500 feet MSL.
B—18,000 feet MSL.
C—60,000 feet MSL.

4530. (Refer to Figure 93.) What is the maximum altitude that Class G airspace will exist? (Does not include airspace less than 1,500 feet AGL.)

A—18,000 feet MSL.
B—14,500 feet MSL.
C—14,000 feet MSL.

4531. (Refer to Figure 93.) What is generally the maximum altitude for Class B airspace?

A—4,000 feet MSL.
B—10,000 feet MSL.
C—14,500 feet MSL.

4532. (Refer to Figure 93.) What are the normal lateral limits for Class D airspace?

A—8 NM.
B—5 NM.
C—4 NM.

4533. (Refer to Figure 93.) What is the floor of Class A airspace?

A—10,000 feet MSL.
B—14,500 feet MSL.
C—18,000 feet MSL.

4534. (Refer to Figure 94.) Mandatory airport instruction signs are designated by having—

A—Yellow lettering with a black background.
B—White lettering with a red background.
C—Black lettering with a yellow background.

4535. (Refer to Figure 94.) What sign is designated by illustration 7?

A—Location sign.
B—Mandatory instruction sign.
C—Direction sign.

4536. (Refer to Figure 94.) What color are runway holding position signs?

A—White with a red background.
B—Red with a white background.
C—Yellow with a black background.

4537. (Refer to Figure 94.) Hold line markings at the intersection of taxiways and runways consist of four lines that extend across the width of the taxiway. These lines are—

A—white and the dashed lines are nearest the runway.
B—yellow and the dashed lines are nearest the runway.
C—yellow and the solid lines are nearest the runway.

4538. When should pilots state their position on the airport when calling the tower for takeoff?

A—When visibility is less than 1 mile.
B—When parallel runways are in use.
C—When departing from a runway intersection.

4539. What minimum aircraft equipment is required for operation within Class C airspace?

A—Two-way communications and Mode C transponder.
B—Two-way communications.
C—Transponder and DME.

4540. What obstacle clearance and navigation signal coverage is a pilot assured with the Minimum Sector Altitudes depicted on the IAP charts?

A—1,000 feet and acceptable navigation signal coverage within a 25 NM radius of the navigation facility.
B—1,000 feet within a 25 NM radius of the navigation facility but not acceptable navigation signal coverage.
C—500 feet and acceptable navigation signal coverage within a 10 NM radius of the navigation facility.

4541. In the case of operations over an area designated as a mountainous area where no other minimum altitude is prescribed, no person may operate an aircraft under IFR below an altitude of

A—500 feet above the highest obstacle.
B—1,000 feet above the highest obstacle.
C—2,000 feet above the highest obstacle.

4542. MEA is an altitude which assures

A—obstacle clearance, accurate navigational signals from more than one VORTAC, and accurate DME mileage.
B—a 1,000-foot obstacle clearance within 2 miles of an airway and assures accurate DME mileage.
C—acceptable navigational signal coverage and meets obstruction clearance requirements.

4543. If, while in Class E airspace, a clearance is received to "maintain VFR conditions on top," the pilot should maintain a VFR cruising altitude based on the direction of the

A—true course.
B—magnetic heading.
C—magnetic course.

4544. Reception of signals from a radio facility, located off the airway being flown, may be inadequate at the designated MEA to identify the fix. In this case, which altitude is designated for the fix?

A—MOCA.
B—MRA.
C—MCA.

4545. ATC may assign the MOCA when certain special conditions exist, and when within

A—22 NM of a VOR.
B—25 NM of a VOR.
C—30 NM of a VOR.

4546. Which aeronautical chart depicts Military Training Routes (MTR) above 1500 feet?

A—IFR Planning Chart.
B—IFR Low Altitude En Route Chart.
C—IFR High Altitude En Route Chart.

4547. Acceptable navigational signal coverage at the MOCA is assured for a distance from the VOR of only

A—12 NM.
B—22 NM.
C—25 NM.

4548. What angular deviation from a VOR course centerline is represented by a full-scale deflection of the CDI?

A—4°.
B—5°.
C—10°.

4549. When using VOR for navigation, which of the following should be considered as station passage?

A—The first movement of the CDI as the airplane enters the zone of confusion.
B—The moment the TO-FROM indicator becomes blank.
C—The first positive, complete reversal of the TO-FROM indicator.

4550. Which of the following should be considered as station passage when using VOR?

A—The first flickering of the TO-FROM indicator and CDI as the station is approached.
B—The first full-scale deflection of the CDI.
C—The first complete reversal of the TO-FROM indicator.

4551. When checking the sensitivity of a VOR receiver, the number of degrees in course change as the OBS is rotated to move the CDI from center to the last dot on either side should be between

A—5° and 6°.
B—8° and 10°.
C—10° and 12°.

4552. A VOR receiver with normal five-dot course sensitivity shows a three-dot deflection at 30 NM from the station. The aircraft would be displaced approximately how far from the course centerline?

A—2 NM.
B—3 NM.
C—5 NM.

4553. An aircraft which is located 30 miles from a VOR station and shows a 1/2 scale deflection on the CDI would be how far from the selected course centerline?

A—1 1/2 miles.
B—2 1/2 miles.
C—3 1/2 miles.

4554. What angular deviation from a VOR course centerline is represented by a 1/2 scale deflection of the CDI?

A—2°.
B—4°.
C—5°.

4555. To comply with ATC instructions for altitude changes of more than 1,000 feet, what rate of climb or descent should be used?

A—As rapidly as practicable to 500 feet above/below the assigned altitude, and then at 500 feet per minute until the assigned altitude is reached.
B—1,000 feet per minute during climb and 500 feet per minute during descents until reaching the assigned altitude.
C—As rapidly as practicable to 1,000 feet above/below the assigned altitude, and then between 500 and 1,500 feet per minute until reaching the assigned altitude.

4556. After passing a VORTAC, the CDI shows 1/2 scale deflection to the right. What is indicated if the deflection remains constant for a period of time?

A—The airplane is getting closer to the radial.
B—The OBS is erroneously set on the reciprocal heading.
C—The airplane is flying away from the radial.

4557. (Refer to Figure 95.) What is the lateral displacement of the aircraft in NM from the radial selected on the No. 1 NAV?

A—5.0 NM.
B—7.5 NM.
C—10.0 NM.

4558. (Refer to Figure 95.) On which radial is the aircraft as indicated by the No. 1 NAV?

A—R-175.
B—R-165.
C—R-345.

4559. (Refer to Figure 95.) Which OBS selection on the No. 1 NAV would center the CDI and change the ambiguity indication to a TO?

A—175°.
B—165°.
C—345°.

4560. (Refer to Figure 95.) What is the lateral displacement in degrees from the desired radial on the No. 2 NAV?

A—1°.
B—2°.
C—4°.

4561. (Refer to Figure 95.) Which OBS selection on the No. 2 NAV would center the CDI?

A—174°.
B—166°.
C—335°.

4562. (Refer to Figure 95.) Which OBS selection on the No. 2 NAV would center the CDI and change the ambiguity indication to a TO?

A—166°.
B—346°.
C—354°.

4563. (Refer to Figures 96 and 97.) To which aircraft position(s) does HSI presentation "A" correspond?

A—9 and 6.
B—9 only.
C—6 only.

4564. (Refer to Figures 96 and 97.) To which aircraft position(s) does HSI presentation "B" correspond?

A—11.
B—5 and 13.
C—7 and 11.

4565. (Refer to Figures 96 and 97.) To which aircraft position does HSI presentation "C" correspond?

A—9.
B—4.
C—12.

4566. (Refer to Figures 96 and 97.) To which aircraft position does HSI presentation "D" correspond?

A—1.
B—10.
C—2.

4567. (Refer to Figures 96 and 97.) To which aircraft position(s) does HSI presentation "E" correspond?

A—8 only.
B—3 only.
C—8 and 3.

4568. (Refer to Figures 96 and 97.) To which aircraft position does HSI presentation "F" correspond?

A—4.
B—11.
C—5.

4569. (Refer to Figures 96 and 97.) To which aircraft position(s) does HSI presentation "G" correspond?

A—7 only.
B—7 and 11.
C—5 and 13.

4570. (Refer to Figures 96 and 97.) To which aircraft position does HSI presentation "H" correspond?

A—8.
B—1.
C—2.

4571. (Refer to Figures 96 and 97.) To which aircraft position does HSI presentation "I" correspond?

A—4.
B—12.
C—11.

4572. (Refer to Figures 98 and 99.) To which aircraft position does HSI presentation "D" correspond?

A—4.
B—15.
C—17.

4573. (Refer to Figures 98 and 99.) To which aircraft position does HSI presentation "E" correspond?

A—5.
B—6.
C—15.

4574. (Refer to Figures 98 and 99.) To which aircraft position does HSI presentation "F" correspond?

A—10.
B—14.
C—16.

4575. (Refer to Figures 98 and 99.) To which aircraft position does HSI presentation "A" correspond?

A—1.
B—8.
C—11.

4576. (Refer to Figures 98 and 99.) To which aircraft position does HSI presentation "B" correspond?

A—9.
B—13.
C—19.

4577. (Refer to Figures 98 and 99.) To which aircraft position does HSI presentation "C" correspond?

A—6.
B—7.
C—12.

4578. (Refer to Figure 101.) What is the magnetic bearing TO the station?

A—060°.
B—260°.
C—270°.

4579. (Refer to Figure 100.) Which RMI illustration indicates the aircraft to be flying outbound on the magnetic bearing of 235° FROM the station? (Wind 050° at 20 knots.)

A—2.
B—3.
C—4.

4580. (Refer to Figure 100.) What is the magnetic bearing TO the station as indicated by illustration 4?

A—285°.
B—055°.
C—235°.

4581. (Refer to Figure 100.) Which RMI illustration indicates the aircraft is southwest of the station and moving closer TO the station?

A—1.
B—2.
C—3.

4582. (Refer to Figure 100.) Which RMI illustration indicates the aircraft is located on the 055° radial of the station and heading away from the station?

A—1.
B—2.
C—3.

4583. (Refer to instruments in Figure 102.) On the basis of this information, the magnetic bearing TO the station would be

A—175°.
B—255°.
C—355°.

4584. (Refer to instruments in Figure 102.) On the basis of this information, the magnetic bearing FROM the station would be

A—175°.
B—255°.
C—355°.

4585. (Refer to instruments in Figure 103.) On the basis of this information, the magnetic bearing FROM the station would be

A—030°.
B—060°.
C—240°.

4586. (Refer to instruments in Figure 103.) On the basis of this information, the magnetic bearing TO the station would be

A—060°.
B—240°.
C—270°.

4587. (Refer to Figure 104.) If the radio magnetic indicator is tuned to a VOR, which illustration indicates the aircraft is on the 115° radial?

A—1.
B—2.
C—3.

4588. (Refer to Figure 104.) If the radio magnetic indicator is tuned to a VOR, which illustration indicates the aircraft is on the 335° radial?

A—2.
B—3.
C—4.

4589. (Refer to Figure 104.) If the radio magnetic indicator is tuned to a VOR, which illustration indicates the aircraft is on the 315° radial?

A—2.
B—3.
C—4.

4590. (Refer to Figure 104.) If the radio magnetic indicator is tuned to a VOR, which illustration indicates the aircraft is on the 010° radial?

A—1.
B—2.
C—3.

4591. (Refer to Figure 105.) If the magnetic heading shown for airplane 7 is maintained, which ADF illustration would indicate the airplane is on the 120° magnetic bearing FROM the station?

A—2.
B—4.
C—5.

4592. (Refer to Figure 105.) If the magnetic heading shown for airplane 5 is maintained, which ADF illustration would indicate the airplane is on the 210° magnetic bearing FROM the station?

A—2.
B—3.
C—4.

4593. (Refer to Figure 105.) If the magnetic heading shown for airplane 3 is maintained, which ADF illustration would indicate the airplane is on the 120° magnetic bearing TO the station?

A—4.
B—5.
C—8.

4594. (Refer to Figure 105.) If the magnetic heading shown for airplane 1 is maintained, which ADF illustration would indicate the airplane is on the 060° magnetic bearing TO the station?

A—2.
B—4.
C—5.

4595. (Refer to Figure 105.) If the magnetic heading shown for airplane 2 is maintained, which ADF illustration would indicate the airplane is on the 255° magnetic bearing TO the station?

A—2.
B—4.
C—5.

4596. (Refer to Figure 105.) If the magnetic heading shown for airplane 4 is maintained, which ADF illustration would indicate the airplane is on the 135° magnetic bearing TO the station?

A—1.
B—4.
C—8.

4597. (Refer to Figure 105.) If the magnetic heading shown for airplane 6 is maintained, which ADF illustration would indicate the airplane is on the 255° magnetic bearing FROM the station?

A—2.
B—4.
C—5.

4598. (Refer to Figure 105.) If the magnetic heading shown for airplane 8 is maintained, which ADF illustration would indicate the airplane is on the 090° magnetic bearing FROM the station?

A—3.
B—4.
C—6.

4599. (Refer to Figure 105.) If the magnetic heading shown for airplane 5 is maintained, which ADF illustration would indicate the airplane is on the 240° magnetic bearing TO the station?

A—2.
B—3.
C—4.

4600. (Refer to Figure 105.) If the magnetic heading shown for airplane 8 is maintained, which ADF illustration would indicate the airplane is on the 315° magnetic bearing TO the station?

A—3.
B—4.
C—1.

4601. (Refer to Figure 106.) The course selector of each aircraft is set on 360°. Which aircraft would have a FROM indication on the ambiguity meter and the CDI pointing left of center?

A—1.
B—2.
C—3.

4602. (Refer to Figure 107.) Where should the bearing pointer be located relative to the wingtip reference to maintain the 16 DME range in a right-hand arc with a right crosswind component?

A—Behind the right wingtip reference for VOR-2.
B—Ahead of the right wingtip reference for VOR-2.
C—Behind the right wingtip reference for VOR-1.

4603. (Refer to Figure 108.) Where should the bearing pointer be located relative to the wingtip reference to maintain the 16 DME range in a left-hand arc with a left crosswind component?

A—Ahead of the left wingtip reference for the VOR-2.
B—Ahead of the right wingtip reference for the VOR-1.
C—Behind the left wingtip reference for the VOR-2.

4604. Determine the approximate time and distance to a station if a 5° wingtip bearing change occurs in 1.5 minutes with a true airspeed of 95 knots.

A—16 minutes and 14.3 NM.
B—18 minutes and 28.5 NM.
C—18 minutes and 33.0 NM.

4605. During the en route phase of an IFR flight, the pilot is advised "Radar service terminated." What action is appropriate?

A—Set transponder to code 1200.
B—Resume normal position reporting.
C—Activate the IDENT feature of the transponder to re-establish radar contact.

4606. (Refer to Figure 109.) In which general direction from the VORTAC is the aircraft located?

A—Northeast.
B—Southeast.
C—Southwest.

4607. (Refer to Figure 110.) In which general direction from the VORTAC is the aircraft located?

A—Southwest.
B—Northwest.
C—Northeast.

4608. (Refer to Figure 111.) In which general direction from the VORTAC is the aircraft located?

A—Northeast.
B—Southeast.
C—Northwest.

4609. (Refer to Figure 112.) You arrive at the 15 DME fix on a heading of 350°. Which holding pattern correctly complies with the ATC clearance below, and what is the recommended entry procedure?

"...HOLD WEST OF THE ONE FIVE DME FIX ON THE ZERO EIGHT SIX RADIAL OF THE ABC VORTAC, FIVE MILE LEGS, LEFT TURNS..."

A—1; teardrop entry.
B—1; direct entry.
C—2; direct entry.

4610. (Refer to Figure 113.) You receive this ATC clearance:

"...HOLD EAST OF THE ABC VORTAC ON THE ZERO NINER ZERO RADIAL, LEFT TURNS..."

What is the recommended procedure to enter the holding pattern?

A—Parallel only.
B—Direct only.
C—Teardrop only.

4611. (Refer to Figure 113.) You receive this ATC clearance:

"...CLEARED TO THE ABC VORTAC. HOLD SOUTH ON THE ONE EIGHT ZERO RADIAL..."

What is the recommended procedure to enter the holding pattern?

A—Teardrop only.
B—Direct only.
C—Parallel only.

4612. (Refer to Figure 113.) You receive this ATC clearance:

"...CLEARED TO THE XYZ VORTAC. HOLD NORTH ON THE THREE SIX ZERO RADIAL, LEFT TURNS..."

What is the recommended procedure to enter the holding pattern.

A—Parallel only.
B—Direct only.
C—Teardrop only.

4613. (Refer to Figure 113.) You receive this ATC clearance:

"...CLEARED TO THE ABC VORTAC. HOLD WEST ON THE TWO SEVEN ZERO RADIAL..."

What is the recommended procedure to enter the holding pattern?

A—Parallel only.
B—Direct only.
C—Teardrop only.

4614. (Refer to Figure 114.) A pilot receives this ATC clearance:

"...CLEARED TO THE ABC VORTAC. HOLD WEST ON THE TWO SEVEN ZERO RADIAL..."

What is the recommended procedure to enter the holding pattern?

A—Parallel or teardrop.
B—Parallel only.
C—Direct only.

4615. (Refer to Figure 114.) A pilot receives this ATC clearance:

"...CLEARED TO THE XYZ VORTAC. HOLD NORTH ON THE THREE SIX ZERO RADIAL, LEFT TURNS..."

What is the recommended procedure to enter the holding pattern?

A—Teardrop only.
B—Parallel only.
C—Direct only.

4616. (Refer to Figure 114.) A pilot receives this ATC clearance:

"...CLEARED TO THE ABC VORTAC. HOLD SOUTH ON THE ONE EIGHT ZERO RADIAL..."

What is the recommended procedure to enter the holding pattern?

A—Teardrop only.
B—Parallel only.
C—Direct only.

4617. To ensure proper airspace protection while in a holding pattern, what is the recommended maximum airspeed above 14,000 feet?

A—220 knots.
B—265 knots.
C—200 knots.

4618. (Refer to Figure 115.) You receive this ATC clearance:

"...HOLD WEST OF THE ONE FIVE DME FIX ON THE ZERO EIGHT SIX RADIAL OF ABC VORTAC, FIVE MILE LEGS, LEFT TURNS..."

You arrive at the 15 DME fix on a heading of 350°. Which holding pattern correctly complies with these instructions, and what is the recommended entry procedure?

A—1; teardrop.
B—2; direct.
C—1; direct.

4619. (Refer to Figure 116.) You arrive over the 15 DME fix on a heading of 350°. Which holding pattern correctly complies with the ATC clearance below, and what is the recommended entry procedure?

"...HOLD WEST OF THE ONE FIVE DME FIX ON THE TWO SIX EIGHT RADIAL OF THE ABC VORTAC, FIVE MILE LEGS, LEFT TURNS..."

A—1; teardrop entry.
B—2; direct entry.
C—1; direct entry.

4620. At what point should the timing begin for the first leg outbound in a nonstandard holding pattern?

A—Abeam the holding fix, or wings level, whichever occurs last.
B—When the wings are level at the completion of the 180° turn outbound.
C—When abeam the holding fix.

4621. (Refer to Figure 117.) You receive this ATC clearance:

"...CLEARED TO THE ABC NDB. HOLD SOUTHEAST ON THE ONE FOUR ZERO DEGREE BEARING FROM THE NDB. LEFT TURNS..."

At station passage you note the indications in Figure 117. What is the recommended procedure to enter the holding pattern?

A—Direct only.
B—Teardrop only.
C—Parallel only

4622. (Refer to Figure 117.) You receive this ATC clearance:

"...CLEARED TO THE XYZ NDB. HOLD NORTHEAST ON THE ZERO FOUR ZERO DEGREE BEARING FROM THE NDB. LEFT TURNS..."

At station passage you note the indications in Figure 117. What is the recommended procedure to enter the holding pattern?

A—Direct only.
B—Teardrop only.
C—Parallel only.

4623. (Refer to Figure 117.) You receive this ATC clearance:

"...CLEARED TO THE ABC NDB. HOLD SOUTHWEST ON THE TWO THREE ZERO DEGREE BEARING FROM THE NDB..."

At station passage you note the indications in Figure 117. What is the recommended procedure to enter the holding pattern?

A—Direct only.
B—Teardrop only.
C—Parallel only.

4624. What timing procedure should be used when performing a holding pattern at a VOR?

A—Timing for the outbound leg begins over or abeam the VOR, whichever occurs later.
B—Timing for the inbound leg begins when initiating the turn inbound.
C—Adjustments in timing of each pattern should be made on the inbound leg.

4625. When holding at an NDB, at what point should the timing begin for the second leg outbound?

A—When the wings are level and the wind drift correction angle is established after completing the turn to the outbound heading.
B—When the wings are level after completing the turn to the outbound heading, or abeam the fix, whichever occurs first.
C—When abeam the holding fix.

4626. To ensure proper airspace protection while holding at 5,000 feet in a civil turbojet airplane, what is the recommended maximum indicated airspeed a pilot should use?

A—230 knots.
B—200 knots.
C—210 knots.

4627. If only one missed approach procedure is available, which of the following conditions is required when conducting "timed approaches from a holding fix"?

A—The pilot must contact the airport control tower prior to departing the holding fix inbound.
B—The reported ceiling and visibility minimums must be equal to or greater than the highest prescribed circling minimums for the IAP.
C—The reported ceiling and visibility minimums must be equal to or greater than the highest prescribed straight-in MDA minimums for the IAP.

4628. Prior to conducting "timed approaches from a holding fix," which one of the following is required?

A—The time required to fly from the primary facility to the field boundary must be determined by a reliable means.
B—The airport where the approach is to be conducted must have a control tower in operation.
C—The pilot must have established two-way communications with the tower before departing the holding fix.

4629. When making a "timed approach" from a holding fix at the outer marker, the pilot should adjust the

A—holding pattern to start the procedure turn at the assigned time.
B—airspeed at the final approach fix in order to arrive at the missed approach point at the assigned time.
C—holding pattern to leave the final approach fix inbound at the assigned time.

4630. If a pilot elects to proceed to the selected alternate, the landing minimums used at that airport should be the

A—minimums specified for the approach procedure selected.
B—alternate minimums shown on the approach chart.
C—minimums shown for that airport in a separate listing of "IFR Alternate Minimums."

4631. If the pilot loses visual reference while circling to land from an instrument approach and ATC radar service is not available, the missed approach action should be to

A—execute a climbing turn to parallel the published final approach course and climb to the initial approach altitude.
B—climb to the published circling minimums then proceed direct to the final approach fix.
C—make a climbing turn toward the landing runway and continue the turn until established on the missed approach course.

4632. When the approach procedure involves a procedure turn, the maximum speed should not be greater than

A—180 knots IAS.
B—200 knots IAS.
C—250 knots IAS.

4633. Under which of the following circumstances will ATC issue a VFR restriction to an IFR flight?

A—Whenever the pilot reports the loss of any navigational aid.
B—When it is necessary to provide separation between IFR and special VFR traffic.
C—When the pilot requests it.

4634. What is expected of you as pilot on an IFR flight plan if you are descending or climbing in VFR conditions?

A—If on an airway, climb or descend to the right of the centerline.
B—Advise ATC you are in visual conditions and will remain a short distance to the right of the centerline while climbing.
C—Execute gentle banks, left and right, at a frequency which permits continuous visual scanning of the airspace about you.

4635. (Refer to Figure 118.) During the ILS RWY 12L procedure at DSM, what altitude minimum applies if the glide slope becomes inoperative?

A—1,420 feet.
B—1,360 feet.
C—1,121 feet.

4636. What does the absence of the procedure turn barb on the planview on an approach chart indicate?

A—A procedure turn is not authorized.
B—Teardrop-type procedure turn is authorized.
C—Racetrack-type procedure turn is authorized.

4637. When making an instrument approach at the selected alternate airport, what landing minimums apply?

A—Standard alternate minimums (600-2 or 800-2).
B—The IFR alternate minimums listed for that airport.
C—The landing minimums published for the type of procedure selected.

4638. Which is true regarding the use of a SID chart?

A—At airfields where SID's have been established, SID usage is mandatory for IFR departures.
B—To use a SID, the pilot must possess at least the textual description of the approved standard departure.
C—To use a SID, the pilot must possess both the textual and graphic form of the approved standard departure.

4639. For airship IFR operations off established airways, the "route of flight" portion of an IFR flight plan should list VOR navigational aids which are no more than

A—40 NM apart.
B—70 NM apart.
C—80 NM apart.

4640. Which is true regarding STAR's?

A—STAR's are used to separate IFR and VFR traffic.
B—STAR's are established to simplify clearance delivery procedures.
C—STAR's are used at certain airports to decrease traffic congestion.

4641. While being radar vectored, an approach clearance is received. The last assigned altitude should be maintained until

A—reaching the FAF.
B—advised to begin descent.
C—established on a segment of a published route or IAP.

4642. (Refer to Figure 119.) The final approach fix for the precision approach is located at

A—DENAY intersection.
B—Glide slope intercept (lightning bolt).
C—ROMEN intersection/locator outer marker.

4643. When operating an airship under IFR with a VFR-On-Top clearance, what altitude should be maintained?

A—The last IFR altitude assigned by ATC.
B—An IFR cruising altitude appropriate to the magnetic course being flown.
C—A VFR cruising altitude appropriate to the magnetic course being flown and as restricted by ATC.

4644. Does the ATC term "cleared to cruise" apply to airship IFR operations?

A—No, this term applies to airplane IFR operations only.
B—Yes, it means that any assigned altitude can be vacated without notifying ATC.
C—Yes, in part, it authorizes the pilot to commence the approach at the destination airport at pilot's discretion.

4645. (Refer to Figure 47.) En route on V112 from BTG VORTAC to DLS VORTAC, the minimum altitude crossing Gymme intersection is

A—6,400 feet.
B—6,500 feet.
C—7,000 feet.

4646. (Refer to Figure 47.) When en route on V448 from YKM VORTAC to BTG VORTAC, what minimum navigation equipment is required to identify ANGOO intersection?

A—One VOR receiver.
B—One VOR receiver and DME.
C—Two VOR receivers.

4647. (Refer to Figure 47.) En route on V468 from BTG VORTAC to YKM VORTAC, the minimum altitude at TROTS intersection is

A—7,100 feet.
B—10,000 feet.
C—11,500 feet.

4648. (Refer to Figure 120.) Refer to the DEN ILS RWY 35R procedure. The FAF intercept altitude is

A—7,488 feet MSL.
B—7,500 feet MSL.
C—9,000 feet MSL.

4649. (Refer to Figure 120.) The symbol on the planview of the ILS RWY 35R procedure at DEN represents a minimum safe sector altitude within 25 NM of

A—Denver VORTAC.
B—Gandi outer marker.
C—Denver/Stapleton International Airport.

4650. (Refer to Figure 121.) During the ILS RWY 30R procedure at DSM, the minimum altitude for glide slope interception is

A—2,365 feet MSL.
B—2,500 feet MSL.
C—3,000 feet MSL.

4651. (Refer to Figure 121.) During the ILS RWY 30R procedure at DSM, what MDA applies should the glide slope become inoperative?

A—1,157 feet.
B—1,320 feet.
C—1,360 feet.

4652. (Refer to Figure 122.) The missed approach point of the ATL S-LOC 8L procedure is located how far from the LOM?

A—4.8 NM.
B—5.1 NM.
C—5.2 NM.

4653. (Refer to Figure 123.) What minimum navigation equipment is required to complete the VOR/DME-A procedure?

A—One VOR receiver.
B—One VOR receiver and DME.
C—Two VOR receivers and DME.

4654. (Refer to Figure 13.) The symbol on the planview of the VOR/DME-A procedure at 7D3 represents a minimum safe sector altitude within 25 NM of

A—DEANI intersection.
B—White Cloud VORTAC.
C—Baldwin Municipal Airport.

4655. (Refer to Figure 124 or 124A.) What options are available concerning the teardrop course reversal for LOC RWY 35 approach to Duncan/Halliburton Field?

A—If a course reversal is required, only the teardrop can be executed.
B—The point where the turn is begun and the type and rate of turn are optional.
C—A normal procedure turn may be made if the 10 DME limit is not exceeded.

4656. (Refer to Figure 124 or 124A.) The point on the teardrop procedure where the turn inbound (LOC RWY 35) Duncan/Halliburton, is initiated is determined by

A—DME and timing to remain within the 10-NM limit.
B—Timing for a 2 minute maximum.
C—Estimating groundspeed and radius of turn.

4657. (Refer to Figure 125 or 125A.) If your aircraft was cleared for the ILS RWY 17R at Lincoln Municipal and crossed the Lincoln VOR at 5,000 feet MSL, at what point in the teardrop could a descent to 3,000 feet commence?

A—As soon as intercepting LOC inbound.
B—Immediately.
C—Only at the point authorized by ATC.

4658. (Refer to Figure 125 or 125A.) If cleared for an S-LOC 17R approach at Lincoln Municipal from over TOUHY, it means the flight should

A—land straight in on runway 17R.
B—comply with straight-in landing minimums.
C—begin final approach without making a procedure turn.

4659. (Refer to Figure 126 or 126A.) What landing minimums apply for a FAR part 91 operator at Dothan, AL using a category C aircraft during a circling LOC 31 approach at 120 knots? (DME available).

A—MDA 860 feet MSL and visibility 2 SM.
B—MDA 860 feet MSL and visibility 1 and 1/2 SM.
C—MDA 720 feet MSL and visibility 3/4 SM.

4660. (Refer to Figure 126 or 126A.) If cleared for a straight-in LOC approach from over OALDY, it means the flight should

A—land straight in on runway 31.
B—comply with straight-in landing minimums.
C—begin final approach without making a procedure turn.

4661. (Refer to Figure 126 or 126A.) What is the ability to identify the RRS 2.5 stepdown fix worth in terms of localizer circle-to-land minimums for a category C aircraft?

A—Decreases MDA by 20 feet.
B—Decreases visibility by 1/2 SM.
C—Without the stepdown fix, a circling approach is not available.

4662. (Refer to Figure 127 or 127A.) If cleared for NDB RWY 28 approach (Lancaster/Fairfield) over ZZV VOR, the flight would be expected to

Category A aircraft
Last assigned altitude 3,000 feet

A—proceed straight in from CRISY to the S-28 minimums of 1620-1.
B—proceed to CRISY, then execute the teardrop procedure as depicted on the approach chart.
C—proceed direct to CASER, then straight in to S-28 minimums of 1620-1.

4663. When a VOR/DME is collocated under frequency pairings and the VOR portion is inoperative, the DME identifier will repeat at an interval of

A—20 second intervals at 1020 Hz.
B—30 second intervals at 1350 Hz.
C—60 second intervals at 1350 Hz.

4664. When installed with the ILS and specified in the approach procedures, DME may be used

A—in lieu of the OM.
B—in lieu of visibility requirements.
C—to determine distance from TDZ.

4665. By which means may a pilot determine if a Loran C equipped aircraft is approved for IFR operations?

A—Not necessary; Loran C is not approved for IFR.
B—Check aircraft logbook.
C—Check the Airplane Flight Manual Supplement.

4666. Full scale deflection of a CDI occurs when the course deviation bar or needle

A—deflects from left side of the scale to right side of the scale.
B—deflects from the center of the scale to either far side of the scale.
C—deflects from half scale left to half scale right.

4667. If an early missed approach is initiated before reaching the MAP, the following procedure should be used unless otherwise cleared by ATC.

A—Proceed to the missed approach point at or above the MDA or DH before executing a turning maneuver.
B—Begin a climbing turn immediately and follow missed approach procedures.
C—Maintain altitude and continue past MAP for 1 minute or 1 mile whichever occurs first.

4668. When more than one circuit of the holding pattern is needed to lose altitude or become better established on course, the additional circuits can be made

A—at pilot's discretion.
B—only in an emergency.
C—only if pilot advises ATC and ATC approves.

4669. How does a pilot determine if DME is available on an ILS/LOC?

A—IAP indicate DME\TACAN channel in LOC frequency box.
B—LOC\DME are indicated on en route low altitude frequency box.
C—LOC\DME frequencies available in the Airman's Information Manual.

4670. When simultaneous approaches are in progress, how does each pilot receive radar advisories?

A—On tower frequency.
B—On approach control frequency.
C—One pilot on tower frequency and the other on approach control frequency.

4671. During an instrument approach, under what conditions, if any, is the holding pattern course reversal not required?

A—When radar vectors are provided.
B—When cleared for the approach.
C—None, since it is always mandatory.

4672. During an instrument precision approach, terrain and obstacle clearance depends on adherence to

A—minimum altitude shown on the IAP.
B—terrain contour information.
C—natural and man-made reference point information.

4673. (Refer to Figure 128 or 128A.) What is the helicopter landing minimum for the VOR RWY 36 approach at Price/Carbon County Airport?

A—500-foot ceiling and 1/2 mile visibility.
B—1-mile visibility.
C—one-half mile visibility.

4674. (Refer to Figure 128 or 128A.) How should a pilot determine when the DME at Price/Carbon County Airport is inoperative?

A—The airborne DME will always indicate "0" mileage.
B—The airborne DME will "search," but will not "lock on."
C—The airborne DME may appear normal, but there will be no code tone.

4675. (Refer to Figure 128 or 128A.) What type entry is recommended for the missed approach holding pattern depicted on the VOR RWY 36 approach chart for Price/Carbon County Airport?

A—Direct only.
B—Teardrop only.
C—Parallel only.

4676. (Refer to Figure 128 or 128A.) What is the helicopter MDA for a straight-in VOR RWY 36 approach at Price/Carbon County Airport (VOR only)?

A—6,090 feet MSL.
B—500 feet MSL.
C—6,400 feet MSL.

4677. (Refer to Figure 128 or 128A.) At which points may you initiate a descent to the next lower minimum altitude when cleared for the VOR RWY 36 approach, from the PUC R-095 IAF (DME operative)?

A—Start descent from 8,000 when established on final, from 7,500 when at the 4 DME fix, and from 6,180 when landing requirements are met.
B—Start descent from 8,000 when established on the PUC R-186, from 6,400 at the 4 DME fix, and from 6,180 when landing requirements are met.
C—Start descent from 8,000 at the R-127, from 6,400 at the LR-127, from 6,180 at the 4 DME fix.

4678. (Refer to Figure 128 or 128A.) What is the purpose of the 10,300 MSA on the Price/Carbon County Airport Approach Chart?

A—It provides safe clearance above the highest obstacle in the defined sector out to 25 NM.
B—It provides an altitude above which navigational course guidance is assured.
C—It is the minimum vector altitude for radar vectors in the sector southeast of PUC between 020° and 290° magnetic bearing to PUC VOR.

4679. (Refer to Figure 129 or 129A.) As you approach LABER during a straight-in RNAV RWY 36 approach in a helicopter, Little Rock Approach Control advises that the ceiling is 400 feet and the visibility is 1/4 mile. Do regulations permit you to continue the approach and land?

A—No, you may not reduce the visibility prescribed for Category A airplanes by more than 50 percent.
B—Yes, only a 1/4 mile visibility or an RVR of 1,200 feet is required for any approach, including RNAV.
C—No, neither the ceiling nor the visibility meet regulatory requirements.

4680. (Refer to Figure 129 or 129A.) What indication should you get when it is time to turn inbound while in the procedure turn at LABER?

A—4 DME miles from LABER.
B—10 DME miles from the MAP.
C—12 DME miles from LIT VORTAC.

4681. (Refer to Figure 129 or 129A.) What type of entry is recommended to the missed approach holding pattern if the inbound heading is 050°?

A—Direct.
B—Parallel.
C—Teardrop.

4682. (Refer to Figure 129 or 129A.) How should the missed approach point be identified when executing the RNAV RWY 36 approach at Adams Field?

A—When the TO-FROM indicator changes.
B—Upon arrival at 760 feet on the glidepath.
C—When time has expired for 5 NM past the FAF.

4683. (Refer to Figure 129 or 129A.) What is the position of LABER relative to the reference facility?

A—316°, 24.3 NM.
B—177°, 10 NM.
C—198°, 8 NM.

4684. (Refer to Figure 129 or 129A.) What minimum airborne equipment is required to be operative for RNAV RWY 36 approach at Adams Field?

A—An approved RNAV receiver that provides both horizontal and vertical guidance.
B—A transponder and an approved RNAV receiver that provides both horizontal and vertical guidance.
C—Any approved RNAV receiver.

4685. (Refer to Figure 130 or 130A.) How does an LDA facility, such as the one at Roanoke Regional, differ from a standard ILS approach facility?

A—The LOC is wider.
B—The LOC is offset from the runway.
C—The GS is unusable beyond the MM.

4686. (Refer to Figure 130 or 130A.) What are the procedure turn restrictions on the LDA RWY 6 approach at Roanoke Regional?

A—Remain within 10 NM of CLAMM INT and on the north side of the approach course.
B—Remain within 10 NM of the airport on the north side of the approach course.
C—Remain within 10 NM of the outer marker on the north side of the approach course.

4687. (Refer to Figure 130 or 130A.) What are the restrictions regarding circle to land procedures for LDA RWY/GS 6 approach at Roanoke Regional?

A—Circling to runway 24 not authorized.
B—Circling not authorized NW of RWY 6-24.
C—Visibility increased 1/2 mile for circling approach.

4688. (Refer to Figure 130 or 130A.) At what minimum altitude should you cross CLAMM intersection during the S-LDA 6 approach at Roanoke Regional?

A—4,200 MSL.
B—4,182 MSL.
C—2,800 MSL.

4689. (Refer to Figure 130 or 130A.) How should the pilot identify the missed approach point for the S-LDA GS 6 approach to Roanoke Regional?

A—Arrival at 1,540 feet on the glide slope.
B—Arrival at 1.0 DME on the LDA course.
C—Time expired for distance from OM to MAP.

4690. (Refer to Figure 131.) What is the landing minimum if you are cleared for the LOC RWY S-18 approach at Stapleton International?

A—5,000 RVR.
B—300-foot ceiling and 1 mile visibility.
C—1 mile visibility.

4691. (Refer to Figure 131.) What is the Category B HAT for the LOC RWY S-18 approach at Stapleton International?

A—300 AGL.
B—547 AGL.
C—319 AGL.

4692. Which of the following statements is true regarding Parallel ILS approaches?

A—Parallel ILS approach runway centerlines are separated by at least 4,300 feet and standard IFR separation is provided on the adjacent runway.
B—Parallel ILS approaches provide aircraft a minimum of two miles radar separation between successive aircraft on the adjacent localizer course.
C—Landing minimums to the adjacent runway will be higher than the minimums to the primary runway, but will normally be lower than the published circling minimums.

4693. (Refer to Figure 131.) Which approach and tower frequencies are appropriate if you are approaching Stapleton International Airport from the north and are cleared for the LOC RWY 18 approach?

A—120.8 and 119.5.
B—127.4 and 118.3.
C—125.6 and 119.5.

4694. (Refer to Figure 131.) What are the alternate minimums for the LOC RWY 18 approach at Stapleton International Airport?

A—800-2.
B—600-2.
C—Nonstandard.

4695. (Refer to Figure 131.) What minimum equipment, in addition to voice communications, is required to make the LOC RWY 18 approach at Stapleton International Airport?

A—VOR/LOC, transponder with altitude/ encoding.
B—VOR/LOC, ADF, and DME.
C—DME, transponder with altitude encoding, VOR/LOC, and Marker Beacon.

4696. (Refer to Figure 133 or 133A.) How should a pilot reverse course to get established on the inbound course of the ILS RWY 9, if radar vectoring or the three IAF's are not utilized?

A—Execute a standard 45° procedure turn toward Seal Beach VORTAC or Pomona VORTAC.
B—Make an appropriate entry to the depicted holding pattern at Swan Lake OM/INT.
C—Use any type of procedure turn, but remain within 10 NM of Riverside VOR.

4697. (Refer to Figure 133 or 133A.) If the Class D airspace is not effective, what is the LOC/VOR minima for a helicopter if cleared for the S-LOC 9 approach at Riverside Municipal?

A—1,200 and 1/4 mile.
B—991 and RVR 24.
C—1,300 and 1/4 mile.

4698. (Refer to Figure 133 or 133A.) What type of entry is recommended for the missed approach holding pattern at Riverside Municipal?

A—Direct.
B—Parallel.
C—Teardrop.

4699. (Refer to Figure 133 or 133A.) What action should the pilot take if the marker beacon receiver becomes inoperative during the S-ILS 9 approach at Riverside Municipal?

A—Substitute SWAN LAKE INT. for the OM and surveillance radar for the MM.
B—Raise the DH 100 feet (50 feet for the OM and 50 feet for the MM).
C—Substitute SWAN LAKE INT. for the OM and use published minimums.

4700. (Refer to Figure 133 or 133A.) Why are two VOR/LOC receivers recommended to obtain an MDA of 1,160 when making an S-LOC 9 approach to Riverside Municipal?

A—To obtain R-327 of PDZ when on the localizer course.
B—In order to identify Riverside VOR.
C—To utilize the published stepdown fix.

4701. (Refer to Figure 133 or 133A.) What is the minimum altitude descent procedure if cleared for the S-ILS 9 approach from Seal Beach VORTAC?

A—Descend and maintain 3,000 to JASER INT, descend to and maintain 2,500 until crossing SWAN LAKE, descend and maintain 1,260 until crossing AGNES, and to 991 (DH) after passing AGNES.
B—Descend and maintain 3,000 to JASER INT, descend to 2,800 when established on the LOC course, intercept and maintain the GS to 991 (DH).
C—Descend and maintain 3,000 to JASER INT, descend to 2,500 while established on the LOC course inbound, intercept and maintain the GS to 991 (DH).

4702. What is a difference between an SDF and an LDA facility?

A—The SDF course width is either 6° or 12° while the LDA course width is approximately 5°.
B—The SDF course has no glide slope guidance while the LDA does.
C—The SDF has no marker beacons while the LDA has at least an OM.

4703. What is the difference between a Localizer-Type Directional Aid (LDA) and the ILS localizer?

A—The LDA is not aligned with the runway.
B—The LDA uses a course width of 6° or 12°, while an ILS uses only 5°.
C—The LDA signal is generated from a VOR-type facility and has no glide slope.

4704. How wide is an SDF course?

A—Either 3° or 6°.
B—Either 6° or 12°.
C—Varies from 5° to 10°.

4705. What are the main differences between the SDF and the localizer of an ILS?

A—The useable off-course indications are limited to 35° for the localizer and up to 90° for the SDF.
B—The SDF course may not be aligned with the runway and the course may be wider.
C—The course width for the localizer will always be 5° while the SDF course will be between 6° and 12°.

4706. A pilot is making an ILS approach and is past the OM to a runway which has a VASI. What action should the pilot take if an electronic glide slope malfunction occurs and the pilot has the VASI in sight?

A—The pilot should inform ATC of the malfunction and then descend immediately to the localizer DH and make a localizer approach.
B—The pilot may continue the approach and use the VASI glide slope in place of the electronic glide slope.
C—The pilot must request an LOC approach, and may descend below the VASI at the pilot's discretion.

4707. What wind condition prolongs the hazards of wake turbulence on a landing runway for the longest period of time?

A—Direct headwind.
B—Direct tailwind.
C—Light quartering tailwind.

4708. Wake turbulence is near maximum behind a jet transport just after takeoff because

A—the engines are at maximum thrust output at slow airspeed.
B—the gear and flap configuration increases the turbulence to maximum.
C—of the high angle of attack and high gross weight.

4709. What effect would a light crosswind of approximately 7 knots have on vortex behavior?

A—The light crosswind would rapidly dissipate vortex strength.
B—The upwind vortex would tend to remain over the runway.
C—The downwind vortex would tend to remain over the runway.

4710. When landing behind a large jet aircraft, at which point on the runway should you plan to land?

A—If any crosswind, land on the windward side of the runway and prior to the jet's touchdown point.
B—At least 1,000 feet beyond the jet's touchdown point.
C—Beyond the jet's touchdown point.

4711. Where may you use a surveillance approach?

A—At any airport that has an approach control.
B—At any airport which has radar service.
C—At airports for which civil radar instrument approach minimums have been published.

4712. You arrive at your destination airport on an IFR flight plan in a helicopter. Which is a prerequisite condition for the performance of a contact approach?

A—Clear of clouds and at least 1 SM flight visibility.
B—A ground visibility of at least 2 SM.
C—A flight visibility of at least 1/2 NM.

4713. During a precision instrument approach (using Category A minimums) a helicopter may not be operated below DH unless

A—the ceiling is forecast to be at or above landing minimums prescribed for that procedure.
B—positioned such that a normal approach to the runway of intended landing can be made.
C—the visibility is forecast to be at or above the landing minimums prescribed for that procedure.

4714. Which procedure should be followed by a pilot who is circling to land in a Category B airplane, but is maintaining a speed 5 knots faster than the maximum specified for that category?

A—Use the approach minimums appropriate for Category C.
B—Use Category B minimums.
C—Use Category D minimums since they apply to all circling approaches.

4715. How can an IAF be identified on a Standard Instrument Approach Procedure (SIAP) Chart?

A—All fixes that are labeled IAF.
B—Any fix illustrated within the 10 mile ring other than the FAF or stepdown fix.
C—The procedure turn and the fixes on the feeder facility ring.

4716. RVR minimums for landing are prescribed in an IAP, but RVR is inoperative and cannot be reported for the intended runway at the time. Which of the following would be an operational consideration?

A—RVR minimums which are specified in the procedures should be converted and applied as ground visibility.
B—RVR minimums may be disregarded, providing the runway has an operative HIRL system.
C—RVR minimums may be disregarded, providing all other components of the ILS system are operative.

4717. Aircraft approach categories are based on

A—certificated approach speed at maximum gross weight.
B—1.3 times the stall speed in landing configuration at maximum gross landing weight.
C—1.3 times the stall speed at maximum gross weight.

4718. What are the main differences between a visual approach and a contact approach?

A—The pilot must request a contact approach; the pilot may be assigned a visual approach and higher weather minimums must exist.
B—The pilot must request a visual approach and report having the field in sight; ATC may assign a contact approach if VFR conditions exist.
C—Any time the pilot reports the field in sight, ATC may clear the pilot for a contact approach; for a visual approach, the pilot must advise that the approach can be made under VFR conditions.

4719. When a pilot elects to proceed to the selected alternate airport, which minimums apply for landing at the alternate?

A—600-1 if the airport has an ILS.
B—Ceiling 200 feet above the published minimum; visibility 2 miles.
C—The landing minimums for the approach to be used.

4720. When passing through an abrupt wind shear which involves a shift from a tailwind to a headwind, what power management would normally be required to maintain a constant indicated airspeed and ILS glide slope?

A—Higher than normal power initially, followed by a further increase as the wind shear is encountered, then a decrease.
B—Lower than normal power initially, followed by a further decrease as the wind shear is encountered, then an increase.
C—Higher than normal power initially, followed by a decrease as the shear is encountered, then an increase.

4721. What effect will a change in wind direction have upon maintaining a 3° glide slope at a constant true airspeed?

A—When groundspeed decreases, rate of descent must increase.
B—When groundspeed increases, rate of descent must increase.
C—Rate of descent must be constant to remain on the glide slope.

4722. Upon what maximum airspeed is the instrument approach criteria for a helicopter based?

A—100 knots.
B—90 knots.
C—80 knots.

4723. All helicopters are considered to be in which approach category for a helicopter IAP?

A—A.
B—A or B, depending upon weight.
C—B.

4724. What reduction, if any, to visibility requirements is authorized when using a fixed-wing IAP for a helicopter instrument approach?

A—All visibility requirements may be reduced by one-half.
B—All visibility requirements may be reduced by one-fourth.
C—The visibility requirements may be reduced by one-half, but in no case lower than 1,200 RVR or 1/4 mile.

4725. What is the pilot in command's responsibility when flying a propeller aircraft within 20 miles of the airport of intended landing and ATC requests the pilot to reduce speed to 160? (Pilot complies with speed adjustment.)

A—Reduce TAS to 160 knots and maintain until advised by ATC.
B—Reduce IAS to 160 MPH and maintain until advised by ATC.
C—Reduce IAS to 160 knots and maintain that speed within 10 knots.

4726. You are being vectored to the ILS approach course, but have not been cleared for the approach. It becomes evident that you will pass through the localizer course. What action should be taken?

A—Turn outbound and make a procedure turn.
B—Continue on the assigned heading and query ATC.
C—Start a turn to the inbound heading and inquire if you are cleared for the approach.

4727. While flying a 3° glide slope, a constant tailwind shears to a calm wind. Which conditions should the pilot expect?

A—Airspeed and pitch attitude decrease and there is a tendency to go below glide slope.
B—Airspeed and pitch attitude increase and there is a tendency to go below glide slope.
C—Airspeed and pitch attitude increase and there is a tendency to go above glide slope.

4728. How is ATC radar used for instrument approaches when the facility is approved for approach control service?

A—Precision approaches, weather surveillance, and as a substitute for any inoperative component of a navigation aid used for approaches.
B—ASR approaches, weather surveillance, and course guidance by approach control.
C—Course guidance to the final approach course, ASR and PAR approaches, and the monitoring of nonradar approaches.

4729. Which range facility associated with the ILS is identified by the last two letters of the localizer identification group?

A—Inner marker.
B—Outer marker.
C—Middle compass locator.

4730. Which range facility associated with the ILS can be identified by a two-letter coded signal?

A—Middle marker.
B—Outer marker.
C—Compass locator.

4731. Which pilot action is appropriate if more than one component of an ILS is unusable?

A—Use the highest minimum required by any single component that is unusable.
B—Request another approach appropriate to the equipment that is useable.
C—Raise the minimums a total of that required by each component that is unusable.

4732. Which substitution is permitted when an ILS component is inoperative?

A—A compass locator or precision radar may be substituted for the ILS outer or middle marker.
B—ADF or VOR bearings which cross either the outer or middle marker sites may be substituted for these markers.
C—DME, when located at the localizer antenna site, should be substituted for the outer or middle marker.

4733. What facilities, if any, may be substituted for an inoperative middle marker during an ILS approach without affecting the straight-in minimums?

A—ASR.
B—Substitution not necessary, minimums do not change.
C—Compass locator, PAR, and ASR.

4734. When being radar vectored for an ILS approach, at what point may you start a descent from your last assigned altitude to a lower minimum altitude if cleared for the approach?

A—When established on a segment of a published route or IAP.
B—You may descend immediately to published glide slope interception altitude.
C—Only after you are established on the final approach unless informed otherwise by ATC.

4735. What are the requirements for a contact approach to an airport that has an approved IAP, if the pilot is on an instrument flight plan and clear of clouds?

A—The controller must determine that the pilot can see the airport at the altitude flown and can remain clear of clouds.
B—The pilot must agree to the approach when given by ATC and the controller must have determined that the visibility was at least 1 mile and be reasonably sure the pilot can remain clear of clouds.
C—The pilot must request the approach, have at least 1-mile visibility, and be reasonably sure of remaining clear of clouds.

4736. When is radar service terminated during a visual approach?

A—Automatically when ATC instructs the pilot to contact the tower.
B—Immediately upon acceptance of the approach by the pilot.
C—When ATC advises, "Radar service terminated; resume own navigation."

4737. When may you obtain a contact approach?

A—ATC may assign a contact approach if VFR conditions exist or you report the runway in sight and are clear of clouds.
B—ATC may assign a contact approach if you are below the clouds and the visibility is at least 1 mile.
C—ATC will assign a contact approach only upon request if the reported visibility is at least 1 mile.

4738. Under which conditions is hydroplaning most likely to occur?

A—When rudder is used for directional control instead of allowing the nosewheel to contact the surface early in the landing roll on a wet runway.
B—During conditions of standing water, slush, high speed, and smooth runway texture.
C—During a landing on any wet runway when brake application is delayed until a wedge of water begins to build ahead of the tires.

4739. Thrust is managed to maintain IAS, and glide slope is being flown. What characteristics should be observed when a headwind shears to be a constant tailwind?

A—PITCH ATTITUDE: Increases; REQUIRED THRUST: Increased, then reduced; VERTICAL SPEED: Increases; IAS: Increases, then decreases to approach speed.
B—PITCH ATTITUDE: Decreases; REQUIRED THRUST: Increased, then reduced; VERTICAL SPEED: Increases; IAS: Decreases, then increases to approach speed.
C—PITCH ATTITUDE: Increases; REQUIRED THRUST: Reduced, then increased; VERTICAL SPEED: Decreases; IAS: Decreases, then increases to approach speed.

4740. When cleared to execute a published sidestep maneuver for a specific approach and landing on the parallel runway, at what point is the pilot expected to commence this maneuver?

A—At the published minimum altitude for a circling approach.
B—As soon as possible after the runway or runway environment is in sight.
C—At the localizer MDA minimum and when the runway is in sight.

4741. Which information, in addition to headings, does the radar controller provide without request during an ASR approach?

A—The recommended altitude for each mile from the runway.
B—When reaching the MDA.
C—When to commence descent to MDA, the aircraft's position each mile on final from the runway, and arrival at the MAP.

4742. Which of these facilities may be substituted for an MM during a complete ILS IAP?

A—Surveillance and precision radar.
B—Compass locator and precision radar.
C—A VOR/DME fix.

4743. What conditions are necessary before ATC can authorize a visual approach?

A—You must have the preceding aircraft in sight, and be able to remain in VFR weather conditions.
B—You must have the airport in sight or the preceding aircraft in sight, and be able to proceed to, and land in IFR conditions.
C—You must have the airport in sight or a preceding aircraft to be followed, and be able to proceed to the airport in VFR conditions.

4744. If all ILS components are operating and the required visual references are not established, the missed approach should be initiated upon

A—arrival at the DH on the glide slope.
B—arrival at the middle marker.
C—expiration of the time listed on the approach chart for missed approach.

4745. The rate of descent required to stay on the ILS glide slope

A—must be increased if the groundspeed is decreased.
B—will remain constant if the indicated airspeed remains constant.
C—must be decreased if the groundspeed is decreased.

4746. Which fixes on the IAP Charts are initial approach fixes?

A—Any fix on the en route facilities ring, the feeder facilities ring, and those at the start of arc approaches.
B—Only the fixes at the start of arc approaches and those on either the feeder facilities ring or en route facilities ring that have a transition course shown to the approach procedure.
C—Any fix that is identified by the letters IAF.

4747. Which indications will a pilot receive where an IM is installed on a front course ILS approach?

A—One dot per second and a steady amber light.
B—Six dots per second and a flashing white light.
C—Alternate dashes and a blue light.

4748. To remain on the ILS glidepath, the rate of descent must be

A—decreased if the airspeed is increased.
B—decreased if the groundspeed is increased.
C—increased if the groundspeed is increased.

4749. When may a pilot make a straight-in landing, if using an IAP having only circling minimums?

A—A straight-in landing may not be made, but the pilot may continue to the runway at MDA and then circle to land on the runway.
B—The pilot may land straight-in if the runway is the active runway and he has been cleared to land.
C—A straight-in landing may be made if the pilot has the runway in sight in sufficient time to make a normal approach for landing, and has been cleared to land.

4750. A contact approach is an approach procedure that may be used

A—in lieu of conducting a SIAP.
B—if assigned by ATC and will facilitate the approach.
C—in lieu of a visual approach.

4751. Under which condition does ATC issue a STAR?

A—To all pilots wherever STAR's are available.
B—Only if the pilot requests a STAR in the "Remarks" section of the flight plan.
C—When ATC deems it appropriate, unless the pilot requests "No STAR."

4752. The rate of descent on the glide slope is dependent upon

A—true airspeed.
B—calibrated airspeed.
C—groundspeed.

4753. Approximately what height is the glide slope centerline at the MM of a typical ILS?

A—100 feet.
B—200 feet.
C—300 feet.

4754. If the RVR is not reported, what meteorological value should you substitute for 2,400 RVR?

A—A ground visibility of 1/2 NM.
B—A slant range visibility of 2,400 feet for the final approach segment of the published approach procedure.
C—A ground visibility of 1/2 SM.

4755. While flying a 3° glide slope, a headwind shears to a tailwind. Which conditions should the pilot expect on the glide slope?

A—Airspeed and pitch attitude decrease and there is a tendency to go below glide slope.
B—Airspeed and pitch attitude increase and there is a tendency to go above glide slope.
C—Airspeed and pitch attitude decrease and there is a tendency to remain on the glide slope.

4756. The glide slope and localizer are centered, but the airspeed is too fast. Which should be adjusted initially?

A—Pitch and power.
B—Power only.
C—Pitch only.

4757. While being vectored, if crossing the ILS final approach course becomes imminent and an approach clearance has not been issued, what action should be taken by the pilot?

A—Turn outbound on the final approach course, execute a procedure turn, and inform ATC.
B—Turn inbound and execute the missed approach procedure at the outer marker if approach clearance has not been received.
C—Maintain the last assigned heading and query ATC.

4758. If during a VFR practice instrument approach, Radar Approach Control assigns an altitude or heading that will cause you to enter the clouds, what action should be taken?

A—Enter the clouds, since ATC authorization for practice approaches is considered an IFR clearance.
B—Avoid the clouds and inform ATC that altitude/heading will not permit VFR.
C—Abandon the approach.

4759. The RVR minimums for takeoff or landing are published in an IAP, but RVR is inoperative and cannot be reported for the runway at the time. Which of the following would apply?

A—RVR minimums which are specified in the procedure should be converted and applied as ground visibility.
B—RVR minimums may be disregarded, providing the runway has an operative HIRL system.
C—RVR minimums may be disregarded, providing all other components of the ILS system are operative.

4760. What are the alternate minimums that must be forecast at the ETA for an airport that has a precision approach procedure?

A—400-foot ceiling and 2 miles visibility.
B—600-foot ceiling and 2 miles visibility.
C—800-foot ceiling and 2 miles visibility.

4761. What point at the destination should be used to compute estimated time en route on an IFR flight plan?

A—The final approach fix on the expected instrument approach.
B—The initial approach fix on the expected instrument approach.
C—The point of first intended landing.

4762. If the RVR equipment is inoperative for an IAP that requires a visibility of 2,400 RVR, how should the pilot expect the visibility requirement to be reported in lieu of the published RVR?

A—As a slant range visibility of 2,400 feet.
B—As an RVR of 2,400 feet.
C—As a ground visibility of 1/2 SM.

4763. If during an ILS approach in IFR conditions, the approach lights are not visible upon arrival at the DH, the pilot is

A—required to immediately execute the missed approach procedure.
B—permitted to continue the approach and descend to the localizer MDA.
C—permitted to continue the approach to the approach threshold of the ILS runway.

4764. Immediately after passing the final approach fix inbound during an ILS approach in IFR conditions, the glide slope warning flag appears. The pilot is

A—permitted to continue the approach and descend to the DH.
B—permitted to continue the approach and descend to the localizer MDA.
C—required to immediately begin the prescribed missed approach procedure.

4765. In the case of operations over an area designated as a mountainous area, no person may operate an aircraft under IFR below 2,000 feet above the highest obstacle within a horizontal distance of

A—3 SM from the course flown.
B—4 SM from the course flown.
C—4 NM from the course flown.

4766. To ensure proper airspace protection while in a holding pattern, what is the recommended maximum indicated airspeed above 14,000 feet?

A—220 knots.
B—265 knots.
C—200 knots.

4767. Where a holding pattern is specified in lieu of a procedure turn, the holding maneuver must be executed within

A—the 1 minute time limitation or published leg length.
B—a radius of 5 miles from the holding fix.
C—10 knots of the specified holding speed.

4768. Which of the following conditions is required before "timed approaches from a holding fix" may be conducted?

A—If more than one missed approach procedure is available, only one may require a course reversal.
B—If more than one missed approach procedure is available, none may require a course reversal.
C—Direct communication between the pilot and the tower must be established prior to beginning the approach.

4769. An airport without an authorized IAP may be included on an IFR flight plan as an alternate, if the current weather forecast indicates that the ceiling and visibility at the ETA will

A—be at least 300 feet and 2 miles.
B—be at least 1,000 feet and 1 mile.
C—allow for a descent from the MEA approach, and a landing under basic VFR.

4770. Which substitution is appropriate during an ILS approach?

A—A VOR radial crossing the outer marker site may be substituted for the outer marker.
B—LOC minimums should be substituted for ILS minimums whenever the glide slope becomes inoperative.
C—DME, when located at the localizer antenna site, should be substituted for either the outer or middle marker.

4771. Assume this clearance is received:

"CLEARED FOR ILS RUNWAY 07 LEFT APPROACH, SIDE-STEP TO RUNWAY 07 RIGHT."

When would the pilot be expected to commence the side-step maneuver?

A—As soon as possible after the runway environment is in sight.
B—Any time after becoming aligned with the final approach course of Runway 07 left, and after passing the final approach fix.
C—After reaching the circling minimums for Runway 07 right.

4772. During a precision radar or ILS approach, the rate of descent required to remain on the glide slope will

A—remain the same regardless of groundspeed.
B—increase as the groundspeed increases.
C—decrease as the groundspeed increases.

4773. When tracking inbound on the localizer, which of the following is the proper procedure regarding drift corrections?

A—Drift corrections should be accurately established before reaching the outer marker and completion of the approach should be accomplished with heading corrections no greater than 2°.
B—Drift corrections should be made in 5° increments after passing the outer marker.
C—Drift corrections should be made in 10° increments after passing the outer marker.

4774. (Refer to Figure 134.) Unless a higher angle is necessary for obstacle clearance, what is the normal glidepath angle for a 2-bar VASI?

A—2.75°.
B—3.00°.
C—3.25°.

4775. Which of the following indications would a pilot see while approaching to land on a runway served by a 2-bar VASI?

A—If on the glidepath, the near bars will appear red, and the far bars will appear white.
B—If departing to the high side of the glidepath, the far bars will change from red to white.
C—If on the glidepath, both near bars and far bars will appear white.

4776. The middle and far bars of a 3-bar VASI will

A—both appear white to the pilot when on the upper glidepath.
B—constitute a 2-bar VASI for using the lower glidepath.
C—constitute a 2-bar VASI for using the upper glidepath.

4777. Tricolor Visual Approach Indicators normally consist of

A—a single unit, projecting a three-color visual approach path.
B—three separate light units, each projecting a different color approach path.
C—three separate light projecting units of very high candle power with a daytime range of approximately 5 miles.

4778. When on the proper glidepath of a 2-bar VASI, the pilot will see the near bar as

A—white and the far bar as red.
B—red and the far bar as white.
C—white and the far bar as white.

4779. If an approach is being made to a runway that has an operating 3-bar VASI and all the VASI lights appear red as the airplane reaches the MDA, the pilot should

A—start a climb to reach the proper glidepath.
B—continue at the same rate of descent if the runway is in sight.
C—level off momentarily to intercept the proper approach path.

4780. Which is a feature of the tricolor VASI?

A—One light projector with three colors: red, green, and amber.
B—Two visual glidepaths for the runway.
C—Three glidepaths, with the center path indicated by a white light.

4781. Which approach and landing objective is assured when the pilot remains on the proper glidepath of the VASI?

A—Continuation of course guidance after transition to VFR.
B—Safe obstruction clearance in the approach area.
C—Course guidance from the visual descent point to touchdown.

4782. (Refer to Figure 135.) Unless a higher angle is required for obstacle clearance, what is the normal glidepath for a 3-bar VASI?

A—2.3°.
B—2.75°.
C—3.0°.

4783. (Refer to Figure 135.) Which illustration would a pilot observe when on the lower glidepath?

A—4.
B—5.
C—6.

4784. (Refer to Figure 135.) Which illustration would a pilot observe if the aircraft is above both glidepaths?

A—5.
B—6.
C—7.

4785. (Refer to Figure 135.) Which illustration would a pilot observe if the aircraft is below both glidepaths?

A—4.
B—5.
C—6.

4786. (Refer to Figure 136.) Which illustration depicts an "on glidepath" indication?

A—8.
B—10.
C—11.

4787. (Refer to Figure 136.) Which illustration depicts a "slightly low" (2.8°) indication?

A—9.
B—10.
C—11.

4788. (Refer to Figure 136.) Which illustration would a pilot observe if the aircraft is on a glidepath higher than 3.5°?

A—8.
B—9.
C—11.

4789. (Refer to Figure 136.) Which illustration would a pilot observe if the aircraft is "slightly high" (3.2°) on the glidepath?

A—8.
B—9.
C—11.

4790. (Refer to Figure 136.) Which illustration would a pilot observe if the aircraft is less than 2.5°?

A—10.
B—11.
C—12.

4791. (Refer to Figure 137.) What is the distance (A) from the beginning of the runway to the fixed distance marker?

A—500 feet.
B—1,000 feet.
C—1,500 feet.

4792. (Refer to Figure 137.) What is the distance (B) from the beginning of the runway to the touchdown zone marker?

A—250 feet.
B—500 feet.
C—750 feet.

4793. (Refer to Figure 137.) What is the distance (C) from the beginning of the touchdown zone marker to the beginning of the fixed distance marker?

A—1,000 feet.
B—500 feet.
C—250 feet.

4794. Which runway marking indicates a displaced threshold on an instrument runway?

A—Arrows leading to the threshold mark.
B—Centerline dashes starting at the threshold.
C—Red chevron marks in the nonlanding portion of the runway.

4795. Which type of runway lighting consists of a pair of synchronized flashing lights, one on each side of the runway threshold?

A—RAIL.
B—HIRL.
C—REIL.

4796. The primary purpose of runway end identifier lights, installed at many airfields, is to provide

A—rapid identification of the approach end of the runway during reduced visibility.
B—a warning of the final 3,000 feet of runway remaining as viewed from the takeoff or approach position.
C—rapid identification of the primary runway during reduced visibility.

4797. (Refer to Figure 138.) What night operations, if any, are authorized between the approach end of the runway and the threshold lights?

A—No aircraft operations are permitted short of the threshold lights.
B—Only taxi operations are permitted in the area short of the threshold lights.
C—Taxi and takeoff operations are permitted, providing the takeoff operations are toward the visible green threshold lights.

4798. What international Morse Code identifier is used to identify a specific interim standard microwave landing system?

A—A two letter Morse Code identifier preceded by the Morse Code for the letters "IM".
B—A three letter Morse Code identifier preceded by the Morse Code for the letter "M".
C—A three letter Morse Code identifier preceded by the Morse Code for the letters "ML".

4799. To at least which altitude AGL is the approach azimuth guidance angle coverage of an MLS?

A—20,000 feet.
B—10,000 feet.
C—8,000 feet.

4800. What are the lateral approach azimuth angle limits, referenced to either side of the landing runway, of an MLS?

A—At least 15°.
B—20°.
C—At least 40°.

4801. What are the respective range limits for the front and back guidance of an MLS?

A—10 NM and 10 NM.
B—15 NM and 10 NM.
C—20 NM and 7 NM.

4802. Without visual aid, a pilot often interprets centrifugal force as a sensation of

A—rising or falling.
B—turning.
C—motion reversal.

4803. Due to visual illusion, when landing on a narrower-than-usual runway, the aircraft will appear to be

A—higher than actual, leading to a lower-than-normal approach.
B—lower than actual, leading to a higher-than-normal approach.
C—higher than actual, leading to a higher-than-normal approach.

4804. What visual illusion creates the same effect as a narrower-than-usual runway?

A—An upsloping runway.
B—A wider-than-usual runway.
C—A downsloping runway.

4805. Abrupt head movement during a prolonged constant rate turn in IMC or simulated instrument conditions can cause

A—pilot disorientation.
B—false horizon.
C—elevator illusion.

4806. A sloping cloud formation, an obscured horizon, and a dark scene spread with ground lights and stars can create an illusion known as

A—elevator illusions.
B—autokinesis.
C—false horizons.

4807. An abrupt change from climb to straight-and-level flight can create the illusion of

A—tumbling backwards.
B—a noseup attitude.
C—a descent with the wings level.

4808. A rapid acceleration during takeoff can create the illusion of

A—spinning in the opposite direction.
B—being in a noseup attitude.
C—diving into the ground.

4809. Why is hypoxia particularly dangerous during flights with one pilot?

A—Night vision may be so impaired that the pilot cannot see other aircraft.
B—Symptoms of hypoxia may be difficult to recognize before the pilot's reactions are affected.
C—The pilot may not be able to control the aircraft even if using oxygen.

4810. The sensations which lead to spatial disorientation during instrument flight conditions

A—are frequently encountered by beginning instrument pilots, but never by pilots with moderate instrument experience.
B—occur, in most instances, during the initial period of transition from visual to instrument flight.
C—must be suppressed and complete reliance placed on the indications of the flight instruments.

4811. How can an instrument pilot best overcome spatial disorientation?

A—Rely on kinesthetic sense.
B—Use a very rapid cross-check.
C—Read and interpret the flight instruments, and act accordingly.

4812. Which statement is correct regarding the use of cockpit lighting for night flight?

A—Reducing the lighting intensity to a minimum level will eliminate blind spots.
B—The use of regular white light, such as a flashlight, will impair night adaptation.
C—Coloration shown on maps is least affected by the use of direct red lighting.

4813. How can an instrument pilot best overcome spatial disorientation?

A—Use a very rapid cross-check.
B—Properly interpret the flight instruments and act accordingly.
C—Avoid banking in excess of 30°.

4814. A pilot is more subject to spatial disorientation if

A—kinesthetic senses are ignored.
B—eyes are moved often in the process of cross-checking the flight instruments.
C—body signals are used to interpret flight attitude.

4815. Which procedure is recommended to prevent or overcome spatial disorientation?

A—Reduce head and eye movements to the extent possible.
B—Rely on the kinesthetic sense.
C—Rely on the indications of the flight instruments.

4816. What action should be taken if hyperventilation is suspected?

A—Breathe at a slower rate by taking very deep breaths.
B—Consciously breathe at a slower rate than normal.
C—Consciously force yourself to take deep breaths and breathe at a faster rate than normal.

4817. Which use of cockpit lighting is correct for night flight?

A—Reducing the interior lighting intensity to a minimum level.
B—The use of regular white light, such as a flashlight, will not impair night adaptation.
C—Coloration shown on maps is least affected by the use of direct red lighting.

4818. Which technique should a pilot use to scan for traffic to the right and left during straight-and-level flight?

A—Systematically focus on different segments of the sky for short intervals.
B—Concentrate on relative movement detected in the peripheral vision area.
C—Continuous sweeping of the windshield from right to left.

4819. What effect does haze have on the ability to see traffic or terrain features during flight?

A—Haze causes the eyes to focus at infinity, making terrain features harder to see.
B—The eyes tend to overwork in haze and do not detect relative movement easily.
C—Haze creates the illusion of being a greater distance than actual from the runway, and causes pilots to fly a lower approach.

4820. As a rule of thumb, altitude corrections of less than 100 feet should be corrected by using a

A—full bar width on the attitude indicator.
B—half bar width on the attitude indicator.
C—two bar width on the attitude indicator.

4821. If both the ram air input and drain hole of the pitot system are blocked, what airspeed indication can be expected?

A—No variation of indicated airspeed in level flight even if large power changes are made.
B—Decrease of indicated airspeed during a climb.
C—Constant indicated airspeed during a descent.

4822. During a "no-gyro" approach and prior to being handed off to the final approach controller, the pilot should make all turns

A—one-half standard rate unless otherwise advised.
B—any rate not exceeding a 30° bank.
C—standard rate unless otherwise advised.

4823. After being handed off to the final approach controller during a "no-gyro" surveillance or precision approach, the pilot should make all turns

A—one-half standard rate.
B—based upon the groundspeed of the aircraft.
C—standard rate.

4824. (Refer to Figures 139 and 140.) Which displacement from the localizer and glide slope at the 1.9 NM point is indicated?

A—710 feet to the left of the localizer centerline and 140 feet below the glide slope.
B—710 feet to the right of the localizer centerline and 140 feet above the glide slope.
C—430 feet to the right of the localizer centerline and 28 feet above the glide slope.

4825. (Refer to Figures 139 and 141.) Which displacement from the localizer centerline and glide slope at the 1,300-foot point from the runway is indicated?

A—21 feet below the glide slope and approximately 320 feet to the right of the runway centerline.
B—28 feet above the glide slope and approximately 250 feet to the left of the runway centerline.
C—21 feet above the glide slope and approximately 320 feet to the left of the runway centerline.

4826. (Refer to Figures 139 and 142.) Which displacement from the localizer and glide slope at the outer marker is indicated?

A—1,550 feet to the left of the localizer centerline and 210 feet below the glide slope.
B—1,550 feet to the right of the localizer centerline and 210 feet above the glide slope.
C—775 feet to the left of the localizer centerline and 420 feet below the glide slope.

4827. (Refer to Figure 143.) The heading on a remote indicating compass is 120° and the magnetic compass indicates 110°. What action is required to correctly align the heading indicator with the magnetic compass?

A—Select the free gyro mode and depress the counterclockwise heading drive button.
B—Select the slaved gyro mode and depress the clockwise heading drive button.
C—Select the free gyro mode and depress the clockwise heading drive button.

4828. (Refer to Figure 143.) When the system is in the free gyro mode, depressing the clockwise manual heading drive button will rotate the remote indicating compass card to the

A—right to eliminate left compass card error.
B—right to eliminate right compass card error.
C—left to eliminate left compass card error.

4829. (Refer to Figure 143.) The heading on a remote indicating compass is 5° to the left of that desired. What action is required to move the desired heading under the heading reference?

A—Select the free gyro mode and depress the clockwise heading drive button.
B—Select the slaved gyro mode and depress the clockwise heading drive button.
C—Select the free gyro mode and depress the counterclockwise heading drive button.

4830. If both the ram air input and the drain hole of the pitot system are blocked, what reaction should you observe on the airspeed indicator when power is applied and a climb is initiated out of severe icing conditions?

A—The indicated airspeed would show a continuous deceleration while climbing.
B—The airspeed would drop to, and remain at, zero.
C—No change until an actual climb rate is established, then indicated airspeed will increase.

4831. What indication should be observed on a turn coordinator during a left turn while taxiing?

A—The miniature aircraft will show a turn to the left and the ball remains centered.
B—The miniature aircraft will show a turn to the left and the ball moves to the right.
C—Both the miniature aircraft and the ball will remain centered.

4832. The gyroscopic heading indicator is inoperative. What is the primary bank instrument in unaccelerated straight-and-level flight?

A—Magnetic compass.
B—Attitude indicator.
C—Miniature aircraft of turn coordinator.

4833. When airspeed is decreased in a turn, what must be done to maintain level flight?

A—Decrease the angle of bank and/or increase the angle of attack.
B—Increase the angle of bank and/or decrease the angle of attack.
C—Increase the angle of attack.

4834. On the taxi check, the magnetic compass should

A—swing opposite to the direction of turn when turning from north.
B—exhibit the same number of degrees of dip as the latitude.
C—swing freely and indicate known headings.

4835. Which condition during taxi is an indication that an attitude indicator is unreliable?

A—The horizon bar tilts more than 5° while making taxi turns.
B—The horizon bar vibrates during warmup.
C—The horizon bar does not align itself with the miniature airplane after warmup.

4836. What instruments are considered supporting bank instruments during a straight, stabilized climb at a constant rate?

A—Attitude indicator and turn coordinator.
B—Heading indicator and attitude indicator.
C—Heading indicator and turn coordinator.

4837. What instruments are primary for pitch, bank, and power, respectively, when transitioning into a constant airspeed climb from straight-and-level flight?

A—Attitude indicator, heading indicator, and manifold pressure gauge or tachometer.
B—Attitude indicator for both pitch and bank; airspeed indicator for power.
C—Vertical speed, attitude indicator, and manifold pressure or tachometer.

4838. What is the primary bank instrument once a standard rate turn is established?

A—Attitude indicator.
B—Turn coordinator.
C—Heading indicator.

4839. What does the miniature aircraft of the turn coordinator directly display?

A—Rate of roll and rate of turn.
B—Angle of bank and rate of turn.
C—Angle of bank.

4840. What is the correct sequence in which to use the three skills used in instrument flying?

A—Aircraft control, cross-check, and instrument interpretation.
B—Instrument interpretation, cross-check, and aircraft control.
C—Cross-check, instrument interpretation, and aircraft control.

4841. Which initial pitch attitude change on the attitude indicator should be made to correct altitude while at normal cruise in a helicopter?

A—Two bar width.
B—One and one-half bar width.
C—One bar width.

4842. What pretakeoff check should be made of the attitude indicator in preparation for an IFR flight?

A—The horizon bar does not vibrate during warmup.
B—The miniature airplane should erect and become stable within 5 minutes.
C—The horizon bar should erect and become stable within 5 minutes.

4843. The rate of turn at any airspeed is dependent upon

A—the horizontal lift component.
B—the vertical lift component.
C—centrifugal force.

4844. During a skidding turn to the right, what is the relationship between the component of lift, centrifugal force, and load factor?

A—Centrifugal force is less than horizontal lift and the load factor is increased.
B—Centrifugal force is greater than horizontal lift and the load factor is increased.
C—Centrifugal force and horizontal lift are equal and the load factor is decreased.

4845. As power is increased to enter a 500 feet per minute rate of climb in straight flight, which instruments are primary for pitch, bank, and power respectively?

A—Airspeed indicator, heading indicator, and manifold pressure gauge or tachometer.
B—VSI, attitude indicator, and airspeed indicator.
C—Airspeed indicator, attitude indicator, and manifold pressure gauge or tachometer.

4846. During the initial acceleration on an instrument takeoff in a helicopter, what flight attitude should be established on the attitude indicator?

A—Level flight attitude.
B—Two bar widths low.
C—One bar width high.

4847. What indications are displayed by the miniature aircraft of a turn coordinator?

A—Rate of roll and rate of turn.
B—Direct indication of bank angle and pitch attitude.
C—Indirect indication of bank angle and pitch attitude.

4848. What is the primary pitch instrument during a stabilized climbing left turn at cruise climb airspeed?

A—Attitude indicator.
B—VSI.
C—Airspeed indicator.

4849. What is the primary pitch instrument during a stabilized autorotation?

A—Altimeter.
B—Airspeed indicator.
C—VSI.

4850. What is the primary pitch instrument when establishing a constant altitude standard rate turn?

A—Altimeter.
B—VSI.
C—Airspeed indicator.

4851. What is the initial primary bank instrument when establishing a level standard rate turn?

A—Turn coordinator.
B—Heading indicator.
C—Attitude indicator.

4852. During a stabilized autorotation, approximately what flight attitude should be established on the attitude indicator?

A—Two bar widths below the artificial horizon.
B—A pitch attitude that will give an established rate of descent of not more than 500 feet per minute.
C—Level flight attitude.

4853. What instrument(s) is(are) supporting bank instrument when entering a constant airspeed climb from straight-and-level flight?

A—Heading indicator.
B—Attitude indicator and turn coordinator.
C—Turn coordinator and heading indicator.

4854. What indication should a pilot observe if an airspeed indicator ram air input and drain hole are blocked?

A—The airspeed indicator will react as an altimeter.
B—The airspeed indicator will show a decrease with an increase in altitude.
C—No airspeed indicator change will occur during climbs or descents.

4855. What are the three fundamental skills involved in attitude instrument flying?

A—Instrument interpretation, trim application, and aircraft control.
B—Cross-check, instrument interpretation, and aircraft control.
C—Cross-check, emphasis, and aircraft control.

4856. What indication is presented by the miniature aircraft of the turn coordinator?

A—Indirect indication of the bank attitude.
B—Direct indication of the bank attitude and the quality of the turn.
C—Quality of the turn.

4857. During normal operation of a vacuum-driven attitude indicator, what attitude indication should you see when rolling out from a 180° skidding turn to straight-and-level coordinated flight?

A—A straight-and-level coordinated flight indication.
B—A nose-high indication relative to level flight.
C—The miniature aircraft shows a turn in the direction opposite the skid.

4858. What is the primary bank instrument while transitioning from straight-and-level flight to a standard rate turn to the left?

A—Attitude indicator.
B—Heading indicator.
C—Turn coordinator (miniature aircraft).

4859. What is the third fundamental skill in attitude instrument flying?

A—Instrument cross-check.
B—Power control.
C—Aircraft control.

4860. During normal coordinated turns, what error due to precession should you observe when rolling out to straight-and-level flight from a 180° steep turn to the right?

A—A straight-and-level coordinated flight indication.
B—The miniature aircraft would show a slight turn indication to the left.
C—The miniature aircraft would show a slight descent and wings-level attitude.

4861. During coordinated turns, which force moves the pendulous vanes of a vacuum-driven attitude indicator resulting in precession of the gyro toward the inside of the turn?

A—Acceleration.
B—Deceleration.
C—Centrifugal.

4862. What is the first fundamental skill in attitude instrument flying?

A—Aircraft control.
B—Instrument cross-check.
C—Instrument interpretation.

4863. As power is reduced to change airspeed from high to low cruise in level flight, which instruments are primary for pitch, bank, and power, respectively?

A—Attitude indicator, heading indicator, and manifold pressure gauge or tachometer.
B—Altimeter, attitude indicator, and airspeed indicator.
C—Altimeter, heading indicator, and manifold pressure gauge or tachometer.

4864. What information does a Mach meter present?

A—The ratio of aircraft true airspeed to the speed of sound.
B—The ratio of aircraft indicated airspeed to the speed of sound.
C—The ratio of aircraft equivalent airspeed, corrected for installation error, to the speed of sound.

4865. Which instrument provides the most pertinent information (primary) for bank control in straight-and-level flight?

A—Turn-and-slip indicator.
B—Attitude indicator.
C—Heading indicator.

4866. Which instruments are considered primary and supporting for bank, respectively, when establishing a level standard rate turn?

A—Turn coordinator and attitude indicator.
B—Attitude indicator and turn coordinator.
C—Turn coordinator and heading indicator.

4867. While recovering from an unusual flight attitude without the aid of the attitude indicator, approximate level pitch attitude is reached when the

A—airspeed and altimeter stop their movement and the VSI reverses its trend.
B—airspeed arrives at cruising speed, the altimeter reverses its trend, and the vertical speed stops its movement.
C—altimeter and vertical speed reverse their trend and the airspeed stops its movement.

4868. What is the relationship between centrifugal force and the horizontal lift component in a coordinated turn?

A—Horizontal lift exceeds centrifugal force.
B—Horizontal lift and centrifugal force are equal.
C—Centrifugal force exceeds horizontal lift.

4869. Which instruments, in addition to the attitude indicator, are pitch instruments?

A—Altimeter and airspeed only.
B—Altimeter and VSI only.
C—Altimeter, airspeed indicator, and vertical speed indicator.

4870. What force causes an airplane to turn?

A—Rudder pressure or force around the vertical axis.
B—Vertical lift component.
C—Horizontal lift component.

4871. Which instrument provides the most pertinent information (primary) for pitch control in straight-and-level flight?

A—Attitude indicator.
B—Airspeed indicator.
C—Altimeter.

4872. Which instruments are considered to be supporting instruments for pitch during change of airspeed in a level turn?

A—Airspeed indicator and VSI.
B—Altimeter and attitude indicator.
C—Attitude indicator and VSI.

4873. If an airplane is in an unusual flight attitude and the attitude indicator has exceeded its limits, which instruments should be relied on to determine pitch attitude before starting recovery?

A—Turn indicator and VSI.
B—Airspeed and altimeter.
C—VSI and airspeed to detect approaching V_{S1} or V_{MO}.

4874. Which instrument is considered primary for power as the airspeed reaches the desired value during change of airspeed in a level turn?

A—Airspeed indicator.
B—Attitude indicator.
C—Altimeter.

4875. Which is the correct sequence for recovery from a spiraling, nose-low, increasing airspeed, unusual flight attitude?

A—Increase pitch attitude, reduce power, and level wings.
B—Reduce power, correct the bank attitude, and raise the nose to a level attitude.
C—Reduce power, raise the nose to level attitude, and correct the bank attitude.

4876. Which instruments should be used to make a pitch correction when you have deviated from your assigned altitude?

A—Altimeter and VSI.
B—Manifold pressure gauge and VSI.
C—Attitude indicator, altimeter, and VSI.

4877. What should be the indication on the magnetic compass as you roll into a standard rate turn to the left from an east heading in the Northern Hemisphere?

A—The compass will initially indicate a turn to the right.
B—The compass will remain on east for a short time, then gradually catch up to the magnetic heading of the aircraft.
C—The compass will indicate the approximate correct magnetic heading if the roll into the turn is smooth.

4878. When airspeed is increased in a turn, what must be done to maintain a constant altitude?

A—Decrease the angle of bank.
B—Increase the angle of bank and/or decrease the angle of attack.
C—Decrease the angle of attack.

4879. What would be the indication on the VSI during entry into a 500 FPM actual descent from level flight if the static ports were iced over?

A—The indication would be in reverse of the actual rate of descent (500 FPM climb).
B—The initial indication would be a climb, then descent at a rate in excess of 500 FPM.
C—The VSI pointer would remain at zero regardless of the actual rate of descent.

4880. How should you preflight check the altimeter prior to an IFR flight?

A—Set the altimeter to the current temperature. With current temperature and the altimeter indication, determine the calibrated altitude to compare with the field elevation.
B—Set the altimeter first with 29.92" Hg and then the current altimeter setting. The change in altitude should correspond to the change in setting.
C—Set the altimeter to the current altimeter setting. The indication should be within 75 feet of the actual elevation for acceptable accuracy.

4881. Which practical test should be made on the electric gyro instruments prior to starting an engine?

A—Check that the electrical connections are secure on the back of the instruments.
B—Check that the attitude of the miniature aircraft is wings level before turning on electrical power.
C—Turn on the electrical power and listen for any unusual or irregular mechanical noise.

4882. Prior to starting an engine, you should check the turn-and-slip indicator to determine if the

A—needle indication properly corresponds to the angle of the wings or rotors with the horizon.
B—needle is approximately centered and the tube is full of fluid.
C—ball will move freely from one end of the tube to the other when the aircraft is rocked.

4883. What indications should you observe on the turn-and-slip indicator during taxi?

A—The ball moves freely opposite the turn, and the needle deflects in the direction of the turn.
B—The needle deflects in the direction of the turn, but the ball remains centered.
C—The ball deflects opposite the turn, but the needle remains centered.

4884. Which instrument indicates the quality of a turn?

A—Attitude indicator.
B—Heading indicator or magnetic compass.
C—Ball of the turn coordinator.

4885. What pretakeoff check should be made of a vacuum-driven heading indicator in preparation for an IFR flight?

A—After 5 minutes, set the indicator to the magnetic heading of the aircraft and check for proper alignment after taxi turns.
B—After 5 minutes, check that the heading indicator card aligns itself with the magnetic heading of the aircraft.
C—Determine that the heading indicator does not precess more than 2° in 5 minutes of ground operation.

4886. What should be the indication on the magnetic compass as you roll into a standard rate turn to the right from an easterly heading in the Northern Hemisphere?

A—The compass will initially indicate a turn to the left.
B—The compass will remain on east for a short time, then gradually catch up to the magnetic heading of the aircraft.
C—The compass will indicate the approximate correct magnetic heading if the roll into the turn is smooth.

4887. What should be the indication on the magnetic compass as you roll into a standard rate turn to the right from a south heading in the Northern Hemisphere?

A—The compass will indicate a turn to the right, but at a faster rate than is actually occurring.
B—The compass will initially indicate a turn to the left.
C—The compass will remain on south for a short time, then gradually catch up to the magnetic heading of the aircraft.

4888. On what headings will the magnetic compass read most accurately during a level 360° turn, with a bank of approximately 15°?

A—135° through 225°.
B—90° and 270°.
C—180° and 0°.

4889. What causes the northerly turning error in a magnetic compass?

A—Coriolis force at the mid-latitudes.
B—Centrifugal force acting on the compass card.
C—The magnetic dip characteristic.

4890. What should be the indication on the magnetic compass when you roll into a standard rate turn to the left from a south heading in the Northern Hemisphere?

A—The compass will indicate a turn to the left, but at a faster rate than is actually occurring.
B—The compass will initially indicate a turn to the right.
C—The compass will remain on south for a short time, then gradually catch up to the magnetic heading of the aircraft.

4891. What should be the indication on the magnetic compass as you roll into a standard rate turn to the right from a westerly heading in the Northern Hemisphere?

A—The compass will initially show a turn in the opposite direction, then turn to a northerly indication but lagging behind the actual heading of the aircraft.
B—The compass will remain on a westerly heading for a short time, then gradually catch up to the actual heading of the aircraft.
C—The compass will indicate the approximate correct magnetic heading if the roll into the turn is smooth.

4892. What should be the indication on the magnetic compass as you roll into a standard rate turn to the right from a northerly heading in the Northern Hemisphere?

A—The compass will indicate a turn to the right, but at a faster rate than is actually occurring.
B—The compass will initially indicate a turn to the left.
C—The compass will remain on north for a short time, then gradually catch up to the magnetic heading of the aircraft.

4893. What should be the indication on the magnetic compass as you roll into a standard rate turn to the left from a west heading in the Northern Hemisphere?

A—The compass will initially indicate a turn to the right.
B—The compass will remain on west for a short time, then gradually catch up to the magnetic heading of the aircraft.
C—The compass will indicate the approximate correct magnetic heading if the roll into the turn is smooth.

4894. What should be the indication on the magnetic compass as you roll into a standard rate turn to the left from a north heading in the Northern Hemisphere?

A—The compass will indicate a turn to the left, but at a faster rate than is actually occurring.
B—The compass will initially indicate a turn to the right.
C—The compass will remain on north for a short time, then gradually catch up to the magnetic heading of the aircraft.

4895. If a half-standard rate turn is maintained, how long would it take to turn 360°?

A—1 minute.
B—2 minutes.
C—4 minutes.

4896. If a standard rate turn is maintained, how long would it take to turn 180°?

A—1 minute.
B—2 minutes.
C—3 minutes.

4897. If a half-standard rate turn is maintained, how much time would be required to turn clockwise from a heading of 090° to a heading of 180°?

A—30 seconds.
B—1 minute.
C—1 minute 30 seconds.

4898. During a constant-bank level turn, what effect would an increase in airspeed have on the rate and radius of turn?

A—Rate of turn would increase, and radius of turn would increase.
B—Rate of turn would decrease, and radius of turn would decrease.
C—Rate of turn would decrease, and radius of turn would increase.

4899. The three conditions which determine pitch attitude required to maintain level flight are

A—flightpath, wind velocity, and angle of attack.
B—airspeed, air density, and aircraft weight.
C—relative wind, pressure altitude, and vertical lift component.

4900. Errors in both pitch and bank indication on an attitude indicator are usually at a maximum as the aircraft rolls out of a

A—180° turn.
B—270° turn.
C—360° turn.

4901. If a 180° steep turn is made to the right and the aircraft is rolled out to straight-and-level flight by visual reference, the miniature aircraft will

A—show a slight climb and turn to the left.
B—show a slight climb and turn to the right.
C—show a slight skid and climb to the right.

4902. One characteristic that a properly functioning gyro depends upon for operation is the

A—ability to resist precession 90° to any applied force.
B—resistance to deflection of the spinning wheel or disc.
C—deflecting force developed from the angular velocity of the spinning wheel.

4903. If a standard rate turn is maintained, how much time would be required to turn to the right from a heading of 090° to a heading of 270°?

A—1 minute.
B—2 minutes.
C—3 minutes.

4904. If a standard rate turn is maintained, how much time would be required to turn to the left from a heading of 090° to a heading of 300°?

A—30 seconds.
B—40 seconds.
C—50 seconds.

4905. If a half-standard rate turn is maintained, how long would it take to turn 135°?

A—1 minute.
B—1 minute 20 seconds.
C—1 minute 30 seconds.

4906. Approximately what percent of the indicated vertical speed should be used to determine the number of feet to lead the level-off from a climb to a specific altitude?

A—10 percent.
B—20 percent.
C—25 percent.

4907. To level off from a descent to a specific altitude, the pilot should lead the level-off by approximately

A—10 percent of the vertical speed.
B—30 percent of the vertical speed.
C—50 percent of the vertical speed.

4908. If, while in level flight, it becomes necessary to use an alternate source of static pressure vented inside the airplane, which of the following should the pilot expect?

A—The altimeter to read lower than normal.
B—The vertical speed to momentarily show a descent.
C—The vertical speed to momentarily show a climb.

4909. During flight, if the pitot tube becomes clogged with ice, which of the following instruments would be affected?

A—The airspeed indicator only.
B—The airspeed indicator and the altimeter.
C—The airspeed indicator, altimeter, and Vertical Speed Indicator.

4910. The local altimeter setting should be used by all pilots in a particular area, primarily to provide for

A—the cancellation of altimeter error due to nonstandard temperatures aloft.
B—better vertical separation of aircraft.
C—more accurate terrain clearance in mountainous areas.

4911. At an altitude of 6,500 feet MSL, the current altimeter setting is 30.42" Hg. The pressure altitude would be approximately

A—7,500 feet.
B—6,000 feet.
C—6,500 feet.

4912. The pressure altitude at a given location is indicated on the altimeter after the altimeter is set to

A—the field elevation.
B—29.92" Hg.
C—the current altimeter setting.

4913. If the outside air temperature increases during a flight at constant power and at a constant indicated altitude, the true airspeed will

A—decrease and true altitude will increase.
B—increase and true altitude will decrease.
C—increase and true altitude will increase.

4914. Rate of turn can be increased and radius of turn decreased by

A—decreasing airspeed and shallowing the bank.
B—decreasing airspeed and increasing the bank.
C—increasing airspeed and increasing the bank.

4915. The primary reason the angle of attack must be increased, to maintain a constant altitude during a coordinated turn, is because the

A—thrust is acting in a different direction, causing a reduction in airspeed and loss of lift.
B—vertical component of lift has decreased as the result of the bank.
C—use of ailerons has increased the drag.

4916. If severe turbulence is encountered during your IFR flight, the airplane should be slowed to the design maneuvering speed because the

A—maneuverability of the airplane will be increased.
B—amount of excess load that can be imposed on the wing will be decreased.
C—airplane will stall at a lower angle of attack, giving an increased margin of safety.

4917. When a climb or descent through an inversion or wind-shear zone is being performed, the pilot should be alert for which of the following change in airplane performance?

A—A fast rate of climb and a slow rate of descent.
B—A sudden change in airspeed.
C—A sudden surge of thrust.

4918. When an airplane is accelerated, some attitude indicators will precess and incorrectly indicate a

A—climb.
B—descent.
C—right turn.

4919. When an airplane is decelerated, some attitude indicators will precess and incorrectly indicate a

A—left turn.
B—climb.
C—descent.

4920. For maintaining level flight at constant thrust, which instrument would be the least appropriate for determining the need for a pitch change?

A—Altimeter.
B—VSI.
C—Attitude indicator.

4921. The displacement of a turn coordinator during a coordinated turn will

A—indicate the angle of bank.
B—remain constant for a given bank regardless of airspeed.
C—increase as angle of bank increases.

4922. Altimeter setting is the value to which the scale of the pressure altimeter is set so the altimeter indicates

A—pressure altitude at sea level.
B—true altitude at field elevation.
C—pressure altitude at field elevation.

4923. The altimeter indicates the aircraft altitude in relation to

A—sea level.
B—the standard datum plane.
C—the pressure level set in the barometric window.

4924. To enter a constant-airspeed descent from level-cruising flight, and maintain cruising airspeed, the pilot should

A—first adjust the pitch attitude to a descent using the attitude indicator as a reference, then adjust the power to maintain the cruising airspeed.
B—first reduce power, then adjust the pitch using the attitude indicator as a reference to establish a specific rate on the VSI.
C—simultaneously reduce power and adjust the pitch using the attitude indicator as a reference to maintain the cruising airspeed.

4925. To level off at an airspeed higher than the descent speed, the addition of power should be made, assuming a 500 FPM rate of descent, at approximately

A—50 to 100 feet above the desired altitude.
B—100 to 150 feet above the desired altitude.
C—150 to 200 feet above the desired altitude.

4926. To level off from a descent maintaining the descending airspeed, the pilot should lead the desired altitude by approximately

A—20 feet.
B—50 feet.
C—60 feet.

4927. During recoveries from unusual attitudes, level flight is attained the instant

A—the horizon bar on the attitude indicator is exactly overlapped with the miniature airplane.
B—a zero rate of climb is indicated on the VSI.
C—the altimeter and airspeed needles stop prior to reversing their direction of movement.

4928. While cruising at 160 knots, you wish to establish a climb at 130 knots. When entering the climb (full panel), it is proper to make the initial pitch change by increasing back elevator pressure until the

A—attitude indicator, airspeed, and vertical speed indicate a climb.
B—vertical speed indication reaches the predetermined rate of climb.
C—attitude indicator shows the approximate pitch attitude appropriate for the 130-knot climb.

4929. While cruising at 190 knots, you wish to establish a climb at 160 knots. When entering the climb (full panel), it would be proper to make the initial pitch change by increasing back elevator pressure until the

A—attitude indicator shows the approximate pitch attitude appropriate for the 160-knot climb.
B—attitude indicator, airspeed, and vertical speed indicate a climb.
C—airspeed indication reaches 160 knots.

4930. If while in level flight, it becomes necessary to use an alternate source of static pressure vented inside the airplane, which of the following variations in instrument indications should the pilot expect?

A—The altimeter will read lower than normal, airspeed lower than normal, and the VSI will momentarily show a descent.
B—The altimeter will read higher than normal, airspeed greater than normal, and the VSI will momentarily show a climb.
C—The altimeter will read lower than normal, airspeed greater than normal, and the VSI will momentarily show a climb and then a descent.

4931. (Refer to Figure 144.) What changes in control displacement should be made so that "2" would result in a coordinated standard rate turn?

A—Increase left rudder and increase rate of turn.
B—Increase left rudder and decrease rate of turn.
C—Decrease left rudder and decrease angle of bank.

4932. (Refer to Figure 144.) Which illustration indicates a coordinated turn?

A—3.
B—1.
C—2.

4933. (Refer to Figure 144.) Which illustration indicates a skidding turn?

A—2.
B—1.
C—3.

4934. (Refer to Figure 144.) What changes in control displacement should be made so that "1" would result in a coordinated standard rate turn?

A—Increase right rudder and decrease rate of turn.
B—Increase right rudder and increase rate of turn.
C—Decrease right rudder and increase angle of bank.

4935. (Refer to Figure 144.) Which illustration indicates a slipping turn?

A—1.
B—3.
C—2.

4936. (Refer to Figure 145.) What is the correct sequence for recovery from the unusual attitude indicated?

A—Reduce power, increase back elevator pressure, and level the wings.
B—Reduce power, level the wings, bring pitch attitude to level flight.
C—Level the wings, raise the nose of the aircraft to level flight attitude, and obtain desired airspeed.

4937. (Refer to Figure 146.) Identify the system that has failed and determine a corrective action to return the airplane to straight-and-level flight.

A—Static/pitot system is blocked; lower the nose and level the wings to level-flight attitude by use of attitude indicator.
B—Vacuum system has failed; reduce power, roll left to level wings, and pitchup to reduce airspeed.
C—Electrical system has failed; reduce power, roll left to level wings, and raise the nose to reduce airspeed.

4938. (Refer to Figure 147.) Which is the correct sequence for recovery from the unusual attitude indicated?

A—Level wings, add power, lower nose, descend to original attitude, and heading.
B—Add power, lower nose, level wings, return to original attitude and heading.
C—Stop turn by raising right wing and add power at the same time, lower the nose, and return to original attitude and heading.

4939. (Refer to Figure 148.) What is the flight attitude? One system which transmits information to the instruments has malfunctioned.

A—Climbing turn to left.
B—Climbing turn to right.
C—Level turn to left.

4940. (Refer to Figure 149.) What is the flight attitude? One system which transmits information to the instruments has malfunctioned.

A—Level turn to the right.
B—Level turn to the left.
C—Straight-and-level flight.

4941. (Refer to Figure 150.) What is the flight attitude? One instrument has malfunctioned.

A—Climbing turn to the right.
B—Climbing turn to the left.
C—Descending turn to the right.

4942. (Refer to Figure 151.) What is the flight attitude? One instrument has malfunctioned.

A—Climbing turn to the right.
B—Level turn to the right.
C—Level turn to the left.

4943. (Refer to Figure 152.) What is the flight attitude? One system which transmits information to the instruments has failed.

A—Climbing turn to right.
B—Level turn to left.
C—Descending turn to right.

Answers, References & Explanations

LTA
4001 [C] — (A20) — FAR §61.57(e)(i)
No pilot may act as PIC under IFR, nor in weather conditions less than the minimums prescribed for VFR, unless that pilot has, within the past 6 calendar months, logged at least 6 hours of instrument time under actual or simulated IFR conditions, at least 3 of which were in the category of aircraft involved, including at least 6 instrument approaches, or passed an instrument competency check in the category of aircraft involved.

AIR
4002 [C] — (A24) — FAR §61.129(a)
The commercial pilot applicant must hold an instrument rating (airplane), or the Commercial Pilot Certificate that is issued is endorsed with a limitation prohibiting the carriage of passengers for hire in airplanes on cross-country flights of more than 50 NM, or at night.

ALL
4003 [C] — (B08) — FAR §91.103
Each pilot-in-command shall, before beginning a flight, become familiar with all available information concerning that flight. This information must include:

1. For a flight under IFR or a flight not in the vicinity of an airport, weather reports and forecasts, fuel requirements, alternatives available if the planned flight cannot be completed, and any known traffic delays of which the pilot has been advised by ATC.
2. For any flight, runway lengths at airports of intended use, and the following takeoff and landing distance information.
3. For civil aircraft for which an approved airplane or rotorcraft flight manual containing takeoff and landing distance data is required, the takeoff and landing distance data contained therein.

LTA
4004 [A] — (B07) — FAR §91.21(a)(2)
With certain exceptions, no person may operate, nor may any operator or pilot in command of an aircraft allow the operation of any portable electronic device on U.S. registered civil aircraft operated in IFR conditions.

AIR
4005 [B] — (B10) — FAR §91.167(a)
No person may operate in IFR conditions unless the aircraft carries enough fuel (considering weather reports, forecasts, and conditions) to complete the flight to the first airport of intended landing, fly from that airport to the alternate airport and fly after that for 45 minutes at normal cruising speed.

ALL
4006 [C] — (B10) — FAR §91.177
Except when necessary for takeoff or landing, no person may operate an aircraft under IFR below an altitude of 2,000 feet above the highest obstacle within 4 NM for mountainous areas, and in any other case, 1,000 feet above.

ALL
4007 [B] — (B11) — FAR §91.215(d)(1)
ATC may authorize deviations from transponder requirements to allow an aircraft with an inoperative transponder to continue to the airport of ultimate destination, including any intermediate stops, or to proceed to a place where suitable repairs can be made.

ALL
4008 [A] — (A20) — FAR §61.51(c)(4)
Each instrument flight entry must include the place and type of each instrument approach completed, and the name of the safety pilot for each simulated instrument flight.

ALL
4009 [B] — (A20) — FAR §61.51(c)(4)
Instrument flight instructors may log as instrument flight time that time during which they act as instrument flight instructor in actual instrument weather conditions.

ALL
4010 [B] — (A20) — FAR §61.51(c)(4)
A pilot may log as instrument flight time, only that time during which the pilot operates the aircraft solely by reference to instruments under actual or simulated instrument flight conditions.

ALL
4011 [A] — (B08) — FAR §91.109(b)
No person may operate a civil aircraft in simulated instrument flight unless the other control seat is occupied by a safety pilot who possesses at least a Private Pilot Certificate with category and class ratings appropriate to the aircraft being flown.

ALL
4012 [A] — (A20) — FAR §61.57(e)(1)(i)
To act as pilot-in-command under IFR, a pilot must have logged, in the past 6 calendar months, at least 6 hours of actual or simulated instrument time. At least 3 of the 6 hours must be flown in the category of aircraft involved, including at least 6 instrument approaches.

ALL
4013 [A] — (A20) — FAR §61.57(e)(2)
A pilot who does not meet the recent instrument experience requirements during the prescribed time, has 6 months thereafter to pass an instrument competency check in the category of aircraft involved.

ALL
4014 [B] — (A20) — FAR §61.57(e)(2)
A pilot who does not meet the recent instrument experience requirements during the prescribed time, has 6 months thereafter to pass an instrument competency check in the category of aircraft involved, given by an approved FAA Examiner, instrument instructor, or FAA inspector.

ALL
4015 [A] — (A20) — FAR §61.57(e)(2)
A pilot who does not meet the recent instrument experience requirements during the prescribed time, has 6 months thereafter to pass an instrument competency check in the category of aircraft involved. If a pilot's IFR currency expires on July 1 of this year, that pilot would have 6 months, or until December 31, this year, to gain the IFR experience before an instrument competency check would be required.

RTC
4016 [B] — (A20) — FAR §61.57(e)(1)(i)
No pilot may act as pilot-in-command under IFR unless that pilot has, within the past 6 months, logged: at least 6 hours of instrument time under actual or simulated IFR conditions, at least 3 of which were in flight in the category of aircraft involved, including at least 6 instrument approaches; or the pilot can pass an instrument competency check in the category of aircraft involved. With the conditions stated in this question, the pilot must do one more approach to meet the total of 6 instrument approaches required.

ALL
4017 [A] — (A20) — FAR §61.57(e)(1)(i)
The minimum conditions necessary for the instrument approaches is to complete 6 within the past 6 months, to be completed in an aircraft, approved instrument ground trainer, or any combination of these.

RTC
4018 [C] — (A20) — FAR §61.57(e)(1)(i)
No person may act as pilot-in-command of a helicopter unless he/she has a valid pilot certificate with rotorcraft category and helicopter class ratings. In addition, to operate IFR, the pilot-in-command must have either a helicopter instrument rating or an Airline Transport Pilot Certificate with rotorcraft category and helicopter class rating not limited to VFR. In addition to having the appropriate ratings, a pilot must have the required recent experience set forth in FAR §61.57.

RTC
4019 [B] — (A20) — FAR §61.57(e)(1)(i)
No pilot may act as pilot-in-command under IFR unless that pilot has, within the past 6 months, logged: at least 6 hours of instrument time under actual or simulated IFR conditions, at least 3 of which were in flight in the category of aircraft involved, including at least 6 instrument approaches, or passed an instrument competency check in the category of aircraft involved. With the conditions stated in this question, the pilot must fly 3 hours in a helicopter of IFR flight (actual or simulated) to meet the total of 6 hours of instrument flight time, of which at least 3 were in the helicopter.

ALL
4020 [A] — (A20) — FAR §61.57(e)(1)(i)
To act as pilot-in-command under IFR, a pilot must have logged at least 6 hours of actual or simulated instrument time within the past 6 months. At least 3 of the 6 hours must be flown in the category of aircraft involved. In addition to the hour requirements, a pilot must have made six instrument approaches within the past 6 months.

ALL
4021 [B] — (A20) — FAR §61.57(e)(1)(i)
A pilot who does not meet the recent instrument experience requirements during the prescribed time, may not serve as pilot-in-command under IFR until that pilot passes an instrument competency check in the category of aircraft involved. If no other instrument flight is conducted, that instrument competency check will remain current for 6 months.

RTC
4022 [A] — (A20) — FAR §61.57(e)(1)(i)
No pilot may act as pilot-in-command under IFR unless that pilot has, within the past 6 months, logged: at least 6 hours of instrument time under actual or simulated IFR conditions, at least 3 of which were in flight in the category of aircraft involved, including at least 6 instrument approaches; or the pilot can pass an instrument competency check in the category of aircraft involved. With the conditions stated in this question, the pilot must fly 3 hours in the category involved, and complete 4 more instrument approaches to total the 6 required.

AIR
4023 [A] — (A20) — FAR §61.57(e)(1)(i)
The minimum conditions necessary for the instrument approaches is to complete 6 within the past 6 months, to be completed in an aircraft, approved instrument ground trainer, or any combination of these.

ALL
4024 [C] — (A20) — FAR §61.3(e) and §91.135
Each person operating an aircraft in Class A airspace must conduct that operation under instrument flight rules (IFR).

ALL
4025 [B] — (A20) — FAR §61.3(e) and §91.135
No person may act as pilot-in-command of a civil aircraft under instrument flight rules, or in weather conditions less than the minimums prescribed for VFR flight unless:

1. In the case of an airplane, the pilot holds an Instrument Rating or an Airline Transport Pilot Certificate with an airplane category rating on it; or
2. In the case of a helicopter, the pilot holds a helicopter instrument rating or an Airline Transport Pilot Certificate with a rotorcraft category and helicopter class rating not limited to VFR.

Also, each person operating an aircraft in Class A airspace must conduct that operation under instrument flight rules (IFR).

AIR
4026 [A] — (A20) — FAR §61.57(e)(1)(i)
No pilot may act as pilot-in-command under IFR unless that pilot has, within the past 6 months, logged: at least 6 hours of instrument time under actual or simulated IFR conditions, at least 3 of which were in flight in the category of aircraft involved, including at least 6 instrument approaches; or the pilot can pass an instrument competency check in the category of aircraft involved. With the conditions stated in this question, the pilot must complete 4 more instrument approaches to total the 6 required.

ALL
4027 [A] — (A20) — FAR §61.57(e)(1)(i)
No pilot may act as pilot-in-command under IFR unless that pilot has, within the past 6 months, logged: at least 6 hours of instrument time under actual or simulated IFR conditions, at least 3 of which were in flight in the category of aircraft involved, including at least 6 instrument approaches.

AIR
4028 [C] — (A20) — FAR §61.129(a)
A Commercial pilot applicant must hold an instrument rating (airplane), or the Commercial Pilot Certificate that is issued is endorsed with a limitation prohibiting the carriage of passengers for hire in airplanes on cross-country flights of more than 50 nautical miles (NM), or at night.

AIR
4029 [B] — (A20) — FAR §61.129
A Commercial Certificate, with single-engine airplane class, is required to carry passengers for hire in that class of aircraft. In addition, the applicant must hold an instrument rating (airplane), or the Commercial Pilot Certificate is endorsed with a limitation prohibiting the carriage of passengers for hire in airplanes on cross-country flights of more than 50 nautical miles (NM), or at night.

RTC
4030 [B] — (A20) — FAR §61.3 and §61.57
No person may act as pilot-in-command of a helicopter unless the pilot has a valid pilot certificate with rotorcraft category and helicopter class ratings. In addition, to operate IFR, the pilot-in-command must have either an instrument-helicopter rating or an Airline Transport Pilot Certificate with rotorcraft category, and helicopter class rating not limited to VFR. Along with the appropriate ratings, a pilot must have the required recent experience set forth in FAR §61.57.

ALL
4031 [B] — (A20) — FAR §61.3
No person may act as pilot-in-command of a civil aircraft under instrument flight rules, or in weather conditions less than the minimums prescribed for VFR flight unless:

1. In the case of an airplane, the pilot holds an Instrument Rating or an Airline Transport Pilot Certificate with an airplane category rating on it; or
2. In the case of a helicopter, the pilot holds a helicopter Instrument Rating or an Airline Transport Pilot Certificate with a rotorcraft category and helicopter class rating not limited to VFR.

Also, each person operating an aircraft in Class A airspace must conduct that operation under instrument flight rules (IFR).

AIR
4032 [B] — (B10) — FAR §91.167 and §91.169
No person may operate a civil aircraft in IFR conditions unless it carries enough fuel (considering weather reports and forecasts, and weather conditions) to:

1. Complete the flight to the first airport of intended landing;
2. Fly from that airport to the alternate airport; and
3. Fly after that for 45 minutes at normal cruising speed.

The above does not apply if:

1. There is a standard instrument approach procedure for the first airport of intended landing; and
2. For at least 1 hour before and 1 hour after the estimated time of arrival at the airport, the weather reports or forecasts or any combination of them, indicate;
    a. The ceiling will be at least 2,000 feet above the airport elevation; and
    b. Visibility will be at least 3 statute miles.

An alternate, and fuel for flight to the alternate, plus 45 minutes reserve is required for this question because the forecast weather (1,500-foot ceiling) does not meet the 2,000-foot ceiling exemption.

ALL
4033 [C] — (B08) — FAR §91.103
Each pilot-in-command shall, before beginning a flight, become familiar with all available information concerning the flight. This information must include:

1. For a flight under IFR or a flight not in the vicinity of an airport, weather reports and forecasts, fuel requirements, alternatives available if the planned flight cannot be completed, and any known traffic delays of which the pilot has been advised by ATC.
2. For any flight, runway lengths at airports of intended use.
3. For civil aircraft for which an approved airplane or rotorcraft flight manual containing takeoff and landing distance data is required, the takeoff and landing distance data contained therein.

AIR
4034 [C] — (A20) — FAR §61.129(a)
The commercial pilot applicant must hold an instrument rating (airplane), or the Commercial Pilot Certificate that is issued is endorsed with a limitation prohibiting the carriage of passengers for hire in airplanes on cross-country flights of more than 50 NM, or at night.

AIR
4035 [C] — (A20) — FAR §61.129(a)
In order to fly for hire, the pilot-in-command must hold a Commercial Pilot Certificate. In addition commercial pilots must hold an instrument rating (airplane), or they will be prohibited to carry passengers for hire in airplanes on cross-country flights of more than 50 NM, or at night.

AIR
4036 [C] — (B10) — FAR §91.171(a)
No person may operate a civil aircraft under IFR using the VOR system of radio navigation unless the VOR equipment of that aircraft:

1. Is maintained, checked, and inspected under an approved procedure; or
2. Has been operationally checked within the preceding 30 days and was found to be within the limits of the permissible indicated bearing error.

ALL
4037 [B] — (B11) — FAR §91.215(b)
A coded transponder with altitude reporting (Mode C) is required for all operations within the 48 contiguous states and the District of Columbia above an altitude of 10,000 feet MSL. Operations in airspace below 2,500 feet AGL are excluded from this rule.

ALL
4038 [A] — (B11) — FAR §91.215(b)
A coded transponder with altitude reporting (Mode C) is required for all operations within the 48 contiguous states and the District of Columbia above an altitude of 10,000 feet MSL. Operations in airspace below 2,500 feet AGL are excluded from this rule.

ALL
4039 [C] — (B07) — FAR §91.3(a)
The pilot-in-command of an aircraft is directly responsible for, and is the final authority as to, the operation of that aircraft.

RTC
4040 [A] — (B10) — FAR §91.167
No person may operate a civil helicopter in IFR conditions unless it carries enough fuel (considering weather reports and forecasts, and weather conditions) to:

1. Complete the flight to the first airport of intended landing;
2. Fly from that airport to the alternate airport; and
3. Fly after that for 30 minutes at normal cruising speed.

RTC
4041 [A] — (B10) — FAR §91.167
No person may operate a civil helicopter in IFR conditions unless it carries enough fuel (considering weather reports and forecasts, and weather conditions) to:

1. Complete the flight to the first airport of intended landing;
2. Fly from that airport to the alternate airport; and
3. Fly after that for 30 minutes at normal cruising speed.

ALL
4042 [C] — (B11) — FAR §91.211(a)
No person may operate a civil aircraft of U.S. registry at cabin pressure altitudes above 12,500 feet MSL up to and including 14,000 feet MSL unless the required minimum flight crew is provided with and uses supplemental oxygen for that part of the flight at those altitudes that is of more than 30 minutes duration.

Solution:

2 hours 20	minutes
– 30	minutes
1 hour 50	minutes

RTC
4043 [C] — (B11) — FAR §91.205(d)
For IFR flight, the following instruments and equipment are required:

1. Instruments and equipment specified for VFR flight (and those required for night flight, if applicable).
2. Two-way radio communications system and navigational equipment appropriate to the ground facilities to be used.
3. Gyroscopic rate-of-turn indicator (with certain exceptions).
4. Slip-skid indicator.
5. Sensitive altimeter adjustable for barometric pressure.
6. A clock displaying hours, minutes and seconds with a sweep-second pointer or digital presentation.

7. Generator or alternator of adequate capacity.
8. Gyroscopic pitch and bank indicator (artificial horizon).
9. Gyroscopic direction indicator (directional gyro or equivalent).

ALL
4044 [B] — (B10) — FAR §91.171(d)
Each person making the VOR operational check shall enter the date, place, bearing error, and sign the aircraft log or other record.

ALL
4045 [C] — (B11) — FAR §91.211(a)(1)
No person may operate a civil aircraft of U.S. registry at cabin pressure altitudes above 12,500 feet MSL up to and including 14,000 feet MSL unless the required minimum flight crew is provided with and uses supplemental oxygen for that part of the flight at those altitudes that is of more than 30 minutes duration.

ALL
4046 [A] — (B10) — FAR §91.171(d)
Each person making the VOR operational check shall enter the date, place, bearing error, and sign the aircraft log or other record.

ALL
4047 [C] — (B13) — FAR §91.411(a)(1)
No person may operate an airplane or helicopter in controlled airspace under IFR unless within the preceding 24 calendar months, each static pressure system, each altimeter instrument, and each automatic pressure altitude reporting system has been tested and inspected.

ALL
4048 [A] — (B10) — FAR §91.171(a)(2), §91.413(a) and §91.411(a)(1)
No person may operate a civil aircraft under IFR using the VOR system of radio navigation unless the VOR equipment of that aircraft has been operationally checked within the preceding 30 days and was found to be within the limits of the permissible indicated bearing error.

No person may operate an airplane in controlled airspace under IFR unless within the preceding 24 calendar months, each static pressure system, each altimeter instrument, and each automatic pressure altitude reporting system has been tested and inspected.

No person may use an ATC transponder unless, within the preceding 24 calendar months, that ATC transponder has been tested and inspected.

ALL
4049 [C] — (B13) — FAR §91.411(a)(1)
No person may operate an airplane or helicopter in controlled airspace under IFR unless the aircraft altimeter system test and inspection has been accomplished within the preceding 24 calendar months.

ALL
4050 [A] — (B11) — FAR §91.205(e)
DME is required if VOR is used for navigation on flights at or above 24,000 feet MSL.

ALL
4051 [C] — (B11) — FAR §91.205(d)
For IFR flight, the following instruments and equipment are required:

1. Instruments and equipment specified for VFR flight (and those required for night flight, if applicable).
2. Two-way radio communications system and navigational equipment appropriate to the ground facilities to be used.
3. Gyroscopic rate-of-turn indicator (with certain exceptions).
4. Slip-skid indicator.
5. Sensitive altimeter adjustable for barometric pressure.
6. A clock displaying hours, minutes and seconds with a sweep-second pointer or digital presentation.
7. Generator or alternator of adequate capacity.
8. Gyroscopic pitch and bank indicator (artificial horizon).
9. Gyroscopic direction indicator (directional gyro or equivalent).

ALL
4052 [C] — (B11) — FAR §91.211(a)(3)
No person may operate a civil aircraft of U.S. registry at cabin pressure altitudes above 15,000 feet (MSL) unless each occupant of the aircraft is provided with supplemental oxygen.

ALL
4053 [C] — (B11) — FAR §91.211(a)(3)
No person may operate a civil aircraft of U.S. registry at cabin pressure altitudes above 15,000 feet (MSL) unless each occupant of the aircraft is provided with supplemental oxygen.

ALL
4054 [A] — (B10) — FAR §91.171(c)
If dual system VOR (units independent of each other except for the antenna) is installed in the aircraft, the pilot shall tune both systems to the same VOR ground facility and note the indicated bearings TO that station. The maximum permissible variation between the two indicated bearings is 4° on the ground or in the air.

ALL
4055 [C] — (B11) — FAR §91.205(d)
Two-way radio communications system and navigational equipment appropriate to the ground facilities to be used is required for IFR flight.

ALL
4056 [B] — (I05) — AC 61-27C, page 33
The needle of the vertical velocity indicator should indicate zero when the aircraft is on the ground or maintaining a constant pressure level in flight. If this adjustment cannot be made, you must allow for the error when interpreting the indications in flight. Since the VSI is not a required instrument, there are no FAR requirements concerning calibration.

RTC
4057 [A] — (B11) — FAR §91.205(d)
For IFR flight, the following instruments and equipment are required:

1. Instruments and equipment specified for VFR flight (and those required for night flight, if applicable)
2. Two-way radio communications system and navigational equipment appropriate to the ground facilities to be used.
3. Gyroscopic rate-of-turn indicator (with certain exceptions).
4. Slip-skid indicator.
5. Sensitive altimeter adjustable for barometric pressure.
6. A clock displaying hours, minutes and seconds with a sweep-second pointer or digital presentation.
7. Generator or alternator of adequate capacity.
8. Gyroscopic pitch and bank indicator (artificial horizon).
9. Gyroscopic direction indicator (directional gyro or equivalent).

ALL
4058 [C] — (J15) — AIM ¶5-13(e),(f)
If operating on an IFR flight plan to an airport where there is no functioning control tower, the pilot must initiate cancellation of the IFR flight plan. This can be done after landing if there is a functioning FSS or other means of direct communications (radio or telephone) with ATC.

ALL
4059 [B] — (J15) — AIM ¶5-6
Composite flight plans can be filed whenever VFR operation will be used for one portion of a flight and IFR for another portion.

ALL
4060 [A] — (J15) — AIM ¶5-7(b)
The flight plan should define the proposed route of flight by using airway designations, transition points between airways and fixes defining off-airway routes. A composite flight should also indicate the point at which the transition from IFR to VFR is proposed.

ALL
4061 [A] — (J15) — AIM ¶5-6
When VFR flight is conducted for the first part of a composite flight, close the VFR portion and request ATC clearance from the FSS nearest the point at which the change from VFR to IFR is proposed.

ALL
4062 [C] — (B10) — FAR §91.173 and §91.135
No person may operate an aircraft in controlled airspace under IFR unless the person has:

1. Filed an IFR flight plan; and
2. Received an appropriate ATC clearance.

No person may operate an aircraft within Class A airspace unless that aircraft is:

1. Operated under IFR at a specific flight level assigned by ATC;
2. Equipped with instruments and equipment required for IFR operations;
3. Flown by a pilot rated for instrument flight.

ALL
4063 [B] — (B10) — FAR §91.173
No person may operate an aircraft in controlled airspace under IFR unless that person has:

1. Filed an IFR flight plan; and
2. Received an appropriate ATC clearance.

ALL
4064 [A] — (B10) — FAR §91.173
No person may operate an aircraft in controlled airspace under IFR unless that person has:

1. Filed an IFR flight plan; and
2. Received an appropriate ATC clearance.

ALL
4065 [C] — (B10) — FAR §91.173
No person may operate an aircraft in controlled airspace under IFR unless that person has:

1. Filed an IFR flight plan; and
2. Received an appropriate ATC clearance.

ALL
4066 [B] — (B10) — FAR §91.135
No person may operate an aircraft within Class A airspace (regardless of the weather conditions) unless that aircraft is:

1. Operated under IFR at a specific flight level assigned by ATC;
2. Equipped with instruments and equipment required for IFR operations;
3. Flown by a pilot rated for instrument flight.

ALL
4067 [B] — (B10) — FAR §91.173, §91.135
No person may operate in controlled airspace under IFR unless that person has:

1. Filed an IFR flight plan; and
2. Received an appropriate ATC clearance.

Each person operating an aircraft within Class A airspace must conduct that operation under instrument flight rules, with an ATC clearance received prior to entering the airspace.

ALL
4068 [C] — (B10) — FAR §91.173
No person may operate an aircraft in controlled airspace under IFR unless that person has:

1. Filed an IFR flight plan; and
2. Received an appropriate ATC clearance.

ALL
4069 [A] — (I07) — AC 61-27C, page 120
Waypoint: A predetermined geographical position used for route/instrument approach definition, or progress reporting purposes, that is defined relative to a VORTAC station, or in terms of latitude/longitude coordinates.

ALL
4070 [B] — (J34) — Airport/Facility Directory
Preferred IFR routes beginning/ending with a fix indicate that aircraft may be routed to/from these fixes via a Standard Instrument Departure (SID) route, radar vectors (RV), or a Standard Terminal Arrival Route (STAR).

ALL
4071 [B] — (J15) — AIM ¶5-32(a)(1)(e)
The pilot of an IFR flight should report changes in the average true airspeed (at cruising altitude) of 5 percent or 10 knots (whichever is greater) from that filed in the flight plan.

ALL
4072 [C] — (J15) — AIM ¶5-7(f), Block 1
For a composite flight, the pilot should check both the VFR and IFR blocks in block 1 of the flight plan.

ALL
4073 [B] — (J15) — AIM ¶5-7(f), Block 12
The pilot should fill in block 12 of the flight plan with the time computed from the departure point, using normal cruising speed and total usable fuel on board.

ALL
4074 [A] — (J15) — AIM ¶5-7(f), Block 7
Enter only the initial requested altitude in this block. When more than one IFR altitude or flight level is desired along the route of flight, it is best to make a subsequent request direct to the controller.

ALL
4075 [C] — (J15) — AIM ¶5-7
When filing an IFR flight plan for flight in an aircraft equipped with a radar beacon transponder, DME equipment, TACAN-only equipment or a combination of both, identify equipment capability by adding a suffix to the aircraft type preceded by a slant. For example, /R means RNAV (which includes DME equipment) and altitude encoding transponder.

ALL
4076 [C] — (J15) — AIM ¶5-13(b)
An IFR flight plan may be canceled at any time the flight is operating in VFR conditions outside Class A airspace by the pilot stating, "Cancel my IFR flight plan" to the controller or air/ground station with which they are communicating.

ALL
4077 [A] — (J15) — AIM ¶9-4(c) and Enroute Low Altitude Chart Legend
All Class B, C, D and E airspace, victor airways, and special use airspace, all up to 18,000 feet MSL, are shown on low altitude enroute charts.

ALL
4078 [A] — (J15) — AIM ¶5-31(c)(2)
When a pilot is on a direct flight, pilots shall report over each reporting point used in the flight plan to define the route of flight.

ALL
4079 [C] — (J34) — AIM ¶5-3(a)
The Airport/Facility Directory provides current airport conditions known. The Distant (D) NOTAMs, and Local (L) NOTAMs provide the latest status of airport conditions, for those items too current or temporary to be published in the A/FD.

ALL
4080 [C] — (J06) — AIM ¶5-3(b)
FDC (Flight Data Center) NOTAMs are regulatory in nature, and contain such things as amendments to published IAPs and other current aeronautical charts. They are also used to advertise temporary flight restrictions caused by such things as natural disasters or large-scale public events that may generate a congestion of air traffic over a site.

ALL

4081 [B] — (B10) — FAR §91.169(c)

No person may include an airport as an alternate in an IFR flight plan unless current weather forecasts indicate that, at the estimated time of arrival at the alternate airport, the ceiling and visibility at that airport will be at or above the following alternate airport weather minimums:

1. Precision approach procedure (ILS): Ceiling 600 feet and visibility 2 statute miles.
2. Nonprecision approach procedure (VOR, NDB, LOC): Ceiling 800 feet and visibility 2 statute miles.

(If an instrument approach procedure has been published for that airport, use the alternate airport minimums specified in that procedure.)

ALL

4082 [C] — (B10) — FAR §91.169(c)

An alternate airport is required on an IFR flight plan unless, from 1 hour before to 1 hour after the ETA at the destination airport, weather reports and forecasts indicate that at the destination:

1. The ceiling will be at least 2,000 feet above the airport elevation; and
2. The visibility will be at least 3 statute miles.

In the forecast for Atlanta, the chance of 1,000-foot ceilings due to overcast is after 2100Z, so an alternate is not required.

ALL

4083 [C] — (B10) — FAR §91.169(c)

An alternate airport is required on an IFR flight plan unless, from 1 hour before to 1 hour after the ETA at the destination airport, weather reports and forecasts indicate that at the destination:

1. The ceiling will be at least 2,000 feet above the airport elevation; and
2. The visibility will be at least 3 statute miles.

RTC

4084 [B] — (B10) — FAR §91.169(c)

An alternate airport is required on an IFR flight plan unless, from 1 hour before to 1 hour after the ETA at the destination airport, weather reports and forecasts indicate that at the destination:

1. The ceiling will be at least 2,000 feet above the airport elevation; and
2. The visibility will be at least 3 statute miles.

ALL

4085 [A] — (B10) — FAR §91.169(c)

An alternate airport with a VOR approach only (a nonprecision approach) can be used only if the current weather conditions indicate that at the ETA, the alternate airport ceiling is at least 800 feet and the visibility is at least 2 statute miles.

ALL

4086 [C] — (B10) — FAR §91.169(c)(2)

If no instrument approach procedure has been published for that airport, the ceiling and visibility minimums are those allowing descent from the MEA, approach, and landing, under basic VFR.

ALL

4087 [A] — (B10) — FAR §91.169(c)

An alternate airport with a precision approach can be used only if the current weather conditions indicate that at the ETA, the alternate airport ceiling is at least 600 feet and the visibility is at least 2 statute miles.

ALL

4088 [C] — (G10) — NTSB Part 830

NTSB Part 830 contains regulations pertaining to notification and reporting of aircraft accidents or incidents and overdue aircraft, and preservation of aircraft wreckage, mail, cargo, and records.

ALL

4089 [A] — (I22) — AC 00-6A, page 19

Density altitude is pressure altitude correct for nonstandard temperature. Under standard atmospheric conditions, each level of air in the atmosphere has a specific density, and under standard conditions, pressure altitude and density altitude identify the same level.

ALL

4090 [B] — (I22) — AC 00-6A, page 19

Pressure altitude will be equal to true altitude when standard atmospheric conditions exist for the same altitude.

ALL

4091 [C] — (I22) — AC 00-6A, page 17

Effects of nonstandard conditions can result in a difference of as much as 2,000 feet between true and indicated altitude. On a warm day, the pressure levels are raised, and the altimeter will indicate a lower altitude than the one actually flown (true altitude).

ALL

4092 [A] — (I31) — AC 00-45D, page 7-1

Radar only detects precipitation. Not all clouds contain enough moisture to be detected by weather radar, which makes it possible to fly into instrument weather conditions despite using radar.

ALL

4093 [B] — (I22) — AC 00-6A, page 12

The altimeter indication changes 1,000 feet for every one inch change in pressure. When the pressure setting is changed to a lower number, the altitude indicated on the altimeter will decrease. The altimeter setting in this question has changed 0.15 (30.11 - 29.96 = 0.15), which will cause a 150-foot change in the altimeter reading (.15 x 1,000).

ALL
4094 [C] — (I21) — AC 00-6A, page 9
An increase in temperature with altitude is defined as an inversion. An inversion often develops near the ground on clear, cool nights when wind is light. The ground radiates and cools much faster than the overlying air. Air in contact with the ground becomes cold while the temperature a few hundred feet above changes very little. Thus, temperature increases with height.

ALL
4095 [C] — (I21) — AC 00-45D, page 4-15
According to the winds and temperatures aloft forecast, the temperature is -10°C at 9,000 feet. Using the average lapse rate of 2°C per 1,000 feet, the temperature change from sea level to 9,000 feet is 18°C. Standard sea level temperature is 15°C. Subtract 18°C from 15°C to get -3°C. Compared to the winds and temperatures aloft forecast for 9,000 feet, the difference is 7°C (10 – 3).

ALL
4096 [A] — (I21) — AC 00-6A, page 7
Every physical process of weather is accompanied by or is the result of a heat exchange. Differences in solar energy create temperature variations. These temperature variations create forces that drive the atmosphere in its endless motion.

ALL
4097 [C] — (I20) — AC 00-6A, page 2
Above the tropopause is the stratosphere. This layer is typified by relatively small changes in temperature with height except for a warming trend near the top.

ALL
4098 [C] — (I27) — AC 00-6A, page 72
Steady precipitation is a characteristic of stable air, which has little or no turbulence. Stratiform clouds are associated with stable air.

ALL
4099 [C] — (I24) — AC 00-6A, page 43
Rain falling through colder air may become supercooled, freezing on impact as freezing rain; or it may freeze during its descent, falling as ice pellets. Ice pellets always indicate freezing rain at higher altitude.

ALL
4100 [C] — (I29) — AC 00-6A, page 41
In order for frost to form, the air temperature must be below the dew point and the dew point of the surrounding air must be colder than freezing. Water vapor will then sublimate directly as ice crystals or frost.

ALL
4101 [A] — (I24) — AC 00-6A, page 38
Dew point is the temperature to which air must be cooled to become saturated by the water vapor already present in the air. Aviation weather reports normally include the air temperature and dew point temperature. Dew point, when related to air temperature, reveals qualitatively how close the air is to saturation.

ALL
4102 [A] — (I24) — AC 00-6A, page 43
Wet snow at your altitude means that the temperature is above freezing since the snow has begun to melt. For snow to form, water vapor must go from the vapor state to the solid state (known as sublimation) with the temperature below freezing. Since melting snow has been encountered, the freezing level must be at a higher altitude.

ALL
4103 [B] — (I24) — AC 00-6A, page 38
Temperature largely determines the maximum amount of water vapor air can hold. Warm air can hold more water vapor than cool air.

ALL
4104 [A] — (I24) — AC 00-6A, page 38
When temperature reaches the dew point, water vapor can no longer remain invisible but is forced to condense, becoming visible on the ground as dew, appearing in the air as fog or clouds, or falling to the earth as rain.

ALL
4105 [B] — (I23) — AC 00-6A, page 30
Friction between the wind and the surface slows the wind. As frictional force slows the wind speed, Coriolis force decreases. However, friction does not affect pressure gradient force. Pressure gradient and Coriolis forces are no longer in balance. The stronger pressure gradient force turns the wind at an angle across the isobars toward lower pressure until the three forces balance. The angle of surface wind to isobars is about 10° over water, increasing with roughness of terrain.

ALL
4106 [B] — (I23) — AC 00-6A, page 30
Surface winds and winds at altitude can differ due to friction. Friction between the wind and the surface slows the wind.

ALL
4107 [B] — (I23) — AC 00-6A, page 30
Close to the earth, wind direction is modified by the contours over which it passes and wind speed is reduced by friction with the surface. Also, the winds at the surface are at an angle across the isobars due to the stronger pressure gradient. At levels 2,000 feet above the surface, the speed is greater and the direction is usually parallel to the isobars.

ALL
4108 [C] — (I23) — AC 00-6A, page 25
Coriolis force is at a right angle to wind direction and directly proportional to wind speed. In the Northern Hemisphere, the air is deflected to the right.

ALL
4109 [B] — (I22) — AC 00-6A, page 17
Effects of nonstandard conditions can result in a difference of as much as 2,000 feet between true and indicated altitude. On a cold day, the pressure levels are lowered, and the altimeter will indicate a higher altitude than the one actually flown (true altitude).

ALL
4110 [B] — (I22) — AC 00-6A, page 19
At or above 18,000 feet MSL, the altimeter is set to 29.92" Hg. You can always determine pressure altitude from your altimeter whether in flight or on the ground. Simply set the altimeter to 29.92 inches, and the altimeter indicates pressure altitude.

ALL
4111 [A] — (I22) — AC 00-6A, page 18
Since the altitude scale is adjustable, you can set the altimeter to read true altitude at some specified height. Takeoff and landing are the most critical phases of flight; therefore, airport elevation is the most desirable altitude for a true reading of the altimeter. Altimeter setting is the value to which the scale of the pressure altimeter is set so the altimeter indicates true altitude at the field elevation.

ALL
4112 [A] — (I21) — AC 00-6A, page 9
An inversion often develops near the ground on clear, cool nights when wind is light. The ground radiates and cools much faster than the overlying air. Air in contact with the ground becomes cold while the temperature a few hundred feet above changes very little. Thus, temperature increases with height.

ALL
4113 [B] — (I21) — AC 00-6A, page 9
Temperature normally decreases with increasing altitude throughout the troposphere. This decrease of temperature with altitude is defined as lapse rate. The average decrease of temperature (average lapse rate) in the troposphere is 2°C per 1,000 feet. An 8°C loss is necessary to reach 0°C, or freezing, in this situation. At 2°/1,000 feet the amount of altitude gain necessary would be:

1. 8°C ÷ 2 = 4 or 4,000 ft
2. 1,350 ft MSL (altitude at +8°C)
   + 4,000 ft (altitude gain necessary to reach 0°C)
   5,350 ft MSL (approximate freezing level)

ALL
4114 [A] — (I21) — AC 00-6A, page 52
A temperature inversion occurs when the temperature increases with altitude. A stable layer of air is characterized by warmer air lying above colder air. With an inversion, the layer is stable and convection is suppressed.

ALL
4115 [C] — (I25) — AC 00-6A, page 51
When stable air is forced upward, the air tends to retain horizontal flow and any cloudiness is flat and stratified. When unstable air is forced upward, the disturbance grows and any resulting cloudiness shows extensive vertical development.

ALL
4116 [A] — (I27) — AC 00-6A, page 64
The stability of an air mass determines its typical weather characteristics. When one type of air mass overlies another, conditions change with height. Characteristics typical of an unstable and a stable air mass are shown in the figure below.

ALL
4117 [B] — (I27) — AC 00-6A, page 64
The stability of an air mass determines its typical weather characteristics. When one type of air mass overlies another, conditions change with height. Characteristics typical of an unstable and a stable air mass are shown in the figure below.

ALL
4118 [C] — (I25) — AC 00-6A, page 51
When stable air is forced upward, the air tends to retain horizontal flow and any cloudiness is flat and stratified. When unstable air is forced upward, the disturbance grows and any resulting cloudiness shows extensive vertical development.

ALL
4119 [C] — (I27) — AC 00-6A, page 64
The stability of an air mass determines its typical weather characteristics. When one type of air mass overlies another, conditions change with height. Characteristics typical of an unstable and a stable air mass are shown in the figure below.

Unstable Air	Stable Air
Cumuliform clouds	Stratiform clouds and fog
Showery precipitation	Continuous precipitation
Rough air (turbulence)	Smooth air
Good visibility, except in blowing obstructions	Fair to poor visibility in haze and smoke

Questions 4116, 4117, 4119, 4120 and 4128

ALL
4120 [C] — (I27) — AC 00-6A, page 64
The stability of an air mass determines its typical weather characteristics. When one type of air mass overlies another, conditions change with height. Characteristics typical of an unstable and a stable air mass are shown in the figure on the previous page.

ALL
4121 [B] — (I25) — AC 00-6A, page 49
A change in ambient temperature lapse rate of an air mass can tip the balance between stable or unstable air. The ambient lapse rate is the rate of decrease in temperature with altitude.

ALL
4122 [B] — (I25) — AC 00-6A, page 50
Cloud type or structure is determined by whether the air is stable or unstable within the layer forced upward. When stable air is forced upward, the air tends to retain horizontal flow and any cloudiness is flat and stratified. When unstable air is forced upward, the disturbance grows, and resulting cloudiness shows "heaped" or cumulus development.

ALL
4123 [C] — (I25) — AC 00-6A, page 48
Unstable air favors convection. A cumulus cloud forms in a convective updraft and builds upward. The initial lifting that triggers a cumuliform cloud can be either orographic (topographical, i.e., mountains) or by surface heating. For convective cumuliform clouds to develop, the air must be unstable after saturation.

ALL
4124 [A] — (I25) — AC 00-6A, page 48
Unsaturated air moving upward and downward cools and warms at about 3°C (5.4°F) per 1,000 feet.

ALL
4125 [A] — (I21) — AC 00-6A, page 9
If the temperature increases with altitude through a layer (an inversion), the layer is stable and convection is suppressed. Air may be unstable beneath the inversion.

ALL
4126 [A] — (I30) — AC 00-6A, page 111
Precipitation beginning to fall from the cloud base signals that a downdraft has developed and a cell has entered the mature stage.

ALL
4127 [A] — (I27) — AC 00-6A, page 66
Frontal waves and cyclones (areas of low pressure) usually form on slow-moving cold fronts or stationary fronts.

ALL
4128 [B] — (I27) — AC 00-6A, page 64
Cool air moving over a warm surface is heated from below, generating instability and increasing the possibility of showers. Stability of an air mass determines its typical weather characteristics. When one type of air mass overlies another, conditions change with height. Characteristics typical of an unstable air mass are cumuliform clouds, turbulence, and good visibility. *See* the figure on the previous page.

ALL
4129 [B] — (I28) — AC 00-6A, page 61
Cumulonimbus clouds are the ultimate manifestation of instability. They are vertically developed clouds of large dimensions with dense boiling tops, often crowned with thick veils of dense cirrus (the anvil). Nearly the entire spectrum of flying hazards are contained in these clouds, including violent turbulence. They should be avoided at all times.

ALL
4130 [C] — (I28) — AC 00-6A, page 85
Standing lenticular altocumulus clouds are formed on the crests of waves created by barriers to the wind flow. The clouds show little movement, hence the name "standing." However, wind can be quite strong blowing through such clouds. The presence of these clouds is a good indication of very strong turbulence and should be avoided.

ALL
4131 [B] — (I26) — AC 00-6A, page 53
The prefix "nimbo-" or the suffix "-nimbus" means raincloud. Thus, stratified clouds from which rain is falling are called nimbostratus. A heavy, swelling cumulus-type cloud which produces precipitation is a cumulonimbus.

ALL
4132 [B] — (I28) — AC 00-6A, page 85
Standing lenticular altocumulus clouds are formed on the crests of waves created by barriers to the wind flow. The clouds show little movement, hence the name "standing." However, wind can be quite strong blowing through such clouds. The presence of these clouds is a good indication of very strong turbulence and they should be avoided.

ALL
4133 [B] — (I29) — AC 00-6A, pages 54 and 92
Two conditions are necessary for structural icing in flight:

1. The aircraft must be flying through visible water such as rain or cloud droplets, and
2. The temperature, at the point where the moisture strikes the aircraft must be 0°C or colder.

The high cloud family is composed almost entirely of ice crystals. Because ice crystals are already frozen, they most likely won't stick to an aircraft.

ALL
4134 [C] — (I26) — AC 00-6A, page 53
For identification purposes, cloud types are divided into four "families." The families are high clouds, middle clouds, low clouds, and clouds with extensive vertical development.

ALL
4135 [C] — (I30) — AC 00-6A, page 115
The winds around a thunderstorm are complex. Wind shear can be found on all sides of a thunderstorm cell and in the downdraft directly under the cell. The wind shift line, or gust front, associated with thunderstorms can precede the actual storm by 15 nautical miles or more.

ALL
4136 [A] — (I27) — AC 00-6A, page 65
Wind always changes across a front. Wind discontinuity may be in direction, in speed, or in both. Temperature and humidity also may change.

ALL
4137 [C] — (I30) — AC 00-6A, page 114
A squall line (instability line) is a nonfrontal, narrow band of active thunderstorms. Often it develops ahead of a cold front in moist, unstable air, but it may develop in unstable air far removed from any front.

ALL
4138 [C] — (I28) — AC 00-6A, page 86
Wind shear may be associated with either a wind shift or a wind speed gradient at any level in the atmosphere.

ALL
4139 [C] — (I28) — AC 00-6A, page 86
Wind shear may be associated with either a wind shift or a wind speed gradient at any level in the atmosphere. Low-level wind shear may result from a frontal passage, thunderstorm activity, or low-level temperature inversion. Wind shear may also be found at higher altitudes in association with a frontal passage due to wind shift through the frontal zone.

ALL
4140 [A] — (I28) — AC 00-6A, page 88
Wind shear occurs with a cold front just after the front passes the airport and for a short period thereafter. If the front is moving 30 knots or more, the frontal surface will usually be 5,000 feet above the airport about three hours after the frontal passage. With a warm front, the most critical period is before the front passes the airport. Warm front wind shear may exist below 5,000 feet for approximately six hours. The problem ceases to exist after the front passes the airport. Data compiled on wind shear indicates that the amount of shear in warm fronts is much greater than that found in cold fronts.

ALL
4141 [C] — (I27) — AC 00-6A, page 72
Usually, thunderstorms are quite visible to the pilot. However, when a thunderstorm is present but not visible to the pilot due to cloud cover, such as a thick stratus layer, the thunderstorm is said to be "embedded."

ALL
4142 [A] — (I43) — AC 00-45D, page 2-9
A squall is a sudden increase in speed of at least 15 knots to a sustained speed of 20 knots or more, lasting for at least one minute.

ALL
4143 [B] — (I30) — AC 00-6A, page 111
A thunderstorm cell during its life cycle progress through three stages: the cumulus, the mature, and the dissipating. Downdrafts characterize the dissipating stage.

ALL
4144 [A] — (I30) — AC 00-6A, page 116
A thunderstorm is a local storm produced by a cumulonimbus cloud. It is always accompanied by lightning and thunder, usually with strong gusts of wind, heavy rain, and sometimes, with hail.

ALL
4145 [B] — (I30) — AC 00-6A, page 114
A squall line often contains severe steady-state thunderstorms and presents the single most intense weather hazard to aircraft. It usually forms rapidly, generally reaching maximum intensity during the late afternoon and the first few hours of darkness. Hail competes with turbulence as the greatest thunderstorm hazard to aircraft.

ALL
4146 [C] — (I30) — AC 00-6A, page 123
Following are some do's and don'ts during thunderstorm penetration:

1. Do keep your eyes on your instruments. Looking outside the cockpit can increase the danger of temporary blindness from lightning.
2. Don't change power settings; maintain settings for reduced airspeed.
3. Do maintain a constant attitude; let the aircraft "ride the waves." Maneuvers that try to maintain constant altitude increase stresses on the aircraft.
4. Don't turn back once you are in the thunderstorm. A straight course through the storm most likely is the quickest way out of the hazards. In addition, turning maneuvers increase stresses on the aircraft.

ALL
4147 [B] — (I30) — AC 00-6A, page 111
Precipitation beginning to fall from the cloud base signals that a downdraft has developed and a cell has entered the mature stage.

ALL
4148 [C] — (I30) — AC 00-6A, page 111
For a thunderstorm to form, the air must have:

1. Sufficient water vapor,
2. An unstable lapse rate, and
3. An initial upward boost (lifting) to start the storm process in motion.

ALL
4149 [A] — (I28) — AC 00-6A, page 81
Fair weather cumulus clouds often indicate bumpy turbulence beneath and in the clouds but good visibility. The cloud tops indicate the approximate upper limit of convection; flight above is usually smooth.

ALL
4150 [C] — (I28) — AC 00-6A, page 86
Wind shear may be associated with either a wind shift or a wind speed gradient at any level in the atmosphere. Low-level wind shear may result from a frontal passage, thunderstorm activity, or low-level temperature inversion. Wind shear may also be found at higher altitudes in association with a frontal passage due to wind shift through the frontal zone.

ALL
4151 [C] — (I29) — AC 00-6A, page 102
Frost does not change the basic aerodynamic shape of the wing, but the roughness of its surface spoils the smooth flow of air, thus causing a slowing of the airflow. This slowing of the air causes early air flow separation over the affected airfoil, resulting in a loss of lift.

ALL
4152 [C] — (I29) — AC 00-6A, page 99
The condition most favorable for very hazardous icing is the presence of many large, supercooled water droplets, also called freezing rain.

ALL
4153 [B] — (I29) — AC 00-6A, page 43
As the rain falls through air that is below freezing, its temperature begins to fall below freezing yet without freezing solid. This is freezing rain. The process requires that the temperature of the rain must be above freezing before it becomes supercooled. Therefore, when freezing rain is encountered, it indicates that warmer temperatures are above.

ALL
4154 [C] — (I20) — AC 00-6A, page 2
The height of the troposphere varies with latitude and seasons. It slopes from about 20,000 feet over the poles, to an average of 37,000 feet over the mid-latitudes, to about 65,000 feet over the Equator, and it is higher in summer than in winter.

ALL
4155 [C] — (I32) — AC 00-6A, page 136
A jetstream occurs in an area of intensified temperature gradients characteristic of the break in the tropopause. The concentrated winds, by arbitrary definition, must be 50 knots or greater to classify as a jetstream.

ALL
4156 [A] — (I31) — AC 00-6A, page 127
Advection fog forms when moist air moves over colder ground or water.

ALL
4157 [C] — (I26) — AC 00-6A, page 54
The high cloud family is cirriform and includes cirrus, cirrocumulus, and cirrostratus. They are composed almost entirely of ice crystals.

ALL
4158 [C] — (I27) — AC 00-6A, page 63
When a body of air comes to rest, or moves slowly over an extensive area having fairly uniform properties of temperature and moisture, the air takes on those properties. Thus, the air over the area becomes one large mass and has fairly uniform horizontal distribution of its properties.

ALL
4159 [B] — (I24) — AC 00-6A, page 43
Cloud particles collide and merge into a larger drop in the more rapid growth process. This process produces larger precipitation particles and does so more rapidly than the simple condensation growth process. Upward currents enhance the growth rate and also support larger drops.

ALL
4160 [B] — (I30) — AC 00-6A, page 123
When flying in severe turbulence, maintaining positive aircraft control may be nearly impossible to do. In attempting to maintain a constant altitude, the stresses applied to the aircraft are greatly increased. Stresses will be lessened if the aircraft is held in a constant attitude and allowed to "ride the waves." Wide fluctuations in airspeed will probably have to be tolerated in order to reduce aircraft stresses.

ALL
4161 [C] — (I24) — AC 00-6A, page 43
Ice pellets always indicate freezing rain at higher altitudes.

ALL
4162 [C] — (I32) — AC 00-6A, page 127
Advection fog forms when moist air moves over colder ground or water.

ALL
4163 [B] — (I31) — AC 00-6A, page 126
Abundant condensation nuclei enhance the formation of fog. Thus, fog is prevalent in industrial areas where byproducts of combustion provide a high concentration of these nuclei.

ALL
4164 [A] — (I31) — AC 00-6A, page 127
Advection fog forms when moist air moves over colder ground or water. During the winter, advection fog over the central and eastern United States results when moist air from the Gulf of Mexico spreads northward over cold ground. The fog may extend as far north as the Great Lakes.

ALL
4165 [A] — (I31) — AC 00-6A, page 127
Advection fog forms when moist air moves over colder ground or water. It is most common along coastal areas but often develops deep into continental areas.

ALL
4166 [C] — (I31) — AC 00-6A, page 127
Advection fog forms when moist air moves over colder ground or water. Upslope fog forms as a result of moist, stable air being cooled adiabatically as it moves up sloping terrain.

ALL
4167 [A] — (I31) — AC 00-6A, page 126
Conditions favorable for radiation fog are clear skies, little or no wind, and small temperature/dew point spread (high relative humidity). The fog forms almost exclusively at night or near daybreak.

ALL
4168 [B] — (I32) — AC 00-6A, page 137
In mid-latitudes, wind speed in the jetstream averages considerably stronger in winter than in summer. Also the jet shifts farther south in winter than in summer.

ALL
4169 [C] — (I31) — AC 00-6A, page 126
Conditions favorable for radiation fog are clear skies, little or no wind, and small temperature/dew point spread (high relative humidity). The fog forms almost exclusively at night or near daybreak.

ALL
4170 [A] — (I43) — AC 00-45D, page 4-1
A Terminal Forecast (FT) is a description of the surface weather expected to occur at an airport. The forecast cloud heights and amounts, visibility, weather and wind relate to flight operations within 5 NM of the center of the runway complex.

ALL
4171 [C] — (I29) — AC 120-58
Test data indicate that ice, snow, or frost formations having a thickness and surface roughness similar to medium or coarse sandpaper on the leading edge and upper surface of a wing can reduce wing lift by as much as 30% and increase drag by 40%.

ALL
4172 [A] — (I48) — AC 00-45D, page 4-15
When the forecast speed is less than 5 knots, the coded group is "9900" and read, LIGHT AND VARIABLE. The "+00" is the outside air temperature for that altitude in degrees Celsius.

ALL
4173 [B] — (I53) — AC 00-45D, page 12-1
Plotted at each reporting station (at the level of the specified pressure) are the observed temperature, temperature/dew point spread, wind, and height of the pressure surface, as well as the height changes over the previous 12-hour period.

ALL
4174 [A] — (I46) — AC 00-45D, page 7-1
Radar does show lines of thunderstorms and thunderstorm cells, all of which are hazardous and are not shown on other weather charts.

ALL
4175 [A] — (I47) — AC 00-45D, page 4-13
The Significant Cloud and Weather Section is a 12-hour forecast (in broad terms), of clouds and weather significant to flight operations plus a 6-hour categorical outlook. The breakdown may be by states, by well-known geographical areas, or in reference to location and movement of a pressure system or front.

ALL
4176 [C] — (I43) — AC 00-45D, pages 4-1, 4-2
A Terminal Forecast (FT) is a prediction of weather conditions to be expected for a specific airport. The area covered in a Terminal Forecast is within a 5-mile radius of the center of the runway complex. Generally, the Terminal Forecast includes expected ceiling and clouds, visibility, weather and obstructions to vision, and surface wind conditions at each terminal. Also included are remarks that more completely describe expected weather. If a change is expected during the forecast period, this change and expected time for change are included.

ALL
4177 [B] — (I43) — AC 00-45D, page 4-2
Omission of a wind entry in the body of an FT implies wind less than 6 knots.

ALL
4178 [B] — (I43) — AC 00-45D, page 4-2
Absence of a visibility entry specifically implies visibility more than 6 statute miles.

ALL
4179 [C] — (I43) — AC 00-45D, page 4-2
The contraction WND or Wind means winds (sustained or gusts) are forecast to be 25 knots or stronger.

ALL
4180 [A] — (I43) — AC 00-45D, page 4-2
The Terminal Forecast for Memphis at 2200Z (22Z 20 SCT.) does not have a wind entry. Omission of a wind entry implies wind less than 6 knots.

ALL
4181 [B] — (I43) — AC 00-45D, page 4-11
A SIGMET advises of weather potentially hazardous to all aircraft other than convective activity. In the continental U.S., items covered are: severe icing, severe or extreme turbulence, duststorms, sandstorms, or volcanic ash lowering visibilities to less than 3 miles.

ALL
4182 [B] — (I42) — AC 00-45D, page 3-2
SK OVC 025/045 OVC 090: The pilot report states the top of the first layer of overcast is 2,500 feet MSL. The base of the second layer is 4,500 feet and the top is 9,000 feet.

ALL
4183 [A] — (I43) — AC 00-45D, page 4-11
A SIGMET advises of weather potentially hazardous to all aircraft other than convective activity. In the continental U.S., items covered are: severe icing, severe or extreme turbulence, duststorms, sandstorms, or volcanic ash lowering visibilities to less than 3 miles.

ALL
4184 [C] — (I43) — AC 00-45D, pages 3-1 and 4-11
Almost all of the listed reports and forecasts will contain information about icing hazards. SIGMETs and AIRMETs are issued when there is an icing hazard. PIREPs (pilot reports) contain information from pilots actually flying in the area forecast to have icing conditions.

ALL
4185 [C] — (I43) — AC 00-45D, page 4-13
The TWEB Route Forecast is similar to the Area Forecast (FA) except more specific information is contained in a route format. Forecast sky cover (height and amount of cloud bases), cloud tops, visibility (including vertical visibility), weather and obstructions to vision are described for a corridor 25 miles either side of the route.

ALL
4186 [A] — (I43) — AC 00-45D, page 4-10
This section contains a 12-hour forecast that identifies and locates aviation weather hazards which meet Inflight Advisory criteria and thunderstorms that are forecast to be at least scattered in area coverage. These hazards include IFR conditions, icing (ICG), turbulence (TURBC), mountain obscurations (MTN OBSCN), and thunderstorms (TSTMS).

ALL
4187 [C] — (I43) — AC 00-45D, page 4-12
AIRMETs are issued with a maximum forecast period of 6 hours. If conditions persist beyond the 6-hour forecast period, the AIRMET must be updated and reissued.

ALL
4188 [B] — (I43) — AC 00-45D, page 4-14
No temperatures are forecast for the 3,000-foot level or for a level within 2,500 feet of station elevation.

ALL
4189 [C] — (I43) — AC 00-45D, page 4-14
No winds are forecast within 1,500 feet of station elevation.

ALL
4190 [B] — (I43) — AC 00-45D, page 4-15
Encoded wind speeds 100 to 199 knots have 50 added to the direction code and 100 subtracted from the speed, so 830558 would be: Wind direction 83 – 50 = 330°; wind speed 05 + 100 = 105 knots; temperatures are negative above 24,000 feet, so 58 becomes -58°C.

ALL
4191 [C] — (I43) — AC 00-45D, page 4-15
The wind direction for all reports and forecasts is always given in relation to true north. The wind speed is always reported and forecast in knots.

ALL
4192 [C] — (I48) — AC 00-45C, page 4-15
To find the conditions at PSB at FL270, you must interpolate between 24,000 and 30,000:

Altitude	Code	Wind Direction	Wind Speed	Temperature
24,000	2368-26	230°	68	-26°C
30,000	781939	280°	119	-39°C
27,000		255°	93	-33°C

To find the temperature relative to ISA, multiply the standard temperature lapse rate (-2°C per 1,000 feet) times the altitude in thousands of feet (in this case 27), then subtract +15°C for standard day conditions:

1. -2°C x 27 = 54°C
2. -54°C – 15 = -39°C (ISA)

Therefore, -33°C (today) is 6°C warmer than -39°C, so the temperature is ISA + 6°C.

ALL
4193 [C] — (I48) — AC 00-45D, page 4-15
To find the conditions at ALB at FL270, you must interpolate between 24,000 and 30,000:

Altitude	Code	Wind Direction	Wind Speed	Temperature
24,000	2777-28	270°	77	-28°C
30,000	781842	280°	118	-42°C
27,000		275°	97	-35°C

To find the temperature relative to ISA, multiply the standard temperature lapse rate (-2°C per 1,000 feet) times the altitude in thousands of feet (in this case 27), then subtract +15°C for standard day conditions:

1. -2°C x 27 = 54°C
2. -54°C – 15 = -39°C (ISA)

Therefore, -35°C (today) is 4°C warmer than -39°C, so the temperature is ISA + 4°C.

ALL
4194 [A] — (I48) — AC 00-45D, page 4-15
To find the conditions at EMI at FL270, you must interpolate between 24,000 and 30,000:

Altitude	Code	Wind Direction	Wind Speed	Temperature
24,000	2891-30	280°	91	-30°C
30,000	751041	250°	110	-41°C
27,000		265°	100	-36°C

To find the temperature relative to ISA, multiply the standard temperature lapse rate (-2°C per 1,000 feet) times the altitude in thousands of feet (in this case 27), then subtract +15°C for standard day conditions:

1. -2°C x 27 = 54°C
2. -54°C – 15 = -39°C (ISA)

Therefore, -36°C (today) is 3°C warmer than -39°C, so the temperature is ISA + 3°C.

ALL
4195 [C] — (I51) — AC 00-45D, page 12-1
Plotted at each reporting station (at the level of the specified pressure) on constant pressure charts are the observed temperature, temperature/dew point spread, wind, height of the pressure surface, as well as height changes over the previous 12-hour period.

ALL
4196 [A] — (I41) — AC 00-45D, page 2-10
OVC/35 in the remarks section means the top of the overcast is 3,500 feet MSL. W8 means the ceiling is indefinite at 800 feet AGL. This means the overcast layer is 2,600 feet MSL. To find the layer thickness:

$$\begin{array}{r} 3{,}500 \\ -2{,}600 \\ \hline 900 \end{array} \text{ feet thick}$$

ALL
4197 [A] — (I50) — AC 00-45D, page 11-1
The single-hatched area indicates possible/probable severe thunderstorms. The line with an arrowhead delineates an area of probable general thunderstorm activity. When you face in the direction of the arrow, activity is expected to the right of the line.

ALL
4198 [A] — (I41) — AC 00-45D, page 3-2
The PIREP interprets: 20 NM south of Atlanta at 1620Z, flight level 5,000 feet, Beech 18 reported moderate rime ice.

ALL
4199 [B] — (I48) — AC 00-45D, page 4-15
Wind speed above 100 knots have 50 added to the direction code (first 2 digits). Wind speeds at 200 knots or greater are coded 99 (middle 2 digits). Temperatures above 24,000 feet are negative so the minus sign is omitted (last 2 digits). The 300° at 200 knots and 54° is a encoded as 809954.

ALL
4200 [A] — (I21) — AC 00-6A, pages 10 and 88
A ground-based inversion favors poor visibility by trapping fog, smoke, and other restrictions into low levels of the atmosphere. Wind just above the inversion may be relatively strong. A wind shear zone develops between the calm and the stronger winds above. Eddies in the shear zone cause airspeed fluctuations as an aircraft climbs or descends through the inversion.

ALL
4201 [C] — (I43) — AC 00-45D, page 4-10
The FA is comprised of five sections that give a forecast of general weather conditions over an area the size of several states:

1. Hazard/Flight Precautions (H)—includes IFR conditions, icing, turbulence, mountain obscuration and thunderstorms.
2. Synopsis (S)—a brief summary of the location and movement of fronts, pressure systems and circulation patterns.
3. Icing (I).
4. Turbulence (T) and low-level wind shear (if applicable).
5. Significant clouds and weather (C).

ALL
4202 [C] — (I41) — AC 00-45D, page 2-1
Ceiling is defined as height of the lowest layer of clouds or obscuring phenomena aloft that is reported as broken or overcast and not classified as thin; or, as vertical visibility into surface-based obscuring phenomena that hides all the sky.

ALL
4203 [A] — (I43) — AC 00-45D, page 2-1
The ceiling is 700 feet AGL measured (M7), and because the field elevation is 1,000 feet MSL (stated in question), that puts the base of the clouds at 1,700 feet MSL. The tops are at 6,500 feet MSL (UA/SK OVC 65). Therefore, 6,500 – 1,700 = 4,800 feet thick.

ALL
4204 [C] — (I41) — AC 00-45D, page 2-7
When obscuring phenomena is surface-based and partially obscures the sky (-X), a remark reports tenths of sky hidden. In this case, F2 means 2/10 of the sky is hidden by fog (F).

ALL
4205 [C] — (I41) — AC 00-45D, page 2-7
RF2 RB12: The "R" is the symbol for rain, and the "F" is the symbol for fog. They both are the obscuring phenomena, and they are obscuring 2/10 of the sky. The next remark (RB12) is the code for weather and obstruction to vision, which means "rain began 12 minutes after the hour" or at 1812.

ALL
4206 [C] — (I45) — AC 00-45D, pages 6-1, 6-2
Northern Illinois and southern Wisconsin are in a shaded area of the chart. The legend to FAA Figure 4 says that shaded areas indicate IFR with ceilings less than 1,000 feet and/or visibilities less than 3 miles. The solid black sky coverage symbols indicate overcast skies. The symbol [• •] means continuous rain slight at time of observation. The symbol [▽̇] means slight rain showers. The number to the left of the sky coverage symbol is visibility; the number underneath is cloud height or ceiling AGL.

ALL
4207 [B] — (I45) — AC 00-45D, pages 6-1, 6-2
The frontal area in Illinois (stationary front) is characterized by the symbols: [ ] (slight or moderate thunderstorms without hail, but with rain at time of observation) and [ ] (slight or moderate thunderstorms without hail but with continuous rain [not freezing] at time of observation). The symbols in the other states that enclose the front indicate either precipitation of lesser intensities or no precipitation at all.

ALL
4208 [C] — (I45) — AC 00-45D, page 6-1
Cloud height is entered under the station circle in hundreds of feet. If total sky cover is few or scattered, the height is the base of the lowest layer. The symbol [◔] means sky coverage is from 1/10 to 5/10 inclusive and is read "scattered." Visibility is not reported unless it is 6 statute miles or less. When it is reported, it will be put on the left side of the sky coverage symbol.

ALL
4209 [C] — (I44) — AC 00-45D, page 5-1
The Surface Analysis Chart provides a ready means of locating pressure systems and fronts and also gives an overview of winds, temperatures and dew points as of the time of the plotted observations. The chart also shows weather and obstruction to vision symbols. It does not show cloud heights or tops or the expected movement of any item depicted. Expected movement would be contained in a forecast.

ALL
4210 [A] — (I53) — AC 00-45D, page 15-3
Light turbulence is defined as turbulence that momentarily causes slight, erratic changes in attitude and/or altitude.

ALL
4211 [A] — (I47) — AC 00-45D, page 8-1
The Low-Level Prog is a four-panel chart. The two lower panels are 12- and 24-hour surface progs. The two upper panels are 12- and 24-hour progs of significant weather from the surface to 400 millibars (24,000 feet). The charts show conditions as they are forecast to be at the valid time of the chart.

ALL
4212 [C] — (I47) — AC 00-45D, page 8-1
The Low-Level Significant Weather Prognostic Chart depicts conditions as they are forecast to be at the valid time of the chart (12- and 24-hour forecasts).

ALL
4213 [A] — (I47) — AC 00-45D, page 8-1
The triangle with a dot over it indicates showery precipitation (i.e., rain showers). The two thick dots indicate continuous rain. The shaded area indicates more than one-half the area is obscured.

ALL
4214 [C] — (I47) — AC 00-45D, page 8-1
The Low-Level Significant Weather Prognostic Chart depicts conditions as they are forecast to be at the valid time of the chart (12- and 24-hour forecasts).

ALL
4215 [B] — (I43) — AC 00-45D, page 4-16
A Convective Outlook (AC) describes prospects for general thunderstorm activity during the following 24 hours. Use the outlook primarily for planning flights later in the day. The charts show conditions as they are forecast to be at the valid time of the chart.

ALL
4216 [B] — (I47) — AC 00-45D, pages 8-1 and 8-5
Use the 12-hour surface prognosis chart (lower left corner), valid at 1800Z. The symbol [ ] indicates a stationary front. The symbol [•] indicates an area of intermittent rain along the front. Tennessee is indicating VFR weather.

ALL
4217 [A] — (I47) — AC 00-45D, page 8-1
Use the 12-hour significant weather surface prognosis chart (upper left corner), valid at 1800Z. The scalloped edged area over northeastern West Virginia indicates marginal VFR (MVFR), with ceilings 1,000 to 3,000 feet inclusive and/or visibility 3 to 5 miles inclusive.

ALL
4218 [A] — (I47) — AC 00-45D, pages 8-1 and 8-3
Use the 24-hour significant weather surface prognosis (right side of figure). The smooth-edged area over southeastern Colorado indicates that the area will have ceilings less than 1,000 feet and/or visibility less than 3 miles. Rain showers (triangle with a dot over it) and thunderstorms (the "R" symbol) are also depicted.

ALL
4219 [A] — (I47) — AC 00-45D, page 8-1
Use the 12-hour significant weather surface prognosis chart (upper left corner), valid at 1800Z. The scalloped edged area over the middle of California's coastline indicates marginal VFR (MVFR), with ceilings 1,000 to 3,000 feet inclusive and/or visibility 3 to 5 miles inclusive.

ALL
4220 [C] — (I42) — AC 00-45D, page 3-2
A pilot reported over (OV) MRB at 6,000 feet (FL060), sky conditions (SK) intermittently between layers (INTMTLY BL), turbulence (TB) of moderate (MDT) intensity, remarks (RM) moderate rain (R), turbulence (TURBC) increasing (INCRS) westward (WWD).

ALL
4221 [C] — (I47) — AC 00-45D, page 8-3
Small-scalloped lines enclose areas of expected cumulonimbus development. The contraction CB denotes cumulonimbus. The contraction FRQ indicates 5/8 to 8/8 coverage cumulonimbus, XXX below the line indicates bases below 24,000 feet (below the scope of the chart), and 370 above the line indicates tops at 37,000 feet. Altitudes are above mean sea level.

ALL
4222 [B] — (I47) — AC 00-45D, page 8-3
The contraction ISOL EMBD indicates embedded cumulonimbus less than 1/8 coverage, XXX below the line indicates bases below 24,000 feet, and 410 above the line indicates tops at 41,000 feet. Altitudes are above mean sea level.

ALL
4223 [A] — (I47) — AC 00-45D, page 8-3
Areas of forecast moderate or greater clear air turbulence (CAT) are bounded by heavy dashed lines. The XXX below the line indicates that the vertical extent of the turbulence extends below 24,000 feet (the lower limit of the chart), and the 320 above the line indicates that the upper extent of the turbulence is at 32,000 feet.

ALL
4224 [B] — (I47) — AC 00-45D, page 8-3
The dashed line indicates clear air turbulence (CAT). The single-hat and double-hat symbols indicate moderate to severe turbulence. The XXX below the line indicates that the vertical extent of the turbulence extends below 24,000 feet (the lower limit of the chart), and the 330 above the line indicates that the upper extent of the turbulence is at 33,000 feet.

ALL
4225 [C] — (I47) — AC 00-45D, page 8-3
Tropopause heights are to be depicted in hundreds of feet MSL and are enclosed in small rectangular blocks.

ALL
4226 [B] — (I43) — AC 00-45D, page 4-16
A Convective Outlook (AC) describes prospects of both severe and general thunderstorms during the following 24 hours.

ALL
4227 [C] — (I20) — AC 00-6A, page 136
Temperature over the tropical tropopause increases with height, but temperatures over the polar tropopause remain almost constant. An abrupt change in temperature lapse rate characterizes the tropopause.

ALL
4228 [C] — (I43) — AC 00-45D, page 4-1
A Terminal Forecast (FT) is a prediction of weather conditions to be expected for a specific airport rather than a larger area. The size of area covered in a Terminal Forecast is within a 5 NM radius of the center of the runway complex. Generally, the Terminal Forecast includes expected ceiling and clouds, visibility, weather and obstructions to vision, and surface wind conditions at each terminal. Also included are remarks that more completely describe expected weather. If a change is expected during the forecast period, this change and expected time of change are included.

ALL
4229 [B] — (I47) — AC 00-45D, page 8-3
The scalloped area indicates cumulonimbus. The contraction OCNL EMBD indicates 1/8 to 4/8 coverage, embedded cumulonimbus. The XXX below the line indicates bases below 24,000 feet (the lower limit of the chart), and 510 above the line indicates tops at 51,000 feet. Altitudes are above mean sea level.

ALL
4230 [C] — (I46) — AC 00-45D, page 7-5
Area A indicates an echo intensity level of 3 to 4, strong to very strong, since it is in the second ring. Echo tops in are at 30,000 feet. Thunderstorms and rain showers are indicated by the contraction "TRW." *See* the figure on the next page.

ALL
4231 [B] — (I46) — AC 00-45D, page 7-5
Area D indicates an echo intensity level of 5 to 6, intense to extreme, since it is in the third ring. Echo tops in are at 29,000 feet. Individual cell movement is indicated by an arrow with the speed in knots entered as a number and the arrow shaft pointing in the direction of movement. In area D an individual cell is moving northeast at 50 knots. *See* the figure on the next page.

ALL
4232 [C] — (I46) — AC 00-45D, page 7-5
Area C indicates an echo intensity level of 3 to 4, strong to very strong, since it is in the second ring. Echo tops in are at 28,000 feet. Line or area movement is indicated by a shaft and barb combination with the shaft indicating the direction and the barbs the speed. A whole barb is 10 knots, a half barb is 5 knots, and a pennant is 50 knots. In area C, the area movement is toward the northeast at 20 knots. *See* the figure on the next page.

ALL
4233 [B] — (I46) — AC 00-45D, page 7-5
Area B indicates an echo intensity of 1 to 2, weak to moderate, since it is in the first ring. The contraction "RW+" indicates rain showers, intensity increasing, or new echo. *See* the figure below.

ALL
4234 [A] — (I46) — AC 00-45D, page 7-5
Area E indicates an echo intensity of 1 to 2, weak to moderate, since it is in the first ring. Echo tops are at 30,000 feet. Individual cell movement is indicated by an arrow with the speed in knots entered as a number and the arrow shaft pointing in the direction of movement. In area E an individual cell is moving northwest at 15 knots. Thunderstorms and rain showers are indicated by the contraction "TRW." *See* the figure below.

ALL
4235 [B] — (I46) — AC 00-45D, page 7-2
The Radar Summary Chart aids in preflight planning by identifying general areas and movement of precipitation and/or thunderstorms. Radar detects only drops or ice particles of precipitation size; it does not detect clouds and fog. Therefore, the absence of echoes does not guarantee clear weather. Furthermore, cloud tops may be higher than precipitation tops detected by radar. The chart must be used in conjunction with other charts, reports and forecasts.

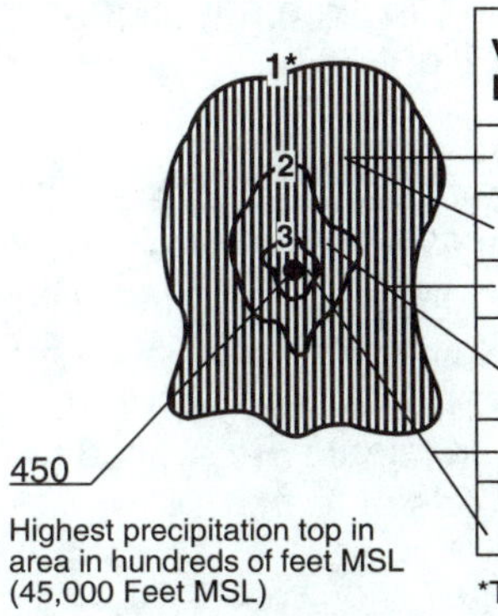

VIP Level	Echo Intensity	Precipitation Intensity	Rainfall Rate in/hr Stratiform	Rainfall Rate in/hr Convective
1	Weak	Light	Less that 0.1	Less than 0.2
2	Moderate	Moderate	0.1 – 0.5	0.2 – 1.1
3	Strong	Heavy	0.5 – 1.0	1.1 – 2.2
4	Very strong	Very heavy	1.0 – 2.0	2.2 – 4.5
5	Intense	Intense	2.0 – 5.0	4.5 – 7.1
6	Extreme	Extreme	More than 5.0	More than 7.1

*The numbers representing the intensity level do not appear on the chart. Beginning from the first contour line, bordering the area, the intensity level is 1–2, second contour is 3–4, and the third contour is 5–6.

Symbols Used on Charts

Symbol	*Meaning*	*Symbol*	*Meaning*	*Symbol*	*Meaning*
R	Rain	**+**	Intensity increasing or new echo	(crescent)	Line of echoes
RW	Rain shower	**–**	Intensity decreasing	**SLD**	8/10 or greater coverage in a line
Hail	Hail	**No Symbol**	No change in intensity	**WS999**	Severe thunderstorm watch
S	Snow	↗ **35**	Cell movement to NE at 35 knots	**WT999**	Tornado watch
IP	Ice pellets	⇒	Line or area movement to East at 20 knots	**LEWP**	Line echo wave pattern
SW	Snow shower	**LM**	Little movement	**HOOK**	Hook echo
L	Drizzle	**MA**	Echoes mostly aloft	**BWER**	Bounded weak echo region
T	Thunderstorm	**PA**	Echoes partly aloft	**PCLL**	Persistent cell
ZR, ZL	Freezing precipitation			**FNLN**	Fine line
NE	No echoes observed				
NA	Observations unavailbale				
OM	Out for Maintenance				
STC	STC ON – all precipitation may not be seen				
ROBEPS	Radar Operating Below Performance Standards				
RHINO	Range Height Indicator Not Operating				

Questions 4230, 4231, 4232, 4233, 4234, 4236 and 4237

ALL
4236 [A] — (I46) — AC 00-45D, page 7-5
Area G indicates an echo intensity of 1 to 2, weak to moderate, since it is in the first ring. Echo bases are at 10,000 feet. Individual cell movement is indicated by an arrow with the speed in knots entered as a number and the arrow shaft pointing in the direction of movement. In area G an individual cell is moving northeast at 15 knots. The contraction "R" indicates rain. *See* the figure on the previous page.

ALL
4237 [A] — (I46) — AC 00-45D, page 7-5
When echoes are reported as a "LINE," a line will be drawn through them on the chart. When there is at least 8/10 coverage in the line, it is labeled solid "SLD" at both ends of the line. The line in area F is not solid, and it is not moving. Echo tops in area F are at 45,000 feet; however, the closest answer is 46,000 feet. Thunderstorms are indicated by the contraction "TRW." *See* the figure on the previous page.

ALL
4238 [C] — (I23) — AC 00-6A, pages 88, 115, 119
You can be relatively certain of a shear zone in a low-level temperature inversion, if you know that the wind at 2,000 to 4,000 feet is 25 knots or more. Wind shear turbulence is also found near the ground outside thunderstorm clouds. *See* the figure below.

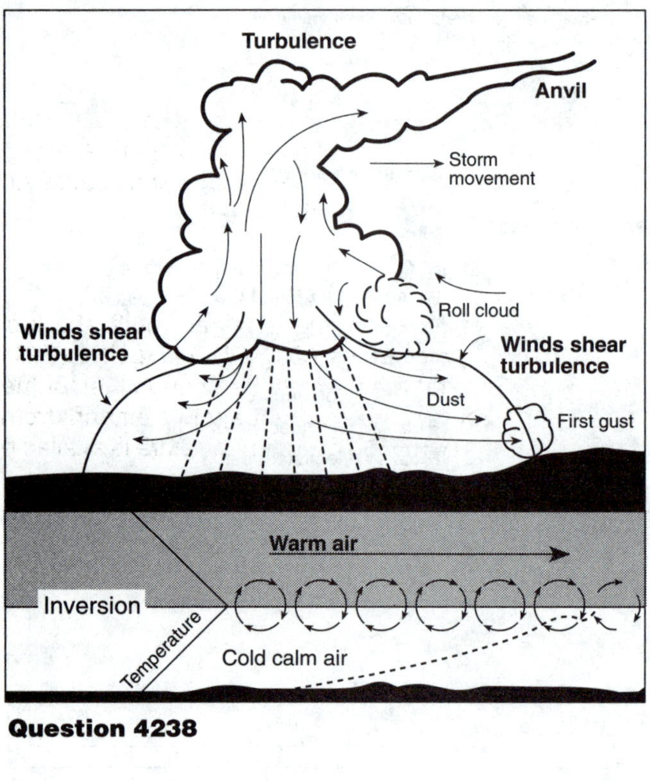

Question 4238

ALL
4239 [C] — (I50) — AC 00-45D, page 11-1
The Severe Weather Outlook Chart is a preliminary 24-hour outlook for thunderstorm activity, tornadoes, and watch areas presented in two panels. The bottom panel covers the 12-hour period 1200Z-0000Z, and the top panel covers the period 0000Z-1200Z.

ALL
4240 [C] — (I50) — AC 00-45D, page 11-1
The significance of the annotations "MDT" and "SLGT" is that they indicate the moderate risk area (6-10% coverage) and the slight risk area (2-5% coverage) for thunderstorm activity.

ALL
4241 [C] — (I43) — AIM ¶7-9(a)(4)
Pilots, upon hearing the alert notice, if they have not received the advisory, or are in doubt, should contact the nearest FSS and ascertain whether or not the advisory is pertinent to their flight.

ALL
4242 [B] — (I52) — AC 00-45D, page 13-1
Tropopause Height/Vertical Wind Shear Prog depicts height of the tropopause in terms of pressure altitude and vertical wind shear in knots per 1,000 feet. Solid lines trace intersections of the tropopause with standard constant pressure surfaces. (Heights are preceded by the letter F and are in hundreds of feet).

ALL
4243 [C] — (I52) — AC 00-45D, page 13-2
Decode the plotted data at Albuquerque, New Mexico as follows: Tropopause wind, 240° (the leg extends to the southwest with a 4 at the end), at 115 knots (pennants are 50 knots, full barbs are 10 knots, and half barbs are 5 knots); tropopause temperature, -61°C; tropopause pressure, 200 millibars.

ALL
4244 [C] — (I52) — AC 00-45D, page 13-2
Decode the plotted data at Albuquerque, New Mexico as follows: Tropopause wind, 240° (the leg extends to the southwest with a 4 at the end), at 115 knots (pennants are 50 knots, full barbs are 10 knots, and half barbs are 5 knots); tropopause temperature, -61°C; tropopause pressure, 200 millibars.

ALL
4245 [A] — (I52) — AC 00-45D, page 13-1
Tropopause Height/Vertical Wind Shear Prog depicts height of the tropopause in terms of pressure altitude and vertical wind shear in knots per 1,000 feet. Vertical wind shear is in knots per 1,000 feet depicted by dashed lines at 2-knot intervals. The symbol "6K" on the dashed line indicates a 6-knot vertical wind shear per 1,000 feet.

ALL
4246 [A] — (I48) — AC 00-45D, page 9-1

Arrow C (in Oregon):

Line extends to the northwest with a 9 at the end of it = 290°

Pennant = 50 knots

ALL
4247 [C] — (I48) — AC 00-45D, page 9-1

Arrow A (in Indiana):

Line extends to the southwest with a 3 at the end of it = 230°

3 full barbs (10 knots each) and 1 half barb (5 knots) = 35 knots

ALL
4248 [B] — (I50) — AC 00-45D, page 11-1

Tornado watches are plotted only if a tornado watch is in effect at chart time. The watch area is cross-hatched. No coverage time is specified. The figure shows a tornado watch in effect for eastern North Dakota and northern Minnesota at the time that the chart was issued.

ALL
4249 [C] — (I48) — AC 00-45D, page 9-1

Arrow A (in Indiana):

Line extends to the southwest with a 3 at the end of it = 230°

3 full barbs (10 knots each) and 1 half barb (5 knots) = 35 knots

ALL
4250 [B] — (I48) — AC 00-45D, page 9-1

Arrow B (in Idaho):

Line extends to the southwest with a 5 at the end = 250°

Pennant (50 knots), 4 full barbs (10 knots each), 1 half barb (5 knots) = 95 knots

ALL
4251 [C] — (J25) — AIM ¶7-24(d)4

An individual microburst will seldom last longer than 15 minutes from the time it strikes the ground until dissipation. However, there may be multiple microbursts in the area.

ALL
4252 [C] — (J25) — AIM ¶7-24(d)2

The downdrafts can be as strong as 6,000 feet per minute in a microburst encounter.

ALL
4253 [C] — (J25) — AIM ¶7-24(d)2

Horizontal winds near the surface can be as strong as 45 knots resulting in a 90-knot shear (headwind to tailwind) across the microburst. These strong horizontal winds occur within a few hundred feet of the ground.

ALL
4254 [C] — (J25) — AIM ¶7-24(e)

The strongest downdraft occurs at point 3. Although there is also downdraft at points 2 and 4, the correct answer is C (3 and 4) because the combination of downdraft and tailwind at 4 has a more negative effect on performance than the combination of downdraft and headwind at 2.

ALL
4255 [C] — (J25) — AIM ¶7-24(e)

At position 1 the aircraft is in the outflow, initially experiencing a headwind with little or no downdraft resulting in a temporary performance increase.

ALL
4256 [C] — (J25) — AIM ¶7-24(e)

Answer C is the best answer. Although all three answer choices are occurring, the strong downdraft is the most severe effect at point 3.

ALL
4257 [A] — (J25) — AIM ¶7-24(e)

In position 4, performance decreases from the strong tailwind and the strong downdraft.

ALL
4258 [B] — (J25) — AIM ¶7-24(e)

At position 4, the aircraft experiences a severe performance decrease. The wind contours indicate a strong downdraft and increasing tailwind which may cause an aircraft to sink toward ground impact.

ALL
4259 [A] — (B07) — FAR §91.167

To answer this question, the times en route and fuel consumption for both trips (GJT-DRO and DRO-GJT) must be calculated. There must be enough fuel for the round trip, plus 45 minutes. Subtract this amount from the total fuel on board, 68 gallons. Any extra is available for holding at DRO.

Continued

Convert wind to magnetic: 230° – 14°E = 216°. Perform standard cross-country calculations, and complete the flight logs for each trip (FAA Figures 22 and 22A):

FROM	TO	CRS	NM	GS	ETE	Fuel
GJT	Herrm	—	(given)	—	24:00	6.0
Herrm	Manca	151°	75	171.5	26:14	6.6
Manca	DRO	—	(given)	—	18:30	4.6
					1:08:44	17.2
DRO	Manca	—	(given)	—	14:30	3.6
Manca	Herrm	333°	75	177.5	25:21	6.3
Herrm	JNC	331°	35	177.2	11:51	3.0
JNC	GJT	—	(given)	—	12:00	3.0
					1:03:42	15.9

68 gal @ 15 GPH = 4:32:00
4:32:00 – (2:12:26 + 0:45:00) = 1 hr 34 minutes

ALL
4260 [A] — (H06) — AC 61-23B, page 161
To answer this question, the flight log (FAA Figure 22) must be completed. Interpolate the 15,000-foot winds aloft as 230° at 8 knots. Convert wind to magnetic: 230° – 14°E = 216°. Perform standard cross country calculations:

FROM	TO	CRS	NM	GS	ETE
GJT	Herrm	—	(given)	—	24:00
Herrm	Manca	152°	75	171.345	26:16
Manca	DRO	—	(given)	—	18:30
					1:08:46

ALL
4261 [A] — (J17) — Enroute Low Altitude Chart Legend
Air Route Traffic Control Center names and frequencies are shown on low altitude charts enclosed in a box of scalloped lines. Just northwest of the Cortez VOR symbol is the box showing Denver Center available at Cortez, using frequency 133.425.

ALL
4262 [B] — (H06) — Enroute Low Altitude Chart Legend
The highest published MEA along the route is 15,000 feet MSL, on V187 between HERRM and MANCA intersections.

ALL
4263 [B] — (J35) — Enroute Low Altitude Chart Legend
The changeover should be made at the changeover symbol if depicted, where there is a change in the direction of the airway, or in the absence of these, at the halfway point between the VORs. In this case, the changeover point (COP) is indicated with a symbol, and should be made 52 miles from JNC and 90 miles from the next VOR.

ALL
4264 [C] — (J35) — Enroute Low Altitude Chart Legend
If an altitude is written under the MEA with an asterisk [*], this altitude represents the MOCA (minimum obstacle clearance altitude). In this case, it is 13,700 feet MSL is the MOCA on V187.

ALL
4265 [A] — (H06) — AC 61-23B, page 162
Use the following steps to complete the flight log and find the fuel consumption at 15 GPH:

1. Find the ground speed. The winds are given 230° (True) at 8 knots. The true course is 165° (151° magnetic + 14° variation), and the true airspeed is 175. Using a flight computer you have an estimated ground speed of 171 knots.
2. Find the time en route for the flight. Looking at FAA Figure 24, the Enroute Low Altitude Chart shows the leg from HERRM to MANLA to be 75 miles. Using a flight computer calculate the time to the second leg to be 0:26:20. Adding the time it takes for each leg (the first leg is given as 00:24:00, and the last leg is given as 00:18:30), the total time en route is 01:08:50.
3. Find the total fuel burn. Using a flight computer, set fuel burn at 15 GPH and find the total fuel burn for 01:09:00 as 17.2 gallons.

ALL
4266 [C] — (J15) — AIM ¶5-7
Look at the Aircraft Equipment/Status at the bottom of FAA Figure 27. It indicates the aircraft has a transponder with Mode C and a DME. The code for DME and transponder with altitude encoding capability is /A. It is recommended that pilots file the maximum transponder or navigation capability of their aircraft in the equipment suffix. This will provide ATC with the necessary information to utilize all facets of navigational equipment and transponder capabilities available. *See* FAA Legend 25.

ALL
4267 [B] — (H06) — AC 61-23B, page 162
To find CAS when TAS is known, the correct OAT and the pressure altitude is required. Since neither are available for the problem, substitute the uncorrected OAT and the planned flight altitude. Using a flight computer and the information from the flight plan in FAA Figure 27:

Altitude 8,000 feet (Block 7 of flight plan)
Temperature -5°C (Given in question)
TAS 155 Knots (Block 4 of flight plan)

Therefore, CAS is 139 KCAS.

ALL

4268 [C] — (H06) — AC 61-23B, page 161

To answer this question, complete the flight log in FAA Figure 28, using the information given in the question:

1. Change the winds aloft for 9,000 feet at OTH (North Bend, OR) from True to Magnetic, using the variation from the flight log (FAA Figure 28):

   Winds at OTH at 9,000 are 2215-05 = 220 true north / 15 knots / -05°C.

   220° true
   – 20° East variation
   200° magnetic

2. Find the distances for the leg from MERLI INT to MOURN INT using the following DME formula.

   a. Miles Flown = $\frac{\text{DME Distance x Number of Degrees}}{60}$

   b. Miles Flown = $\frac{15 \times (\text{R-333°} - \text{R-251°})}{60}$

   c. $\frac{15 \times 82}{60}$ = 20.5 miles

   d. 20.5 miles from the 251° radial to the 333° radial of the arc
   +16.0 miles from the arc to MOURN INT
   36.5 NM

3. Find time en route for the MERLI INT to MOURN INT leg using a flight computer:

   Distance 36.5 NM (calculated in step 2)
   Ground speed 135 Knots (given in flight log FAA Figure 28)

   Time en route is 00:16:13

4. Find the distance for the leg from MOURN INT to the Roseburg VOR on V121. (Remember the distance value that is not boxed is the mileage between reporting points, radio aids, and/or mileage breakdown points.) This distance is 19 miles.

5. Compute the ground speed for the leg from MOURN INT to the Roseburg VOR on V121:

   Wind direction 200 (found in Step 1)
   Wind speed 15 knots (found in Step 1)
   Course 287 (found in FAA Figure 31)
   TAS 155 (found in FAA Figure 28)

   Therefore, ground speed is 153.5 Knots

6. Compute the time en route for the leg from MOURN INT to the Roseburg VOR on V121:

   Distance 19 miles (found in Step 4)
   Ground speed 153.5 (found in Step 5)

   Time en route is 00:07:26

7. Repeat the first six steps to fill in the flight log:

FROM	TO	CRS	NM	GS	ETE
MFR	MERLI	—	—	—	:11:00 (given)
MERLI	MOURN	—	37	135	:16:13
MOURN	RBG	287	19	153	:07:26
RBG	OTH	271	38	149	:15:15
OTH	EUG	024	59	170	:20:50
EUG	App/Ldg	—	—	—	:10:00 (given)
					1:20:44

ALL

4269 [A] — (I08) — AC 61-27C, page 144

If a crosswind is drifting you toward the facility, turn until the RMI bearing pointer is behind the wing tip. With a crosswind from the left on a right-hand arc, the right wingtip reference point will be behind the wing tip.

ALL

4270 [C] — (H06) — AC 61-23B, page 94

Using FAA Legend 18, enter the table in the required rate of climb column at the rate of climb required by the NOTE on the plan view of the SID (400 feet per NM). Move directly to the right in that row until intercepting the 120-knot ground speed column and read 800 fpm. Or you can use the formula:

1. $\frac{\text{Ground speed x Climb Rate Required}}{60}$ = Rate of Climb Needed

2. $\frac{120 \times 400 \text{ feet/NM}}{60}$ = 800 feet per minute

ALL

4271 [A] — (J16) — AIM ¶1-8

In FAA Figure 30, the A/FD lists restrictions to the use of radio aids under the name of that radio aid, found under Radio Aids to Navigation. In this case, 333° falls within the unusable ranges between 280° and 345°.

ALL

4272 [B] — (I05) — AC 61-27C, page 52

On the RMI in FAA Figure 30A, the thin needle is tuned to VIOLE LMM (356). The tail of the needle is on 280, which shows that the aircraft is on the 280° bearing from or 10° off course. This places the aircraft to the north (right) of the published route. The larger needle is set to indicate position from the MEDFORD (OED) VORTAC (113.6). The tail of the needle is on 224, which indicates that the aircraft is on the 225° radial of OED, therefore, west of and past GNATS INT.

ALL

4273 [C] — (J01) — AIM ¶1-8

Looking at FAA Figure 30, the OED (Medford) VORTAC class is written in parenthesis in front of the facility in the A/FD. In this case it is (H), or high. By referring to the radio class designations in FAA Legend 26, note that at 8,000 feet (flight planned altitude shown in FAA Figure 27), the High altitude VORTAC normal usable range is 40 NM.

ALL

4274 [A] — (I10) — 61-27C, page 187

The Touchdown Zone Elevation (TDZE) of 363 feet for RWY 16 is shown on the airport sketch of FAA Figures 29 and 29A.

ALL

4275 [B] — (I10) — A/FD Legend

The hours of operation at Eugene/Mahlon Sweet Field are listed on FAA Figure 29 under Eugene Tower. Hours of operation are expressed in UTC (Coordinated Universal Time) and is shown as "Z" time or Zulu. The top line of the A/FD indicates the number of hours to be subtracted from UTC to obtain local standard time [GMT -8 (-7DT).]:

Published time of operation	1400 Z	to	0800 Z
Time Zone correction	– 800		– 800
Local Standard Time hours of operations	0600	to	0000

ALL

4276 [B] — (I10) — AC 61-27C, page 185

On FAA Figure 29, find the glide slope angle of descent on the profile view, in this case, 3°. Enter the rate of descent table (FAA Legend 19) at the 3° point in the angle of descent column. Move directly to the right until intercepting the 90-knot ground speed column and read the rate of descent, in this case, 480 fpm.

ALL

4277 [C] — (J15) — AIM ¶5-7

Look at the Aircraft Equipment/Status at the bottom of FAA Figure 32. It indicates the aircraft is equipped with transponder with Mode C and RNAV. The code for RNAV and transponder with altitude encoding capability is /R. It is recommended that pilots file the maximum transponder or navigation capability of their aircraft in the equipment suffix. This will provide ATC with the necessary information to utilize all facets of navigational equipment and transponder capabilities available. *See* FAA Legend 25.

ALL

4278 [B] — (H06) — AC 61-23B, page 162

To find CAS when TAS is known, the correct OAT and the pressure altitude is required. Since neither are available for the problem, substitute the uncorrected OAT and the planned flight altitude. Using a flight computer and the information from the flight plan in FAA Figure 32:

Altitude 8,000 feet (Block 7 of flight plan)
Temperature +8°C (Given in question)
TAS 180 Knots (Block 4 of flight plan)

Therefore, CAS is 157.4 KCAS.

ALL

4279 [A] — (H06) — AC 61-23B, page 161

To answer this question, the flight log in FAA Figure 33 must be completed. The related BUJ.BUJ3 STAR (FAA Figure 35) and the enroute chart (FAA Figure 34) provide some alterations to the flight log. On V573 the bend in the airway 10 NM southwest of MARKI intersection requires an additional leg. The check points in the STAR must be accounted for, as well.

1. Interpolate winds aloft, convert to magnetic:

6000	8000	9000
220°	220° True (216° Mag)	220°
39K	39.66K	40K

2. Using a flight computer, calculate the ground speeds and ETE to complete the flight log:

FROM	TO	CRS	DIST	GS	ETE
HOT	Marki	—	—	—	00:12:00 (given)
Marki	VOR COP	221°	10	140.5	00:04:16
COP	TXK	210°	45	140.5	00:19:13
TXK	Conny	272°	61	154.8	00:22:52
Conny	BUJ3	239°	59	142.8	00:24:47
BUJ3	D/A	—	—	—	00:10:00 (given)
					1:33:08

ALL

4280 [A] — (J35) — Enroute Low Altitude Chart Legend

The navigational signal at APINE intersection would be unreliable below 3,500 feet because the MEA and MOCA are 3,500 and 2,600 feet, respectively. Since APINE is further than 22 NM from the HOT VOR, navigational signal is only assured with the MEA.

ALL

4281 [A] — (I08) — AC 61-27C, page 138

FAA Figure 37 has an RMI tuned to 116.3 (TXK VORTAC), with indications that show the airplane on the 270° radial (thick needle tail pointing to 270°) which is left of the 272°. FAA Figure 37 also has an HSI tuned to 114.9 (BUJ VORTAC) and set to 239°. With the bar deflected to the northwest, the airplane is southeast of the BUJ R-059 and approaching it. Thus, the airplane is left of the TXK R-272 and approaching the BUJ R-059.

ALL

4282 [B] — (I07) — AC 61-27C, page 121

The missed approach point on the RNAV RWY 33 approach is arrival at the indicated MAP waypoint. Arrival at a waypoint is shown on the TO-FROM ambiguity indicator.

ALL

4283 [A] — (I07) — IAP Chart Legend

First calculate the aircraft approach category to be used. FAA Figure 32 indicates the V_{SO} is 74 knots. Multiply the stall speed by 1.3 to find the lowest approach category for the aircraft (74 x 1.3 = 96 knots, which is category B). Since we are not told otherwise, assume we have ATIS and the Addison altimeter. FAA Figures 36A and 36B indicate the MDA and visibility criteria for S33 approach to be 1,240 feet and 1 SM with the Addison altimeter.

ALL
4284 [A] — (K24) — AC 20-138, page 21
An appropriate Airplane or Rotorcraft Flight Manual Supplement (or, for aircraft without an FAA Approved Flight Manual, a Supplemental Flight Manual) containing the limitations and operating procedures applicable to the equipment installed, should be provided for each installation of GPS navigation equipment for IFR approval. A Flight Manual Supplement may be necessary for installations limited to VFR use only, depending upon the complexity of the installation and the need to identify necessary limitations and operating procedures.

ALL
4285 [B] — (I10) — IAP Chart Legend
Two waypoints are required for the complete RNAV RWY 33 approach: ADDIS and the MAP.

ALL
4286 [C] — (J18) — STAR Chart
The heavy black lines with arrows portray the Arrival Route, which, in this case, begins at the Blue Ridge VOR (BUJ). Note also, in the textual description, all the transitions end with the following statement, "To BUJ VORTAC thence…"

ALL
4287 [A] — (J35) — Enroute Low Altitude Chart
The Enroute Low Altitude U.S. IFR Chart is for use up to, but not including, 18,000 feet. In controlled airspace, each person operating an aircraft under IFR in level cruising flight shall maintain the altitude assigned by ATC or the regulatory IFR altitudes. Flying on V573 from HOT VORTAC to the TXK VORTAC is a westerly course which requires an even-thousands altitude.

ALL
4288 [B] — (J15) — AIM ¶5-7
Look at the Aircraft Equipment/Status of the bottom of FAA Figure 38. The aircraft is equipped with transponder with Mode C, RNAV, and DME. The code for RNAV and transponder with altitude encoding capability is /R. It is recommended that pilots file the maximum transponder or navigation capability of their aircraft in the equipment suffix. This will provide ATC with the necessary information to utilize all facets of navigational equipment and transponder capabilities available. *See* FAA Legend 25.

ALL
4289 [A] — (H06) — AC 61-23B, page 162
To find CAS when TAS is known, the correct OAT and the pressure altitude is required. Since neither are available for the problem, substitute the uncorrected OAT and the planned flight altitude. Using a flight computer and the information from the flight plan in FAA Figure 38:

Altitude 11,000 feet (Block 7 of flight plan)
Temperature +5°C (Given in question)
TAS 156 Knots (Block 4 of flight plan)

Therefore, CAS is 129.4 KCAS.

ALL
4290 [C] — (H06) — AC 61-23B, page 161
To answer this question, complete the flight log in FAA Figure 39, using the information given in the problem:

1. Change the winds aloft for 12,000 feet at ABI (Abilene) from True to Magnetic, using the variation from the flight log (FAA Figure 39):

   Winds at ABI at 12,000 are 2142+05 = 210 true north / 42 knots / +5°C.

   210° true
   – 11° East variation
   199° magnetic

2. Find the distances for the leg from Big Spring VORTAC to Loran INT on V16. The distance that is not boxed is the mileage between reporting points, radio aids, and/or mileage breakdown points. The distance is 42 NM (22 + 20).

3. Find the ground speed for the leg from Big Spring VORTAC to Loran INT on V16 using a flight computer:

   Wind direction 199 (calculated in step 1)
   Wind speed 42 (found in step 1)
   Course 075 (V16 is the 075 radial from BGS)
   TAS 156 (FAA Figure 38, box 4)

   Therefore, ground speed is 175.6 knots.

4. Compute the time en route for the leg from Big Spring VORTAC to Loran INT on V16:

   Distance 42 miles (found in Step 2)
   Ground speed 175.6 (found in Step 3)

   Therefore, time en route is 00:14:21.

5. Repeat the first four steps to fill in the flight log:

FROM	TO	CRS	NM	GS	ETE
21XS	BGS	—	—	—	:06:00 (given)
BGS	Loran	075	42	176	:14:21
Loran	ABI	076	40	175	:13:44
ABI	COTTN	087	63	167	:22:40
COTTN	AQN	075	50	176	:17:05
AQN	Creek	040	32	195	:09:52
Creek	App/Ldg	—	—	—	:08:00 (given)
					1:31:42

ALL
4291 [A] — (J35) — Enroute Low Altitude Chart
The Enroute Low Altitude U.S. IFR Chart is for use up to, but not including, 18,000 feet.

ALL
4292 [B] — (J41) — U.S. Terminal Procedures, STAR Legend
Heavy lines ending in arrows are arrival routes. Also, arrival descriptions in FAA Figure 41A state that transitions end and arrivals begin ("Thence...") at AQN VORTAC.

ALL
4293 [C] — (I12) — U.S. Terminal Procedures, STAR Legend
Frequencies are in the upper left corner of FAA Figure 41 (with the procedure turned to read correctly). Approaching from the southwest, anticipate using 125.8.

ALL
4294 [C] — (J41) — U.S. Terminal Procedures, STAR Legend
Turbojets landing south at the major area airports will follow the short, thick 350° arrow from CREEK and receive vectors to appropriate final approach courses. (See arrival descriptions in FAA Figure 41A.)

ALL
4295 [C] — (I12) — U.S. Terminal Procedures
Looking at the A/FD excerpt in FAA Figure 42, the Communications section, aircraft approaching from the west would anticipate using 125.8 or 132.1 for approach control, 124.15 for tower control, and 121.65 or 121.8 for ground control.

ALL
4296 [C] — (J42) — U.S. Terminal Procedures, IAP Chart
ILS-1 RWY 36L provides localizer and glide slope (since an ILS approach requires these components). DME is available, as indicated by the channel numbers. "No voice capability" is indicated by the underline on the I-BXN 111.9 frequency.

ALL
4297 [A] — (J42) — U.S. Terminal Procedures, IAP Chart
The airport elevation is 603 feet MSL (see the upper left corner of the airport diagram). The TDZE (Touchdown Zone Elevation) for RWY 36L is 588 feet MSL (see the left center of the airport diagram). The difference is 603 – 588 = 15 feet.

ALL
4298 [C] — (J42) — U.S. Terminal Procedures, IAP Chart
The bottom-right corner of the profile view indicates a 3° glide slope. Enter the Rate-of-Descent Table (FAA Legend 19) in the left-hand column at the given glide slope angle of descent (3°). Move horizontally to the right to the 120-knot Ground speed column and read the recommended rate of descent (635 fpm).

ALL
4299 [A] — (I08) — AC 61-27C, page 135
The NAV-1 is on the I-BXN localizer (111.9) indicating right of course (left deviation). The aircraft is approaching CHARR intersection (DME readout is 7.5, and CHARR is 7.2). The aircraft is below and approaching the glide slope (the aircraft is shown above the glide slope). The NAV-2 is on the LOVE VOR (114.3) indicating the aircraft is south of CHARR (it is indicating R-230 and CHARR is R-233). Therefore, the aircraft is right of the localizer course approaching CHARR intersection and approaching the glide slope.

ALL
4300 [C] — (J15) — AIM ¶5-7
Look at the Aircraft/Equipment Status at the bottom of FAA Figure 44. The aircraft is equipped with transponder with Mode C, RNAV, and DME. The code for RNAV and transponder with altitude encoding capability is /R. It is recommended that pilots file the maximum transponder or navigation capability of their aircraft in the equipment suffix. This will provide ATC with the necessary information to utilize all facets of navigational equipment and transponder capabilities available. *See* FAA Legend 25.

ALL
4301 [A] — (H06) — AC 61-23B, page 162
To find CAS when TAS is known, the correct OAT and the pressure altitude is required. Since neither are available for the problem, substitute the uncorrected OAT and the planned flight altitude. Using a flight computer and the information from the flight plan in FAA Figure 44:

Altitude 12,000 feet (Block 7 of flight plan)
Temperature +5°C (Given in question)
TAS 180 Knots (Block 4 of flight plan)

Therefore, CAS is 146.6 KCAS.

ALL
4302 [B] — (H06) — AC 61-23B, page 161
To answer this question, complete the flight log in FAA Figure 45, using the information given in the problem:

1. Change the winds aloft for 12,000 feet at the VOR COP from True to Magnetic, using the variation from the flight log (FAA Figure 45):

   Winds at YKM at 12,000 are 2035+05 = 200 true north / 35 knots / 5°C.

   200° true
   – 20° East variation
   180° magnetic

2. Find the distances for the leg from Hitch to COP; 20 + 17 = 37 NM. The distance value that is not boxed is the mileage between reporting points, radio aids, and/or mileage breakdown points.

3. Compute the ground speed for the leg from Hitch to COP:

   Wind direction 180 (found in Step 1)
   Wind speed 35 knots (found in Step 1)
   Course 206 (found in FAA Figure 45)
   TAS 180 (found in FAA Figure 44, block 4)

   Therefore, ground speed is 147.9 Knots.

4. Compute the time en route for the leg from Hitch to COP:

   Distance 37 miles (found in Step 2)
   Ground speed 147.9 (found in Step 3)

   Therefore, time en route is 00:15:01.

5. Repeat the first six steps to fill in the flight log:

FROM	TO	CRS	NM	GS	ETE
YKM	Hitch	—	—	—	:10:00 (given)
Hitch	COP	206	37	147.9	:15:00
COP	BTG	234	53	157.2	:20:00
BTG	PDX	160	10	146.7	:04:00
PDX	App/Ldg	—	—	—	:13:00 (given)
					1:02:00

ALL
4303 [C] — (I10) — U.S. Terminal Procedures, IAP Chart
The note on the plan view of the SID states that a climb of 350 feet per NM is required to 6,300 feet. To convert this to a rate of climb, enter the rate-of-climb table (FAA Legend 18) at 350 feet per NM. Move directly to the right in that row until intercepting the 140-knot ground speed column and read 816 fpm.

ALL
4304 [A] — (I08) — AC 61-27C, page 53
The DME readout shows the aircraft on the 9-mile DME arc. The HSI shows that if the aircraft turns to 206°, then the 206° radial will be to the pilot's left. The aircraft is heading 130°, so it is approaching R-206.

ALL
4305 [B] — (J34) — Airport/Facility Directory
The top line of the A/FD indicates the number of hours to be subtracted from UTC to obtain local standard time: GMT -8 (-7DT). The ATIS hours of operation can be found in the "Communications" line of the A/FD: ATIS 125.25 (1400-0600Z‡). The symbol (‡) indicates that during the periods of Daylight Savings Time, effective hours will be 1 hour earlier than shown. In this case, for daylight savings time:

1300Z	to	0500Z	
– 0700		– 0700	
0600	to	2200	Local daylight savings time

ALL
4306 [A] — (I12) — U.S. Terminal Procedures
The missed approach point as shown on the chart is PONCO waypoint. The note above the Maltese cross states "6 NM from MAP WPT."

ALL
4307 [B] — (I10) — U.S. Terminal Procedures, IAP Legend
To find out what category is needed, look at FAA Figure 44. Multiply 77 (V_{SO}) by 1.3 to equal 100.1 kts, which is Category B (90 to 120 knots). Therefore, looking at FAA Figures 49 or 49A, the straight-in minimums for Rwy 10R, Category B is 900 feet MSL and RVR 40 or 3/4 miles visibility.

ALL
4308 [C] — (I10) — U.S. Terminal Procedures, IAP Chart Legend
The Loran RNAV RWY 10R approach requires 4 waypoints. They are SCAPO, WETTR, PONCO, and BATTLE GROUND.

ALL
4309 [C] — (I10) — U.S. Terminal Procedures, IAP Chart Legend
The Touchdown Zone Elevation (TDZE) for RWY 10R is 20 feet MSL, and can be found in the airport sketch portion of the IAP.

ALL
4310 [C] — (I10) — U.S. Terminal Procedures, IAP Legend
The profile view of the IAP shows the Final Approach Angle for Vertical Path Computers (RNAV Descent) to be 3.20°.

ALL
4311 [C] — (I10) — U.S. Terminal Procedures, IAP Legend
On nonprecision approaches, descent to the MDA can begin at the final approach fix (FAF). For the Loran RNAV RWY 10R approach this point is identified by the Maltese cross and is 6 NM from MAP WPT, which is PONCO.

ALL
4312 [A] — (J15) — AIM ¶5-7
Look at the Aircraft/Equipment Status at the bottom of FAA Figure 50. The aircraft is equipped with a transponder with Mode C, RNAV, and DME. The code for RNAV and transponder with altitude encoding capability is /R. It is recommended that pilots file the maximum transponder or navigation capability of their aircraft and equipment suffix. This will ensure ATC awareness of the pilot's ability to navigate point-to-point and may be used to expedite the flight. *See* FAA Legend 25.

ALL

4313 [B] — (H06) — AC 61-23B, page 162

To find CAS when TAS is known, the correct OAT and the pressure altitude is required. Since neither are available for the problem, substitute the uncorrected OAT and the planned flight altitude. Using a flight computer and the information from the flight plan in FAA Figure 50:

Altitude 8,000 feet (Block 7 of flight plan)
Temperature +0°C (Given in question)
TAS 158 Knots (Block 4 of flight plan)

Therefore, CAS is 140.1 KCAS.

ALL

4314 [C] — (H06) — AC 61-23B, page 161

To answer this question, complete the flight log in FAA Figure 51, using the information given in the problem:

1. Change the winds aloft for 9,000 feet at SBA (San Luis Obispo) from True to Magnetic, using the variation from the flight log (FAA Figure 51):

   Winds at SBA at 9,000 are 2525+00 = 250 true north / 25 knots / 0°C.

   250° true
   – 16° East variation
   234° magnetic

2. Find the distance for the leg from HABUT INT to Gaviota VORTAC (GVO) on the HABUT one departure (FAA Figure 52). This distance is 6.4 NM.
3. Compute the ground speed for the leg from HABUT INT to Gaviota VORTAC (GVO):

   Wind direction 234 (found in Step 1)
   Wind speed 25 knots (found in Step 1)
   Course 343 (found in FAA Figure 52)
   TAS 158 (found in FAA Figure 50, block 4)

   Therefore, ground speed is 164.4 Knots.

4. Compute the time en route for the leg from HABUT INT to Gaviota VORTAC (GVO):

   Distance 6.4 miles (found in Step 2)
   Ground speed 164.4 (found in Step 3)

   Therefore, time en route is 00:02:20

5. Repeat the first four steps to fill in the flight log:

FROM	TO	CRS	NM	GS	ETE
SBA	HABUT 1	—	—	—	:08:00 (given)
HABUT 1	GVO	343	6.4	164.4	:02:20
GVO	MQO	306	54	148.7	:32:07
MQO	PRB	358	26	170.6	:09:09
PRB	App/Ldg	—	—	—	:10:00 (given)
					00:51:16

ALL

4315 [B] — (I08) — AC 61-27C, page 135

The standard indicator is tuned to the localizer (110.3). The aircraft is tracking "outbound" on the front course (the aircraft heading is 240), so the CDI will have reversed sensing. The pilot would have to fly left to get back on course (fly away from the needle), thus putting the aircraft north of the localizer course. The RMI indicator #2 (or large needle) is set to receive Gaviota (GVO) VORTAC. It is indicating that the aircraft is on the 130° radial of GVO; therefore, it's east of the HABUT INT.

ALL

4316 [C] — (J40) — U.S. Terminal Procedures, IAP Chart

The note on the plan view of the HABUT ONE DEPARTURE indicates this departure requires a minimum climb rate of 385 feet per NM to 6,000. The Rate-of-Climb Table (FAA Legend 18) is used to convert this to a rate of climb (fpm).

1. Enter the Rate-of-Climb Table in the left-hand column at 350 feet per NM. Move directly to the right in that row until intercepting the 100-knot ground speed column and read 583 feet per minute rate of climb.
2. Enter the table in the left-hand column at 400 feet per NM. Move directly to the right in that row until intercepting the 100-knot ground speed column and read 667 fpm rate of climb.
3. Interpolate between 350 and 400 to find the rate of climb for 385: 642 fpm.

ALL

4317 [A] — (J35) — Enroute Low Altitude Chart Legend

The change-over point (COP) is located 20 NM from GVO or 34 NM from SBP.

ALL

4318 [B] — (J35) — Enroute Low Altitude Chart Legend

The solid square in the upper-left corner of the PRB VORTAC identifier box indicates availability of Hazardous Inflight Weather Advisory Service (HIWAS).

ALL

4319 [A] — (I10) — AC 61-23B, page 161

The distance between the 6 DME and the FAF is 3 NM as shown in the profile view. The aircraft will cover that distance in 2 minutes, at 90 knots. Therefore, the aircraft must lose 400 feet in 2 minutes, or 200 feet per minute.

Computation:

1. Apply the following formula:

$$\frac{\text{Distance} \times 60}{\text{Ground speed}} = \text{Time}$$

or:

$$\frac{3 \times 60}{90} = 2 \text{ minutes}$$

2. Compute the altitude to be lost:

   2,400 feet
   – 2,000 feet
   400 feet (altitude to be lost)

3. Compute the time:

   400 feet in 2 minutes = 200 feet per minute.

ALL
4320 [C] — (I08) — AC 61-27C, page 144
As a guide in making range corrections, change the relative bearing 10° to 20° for each 1/2-mile deviation from the desired arc.

ALL
4321 [C] — (J35) — U.S. Terminal Procedures, IAP Chart
The established missed approach point is the PRB VOR. If, upon reaching it, the runway environment is not in sight and the visibility is less than 1 mile, the missed approach procedure should be commenced.

RTC
4322 [C] — (J15) — AIM ¶5-7
Look at the Aircraft/Equipment Status at the bottom of FAA Figure 56. The aircraft is equipped with transponder with Mode C, RNAV, and DME. The code for RNAV and transponder with altitude encoding capability is /R. It is recommended that pilots file the maximum transponder or navigation capability of their aircraft in the equipment suffix. *See* FAA Legend 25.

RTC
4323 [A] — (H06) — AC 61-23B, page 162
To find CAS when TAS is known, the correct OAT and the pressure altitude is required. Since neither are available for the problem, substitute the uncorrected OAT and the planned flight altitude. Using a flight computer and the information from the flight plan in FAA Figure 56:

Altitude 7,000 feet (Block 7 of flight plan)
Temperature +5°C (Given in question)
TAS 110 Knots (Block 4 of flight plan)

Therefore, CAS is 98.4 KCAS.

RTC
4324 [C] — (H06) — AC 61-23B, page 161
To answer this question, complete the flight log in FAA Figure 57, using the information given in the problem:

1. Change the winds aloft for 6,000 feet at HOU (Houston, TX) from True to Magnetic, using the variation from the flight log (FAA Figure 57):

   Winds at HOU at 6,000 are 1015+05 = 100 true north / 15 knots / 5°C.

   100° true
   – 6° East variation
   094° magnetic

2. Find the distance for the leg from CLL to TNV. (Remember, the distance value that is boxed is the mileage between total mileage between compulsory reporting points and/or radio aids.) This distance is 27 NM.

3. Compute the ground speed for the leg from CLL to TNV:

   Wind direction 094 (found in Step 1)
   Wind speed 15 knots (found in Step 1)
   Course 127 (found in FAA Figure 57)
   TAS 110 (found in FAA Figure 56)

   Therefore, ground speed is 97.1 Knots.

4. Compute the time en route for the leg from CLL to TNV:

   Distance 27 miles (found in Step 2)
   Ground speed 97.1 (found in Step 3)

   Therefore, time en route is 00:16:41.

5. Repeat the first four steps to fill in the flight log:

FROM	TO	CRS	NM	GS	ETE
E.W.	CLL	—	—	—	:05:00 (given)
CLL	TNV	127	27	97.1	:16:41
TNV	IAH	110	42	95.5	:26:23
IAH	HUB	161	18	103.3	:10:27
HUB	App/Ldg	—	—	—	:15:00 (given)
					01:13:31

ALL
4325 [B] — (J35) — A/FD Legend
Look at the Communications section of the excerpt from the A/FD. College Station RCO gives the frequencies 122.65 and 122.2.

ALL
4326 [A] — (J01) — AIM ¶1-4
The bottom of FAA Figure 58 gives the VOR receiver checkpoints. Looking at College Station (Easterwood Field), the checkpoint is on the ground (G), the azimuth is on R-097, and the distance from the facility is 3.2 NM. While on R-097, the CDI should be centered with an OBS setting of 097 FROM or 277 TO the station, with an acceptable error of ±4. Therefore the acceptable VOR indications or 097 FROM and 101 FROM the station.

ALL
4327 [A] — (J34) — A/FD
Look at the bottom of FAA Figure 60 to find the Houston EFAS is operated by Montgomery Co. FSS. The hours of operations are given as 1200-0400Z. To convert this to local time, subtract 6 hours (see the top line of the A/FD: UTC–6). Therefore, the local hours of operation are 0600-2400Z.

RTC

4328 [C] — (I10) — FAR §97.3(d)(1)

Helicopters may also use the Category A Minimum Descent Altitude (MDA) or Decision Height (DH). The required visibility minimum may be reduced to one-half the published visibility minimum for Category A aircraft, but in no case may it be reduced to less than one-quarter mile or 1,200 feet RVR. Looking at FAA Figure 60, the RVR is given as 24, so the acceptable visibility may be reduced to RVR 12.

RTC

4329 [A]—(J08)—FAR §91.131(c)(d), §91.205(d)(2)

At HUMBLE VORTAC the helicopter would be in Class B airspace. Unless otherwise authorized by ATC, no person may operate an aircraft within a Class B airspace unless that aircraft is equipped with an operable two-way radio, and the applicable operating transponder and automatic altitude reporting equipment. Because the question states "on an IFR cross-country flight," VOR is also required.

RTC

4330 [B] — (I10) — AC 61-27C, page 170

Since helicopters can decrease RVR by 50% but not less than 1,200, and no adjustment is necessary to DH (no adjustment is necessary for MM out), the correct choice is 246/12.

ALL

4331 [C] — (I08) — AC 61-27C, page 141

The standard LOC/GS indicator on the right side of FAA Figure 61 shows the aircraft to the right of course and high. The RMI shows a heading of 030° direct to the LOM (TUTTE). The large arrow shows the aircraft on the 300° radial from Scholes VORTAC. Therefore, the aircraft is past PLATS intersection.

ALL

4332 [C] — (I10) — AC 61-27C, pages 187, 188

Touchdown Zone Elevation (TDZE) is the highest elevation in the first 3,000 feet of the landing surface. TDZE is indicated on the Instrument Approach Procedure Chart when straight-in landing minimums are authorized. The TDZE is 46 feet MSL, and can be found at the approach end of the landing runway.

RTC

4333 [C] — (J15) — AIM ¶5-7

Look at the Aircraft Equipment Status at the bottom of FAA Figure 62. The code for RNAV and transponder with altitude encoding capability is /R. The aircraft is also equipped with DME, but it is recommended that pilots file the maximum transponder or navigation capability of their aircraft in the equipment suffix. This will provide ATC with the necessary information to utilize all facets of navigational equipment and transponder capabilities available. *See* FAA Legend 25.

RTC

4334 [B] — (H06) — AC 61-23B, page 162

To find CAS when TAS is known, the correct OAT and the pressure altitude is required. Since neither are available for the problem, substitute the uncorrected OAT and the planned flight altitude. Using a flight computer and the information from the flight plan in FAA Figure 62:

Altitude 5,000 feet (Block 7 of flight plan)
Temperature +15°C (Given in question)
TAS 105 Knots (Block 4 of flight plan)

Therefore, CAS is 95.8 KCAS.

RTC

4335 [B] — (H06) — AC 61-23B, page 161

To answer this question, complete the flight log in FAA Figure 63, using the information given in the problem:

1. Change the winds aloft for 6,000 feet at MSY from True to Magnetic, using the variation from the flight log (FAA Figure 63):

   Winds at MSY at 6,000 are 1626+15 = 160 true north / 26 knots / 15°C.

   160° true
   – 6° East variation
   154° magnetic

2. Find the distance for the leg from Lafayette VORTAC to the Hatch INT. This distance is 28 miles (found in the D with the arrow).

3. Compute the ground speed for the leg from Lafayette VORTAC to the Hatch INT:

   Wind direction 154 (found in Step 1)
   Wind speed 26 knots (found in Step 1)
   Course 109 (found in FAA Figure 65, V552 is R109 from LKM)
   TAS 105 (found in FAA Figure 62)

   Therefore, ground speed is 85.0 Knots.

4. Compute the time en route for the leg from Lafayette VORTAC to the Hatch INT:

   Distance 28 miles (found in Step 2)
   Ground speed 85.0 (found in Step 3)

   Therefore, time en route is 00:19:46.

5. Repeat the first four steps to fill in the flight log:

FROM	TO	CRS	NM	GS	ETE
LFT A/P	LFT VOR	—	—	—	:05:00 (given)
LFT VOR	HATCH	109	28	85	:19:46
HATCH	GRICE	110	18	85	:12:42
GRICE	TBD	110	21	85	:14:53
TBD	App/Ldg	117	—	—	:10:00 (given)
					1:02:21

ALL
4336 [C] — (J35) — Enroute Low Altitude Chart Legend
The total distance between LFT and TBD is 68 NM (shown in the box between NAVAIDs). The changeover point (COP) is half of 68, or 34 NM from LFT. VOR changeover points are where the pilot should change navigation receiver frequency from the station behind the aircraft to the station ahead and are:

1. Halfway between the VORs, if no other designation is given;
2. At a course change; or
3. At the COP symbol.

ALL
4337 [A] — (J01) — AIM ¶1-4
The top-half of the A/FD excerpt for LFT contains the VOR receiver check points. The Lafayette checkpoint is airborne at 1,000 feet over the rotating beacon and the azimuth from the VORTAC is on the R-340. Once established on R-340, the CDI needle should be centered with an OBS setting of 160° TO or 340° FROM, with an acceptable error being ±6°.

ALL
4338 [A] — (I08) — AC 61-27C, pages 129 and 136
The #1 VOR (set to TIBBY VOR) shows you should fly left to get back on course (heading eastbound on V552). Consequently, you are right of course. The #2 NAV indicator is tuned to a localizer, which is why there is no TO/FROM indication. The CDI position indicates that if you were inbound on the localizer, the course centerline would be to your left. This corresponds to a position approaching GRICE intersection.

ALL
4339 [B] — (I10) — Enroute Low Altitude Chart Legend
The localizer symbol at GRICE INT indicates that the localizer has an ATC function in addition to course guidance. *See* FAA Legend 22.

RTC
4340 [B] — (I10) — FAR §97.3
The approach criteria for helicopters are based on airspeeds not exceeding 90 knots, regardless of weight.

RTC
4341 [B] — (I10) — FAR §97.3
The approach minimums for the "helicopter only" approach procedure may differ from the adjusted approach category "A" criteria for the fixed-wing aircraft even though the approach is made to the same general landing area or runway environment. The minimum visibility is 1/2 mile.

RTC
4342 [A] — (I10) — A/FD Legend
The VASI approach slope angle is found at the top of the A/FD excerpt for "RWY 12: REIL. PVASI (PSIL)—GA 3.0°TCH 42'."

RTC
4343 [C] — (I10) — U.S. Terminal Procedures, IAP Legend
In the Remarks section of the IAP, it states that when the local altimeter is not available, use the New Orleans International altimeter setting and increase MDA 140 feet:

440	feet and 1/2 mile
+ 140	feet
580	feet and 1/2 mile

ALL
4344 [A] — (J15) — AIM ¶5-7
Look at the Aircraft Equipment/Status at the bottom of FAA Figure 69. The aircraft is equipped with transponder with Mode C, and DME. The code for DME and transponder with altitude encoding capability is /A. It is recommended that pilots file the maximum transponder or navigation capability of their aircraft in the equipment suffix. This will provide ATC with the necessary information to utilize all facets of navigational equipment and transponder capabilities available. *See* FAA Legend 25.

ALL
4345 [A] — (H06) — AC 61-23B, page 162
To find CAS when TAS is known, the correct OAT and the pressure altitude is required. Since neither are available for the problem, substitute the uncorrected OAT and the planned flight altitude. Using a flight computer and the information from the flight plan in FAA Figure 69:

Altitude 5,000 feet (Block 7 of flight plan)
Temperature +5°C (Given in question)
TAS 128 Knots (Block 4 of flight plan)

Therefore, CAS is 118.9 KCAS.

ALL
4346 [B] — (H06) — AC 61-23B, page 161
To answer this question, complete the flight log in FAA Figure 70, using the information given in the problem:

1. Change the winds aloft for 3,000 feet from True to Magnetic, using the variation from the flight log (FAA Figure 70) and use the winds at 6,000 feet for the planned altitude of 5,000 feet:

   Winds at 3,000 feet are 3320 = 330 true north/20 knots

   Winds at 5,000 feet will be around = 340 true north/25 knots

3,000 feet:	5,000 feet:
330	340
+14 W	+14 W
344	354

Continued

2. Using a flight computer, calculate the ground speeds and ETE to complete the flight log:

FROM	TO	Alt/Climb	CRS	DIST	GS	ETE
GrLake	Shaff	Climb	—	—	—	00:08:00 (given)
Shaff	Helon	5,000	029	24	107	00:13:27
Helon	IGN	5,000	102	21	133	00:09:28
IGN	JD2X	3,000	112	15	139	00:06:28
JD2X	Judds	3,000	100	17	135	00:07:33
Judds	Briss	3,000	057	6	121	00:02:59
Briss	D/A	—	—	—	—	00:12:00 (given)
						00:59:55

ALL

4347 [A] — (I08) — AC 61-27C, page 135

The #1 VOR is tuned to Kingston VORTAC (IGN). With the OBS set to 265, a FROM indication, and the CDI deflected to the right, the aircraft is south of the 265 radial and approaching the Flosi intersection. The #2 VOR is tuned to Sparta VORTAC (SAX). With the OBS set to 029, a FROM indication, and the CDI deflected to the right, the aircraft is west of the 029 radial.

ALL

4348 [C] — (J35) — Enroute Low Altitude Chart Legend

Kingston VORTAC has a solid square in the upper-left corner of the information box. This indicates HIWAS availability.

ALL

4349 [B] — (J41) — U.S. Terminal Procedures, STAR Chart Legend

The heavy black lines with arrows portray the Arrival Route, which, in this case, begins at the Kingston VORTAC (IGN). Note also, the textual description of the transition: "From over IGN VORTAC..."

ALL

4350 [B] — (J42) — A/FD Legend

The bottom of the A/FD excerpt lists the available precision approaches: ILS/DME for runways 06, 33, and 24.

ALL

4351 [B] — (J42) — U.S. Terminal Procedures, IAP Legend; AC 61-27C, pages 161 through 190

The glide slope intercept and final approach fix for a precision approach is identified with a lightning bolt (on the NOS chart) on the profile view of the approach. The minimum altitude to intercept the glide slope on the ILS RWY 6 approach is 1,800 feet MSL.

ALL

4352 [A] — (J42) — U.S. Terminal Procedures, IAP Chart Legend

With respect to the operation of aircraft, Decision Height (DH) means the height at which a decision must be made during an ILS, MLS, or PAR instrument approach to either continue the approach or to execute a missed approach. The DH for the ILS RWY 6 approach is 374 feet MSL.

ALL

4353 [C] — (J42) — AC 61-27C, page 127

The OM is modulated at 400 Hz and identified with continuous dashes at the rate of two dashes per second and a blue marker beacon light. The MM is modulated at 1300 Hz and identified with alternate dots and dashes keyed at the rate of 95 dot/dash combinations per minute and an amber marker beacon light. The IM is modulated at 3000 Hz and identified with continuous dots keyed at the rate of six dots per second and a white marker beacon light (but in this case, the IM is passed after the MAP).

ALL

4354 [B] — (J42) — U.S. Terminal Procedures, IAP Chart

Enter the Rate-of-Descent Table (FAA Legend 19) in the left-hand column at the given glide slope angle of descent (3.0°, found on the profile view of the approach). Move horizontally to the right to the 90-knot ground speed column and read the recommended rate of descent (480 fpm).

ALL

4355 [A] — (J42) — U.S. Terminal Procedures, IAP Chart Legend

The airport sketch shows the Touchdown Zone Elevation (TDZE) for RWY 6 near the approach end of the landing runway (on the NOS IAP) as 174 feet MSL.

ALL

4356 [B] — (J42) —U.S. Terminal Procedures, IAP Chart

FAA Legend 20 indicates no adjustments are necessary to DH or RVR for MM out.

ALL

4357 [C] — (J42) — U.S. Terminal Procedures, IAP Legend

Looking at the airport sketch in the IAP, there is an "A" in a circle, with a dot above it, near the approach end of RWY 6. FAA Legend 21 indicates this means ALSF-2 approach lighting. Also, at the bottom of the airport sketch, it indicates RWY 6 has HIRL (high intensity runway lights).

ALL

4358 [C] — (J15) — AIM ¶5-7

Look at the Aircraft Equipment/Status at the bottom of FAA Figure 74. The aircraft is equipped with transponder with Mode C, and DME. The code for DME and transponder with altitude encoding capability is /A. It is recommended that pilots file the maximum transponder or navigation capability of their aircraft in the equipment suffix. This will provide ATC with the necessary information to utilize all facets of navigational equipment and transponder capabilities available. *See* FAA Legend 25.

ALL
4359 [B] — (H06) — AC 61-23B, page 162
To find CAS when TAS is known, the correct OAT and the pressure altitude is required. Since neither are available for the problem, substitute the uncorrected OAT and the planned flight altitude. Using a flight computer and the information from the flight plan in FAA Figure 74:

Altitude 11,000 feet (Block 7 of flight plan)
Temperature +5°C (Given in question)
TAS 160 Knots (Block 4 of flight plan)

Therefore, CAS is 132.8 KCAS.

ALL
4360 [C] — (H06) — AC 61-23B, page 161
To answer this question, complete the flight log in FAA Figure 75, using the information given in the problem:

1. Change the winds aloft for 12,000 feet at BIL (Logan Intl Airport) from True to Magnetic, using the variation from the flight log (FAA Figure 75):

   Winds at BIL at 12,000 are 2324+05 = 230 true north / 24 knots / 5°C.

   230° true
   – 18° East variation
   212° magnetic

2. Find the distance for the leg from Vests to BZN using FAA Figure 77. The distance is 44 NM.
3. Compute the ground speed for the leg from Vests to BZN:

   Wind direction 212 (found in Step 1)
   Wind speed 24 knots (found in Step 1)
   Course 140 (found in FAA Figure 75)
   TAS 160 (found in FAA Figure 74)

   Therefore, ground speed is 150.9 Knots.

4. Compute the time en route for the leg from Vests to BZN:

   Distance 44 miles (found in Step 2)
   Ground speed 150.9 (found in Step 3)

   Therefore, time en route is 00:17:29.

5. Repeat the first four steps to fill in the flight log:

FROM	TO	CRS	NM	GS	ETE
HLN	Vests	—	—	—	:15:00 (given)
Vests	BZN	140	44	150.9	:17:29
BZN	"x"	110	13	163.3	:04:47
"x"	LVM	063	20	180.1	:06:40
LVM	Reepo	067	39	179.1	:13:04
Reepo	BIL	069	38	178.5	:12:46
BIL	App/Ldg	—	—	—	:15:00 (given)
					01:24:46

ALL
4361 [B] — (J40) — U.S. Terminal Procedures, SID Chart Legend
The heavy black lines with arrows portray the Departure Route, which, in this case, terminates at STAKK INT. Note also in the textual description all the transitions are indicated with the following statement, "...localizer course to STAKK. INT at or above 8,400 feet. Thence..."

ALL
4362 [C] — (J01) — AIM ¶1-4
If a test signal is not available at the airport of intended departure, use a point on an airport surface designated as a VOR system checkpoint by the Administrator, the maximum permissible bearing error for a ground check is ±4°. The tail of the RMI will show which radial that the aircraft is on. The ground check point for Helena is located on the 237° radial from the HLN VOR, therefore, both RMI tails should be indicating 237° ±4°.

ALL
4363 [C] — (J10) — U.S. Terminal Procedures, SID Chart Legend
The departure route descriptions for the STAKK TWO DEPARTURE state "... cross STAKK INT at or above 10,200'..." for either runway.

ALL
4364 [C] — (J40) — U.S. Terminal Procedures
Enter the Rate-of-Climb Table (FAA Legend 18) in the left-hand column at the required climb rate of 300 feet (see the Note on the left side of the plan view of FAA Figure 77). Move horizontally to the right to the 140-knot ground speed column and read the required rate of climb (700 fpm).

ALL
4365 [C] — (J35) — U.S. Terminal Procedures, SID Chart Legend
The name and letter numbers in the top right corner of an airport/facility directory are the charts on which the airport and surrounding area are found. FAA Figure 76 indicates L-9 is the chart to cover the BOZEMAN area.

ALL
4366 [B] — (J35) — Enroute Low Altitude Chart Legend
The Enroute Low Altitude U.S. IFR Chart is for use up to, but not including, 18,000 feet.

ALL
4367 [C] — (I08) — AC 61-27C, pages 52 and 141
The aircraft is proceeding SE on a heading of 130° (indicated under the index at the top of the instrument). The #1 needle (thin needle) is tuned to Livingston VORTAC (116.1) and on the 239° radial (indicated by the tail of the needle). This means the aircraft is past the 246° radial of LVM. The #2 needle (double needle) is tuned to Bozeman (112.2) and the aircraft is on the 102° radial (indicated by the tail of the needle). This means the aircraft is to the left (east) of the 110° radial of BZN.

ALL
4368 [B] — (J42) — U.S. Terminal Procedures, IAP Legend
The approach category is based on 1.3 times V_{SO}. V_{SO} can be found in FAA Figure 74. For this problem, 1.3 x 72 = 93.6 knots, so the aircraft would be Category B (91-120 knots).

ALL
4369 [B] — (J42) — U.S. Terminal Procedures, IAP Legend
The initial approach fixes (identified by "IAF") are: (1) R-040 and DME arc, (2) R157 and DME arc, (3) MUSTY intersection and (4) BILLINGS VORTAC.

ALL
4370 [B] — (J35) — Enroute Low Altitude Chart Legend
Bozeman (BZN) has a flag with an X in it, which indicates it has a minimum crossing altitude (MCA). Above the Bozeman communications box is the list of MCAs. For a flight southeast bound on V86, the MCA is 9,300 feet MSL.

ALL
4371 [B] — (J42) — Pilot/Controller Glossary
The instrument approach plate shows the touchdown zone elevation (TDZE) in the airport diagram (bottom, right-hand quadrant) near the approach end of runway 27R as 3,514 feet MSL.

ALL
4372 [A] — (J01) — AIM ¶1-4
The maximum tolerance allowed for an operational VOR equipment check when using a VOT is ±4°.

ALL
4373 [B] — (J01) — FAR §91.113(b)
When weather conditions permit, regardless of whether an operation is conducted under instrument flight rules or visual flight rules, vigilance shall be maintained by each person operating an aircraft so as to see and avoid other aircraft in compliance with this section.

LTA
4374 [A] — (J24) — AIM ¶6-30
If the failure occurs in VFR conditions, or if VFR conditions are encountered after the failure, each pilot shall continue the flight under VFR and land as soon as practicable.

ALL
4375 [B] — (B13) — FAR §91.215
ATC may authorize deviations from transponder requirements immediately to allow an aircraft with an inoperative transponder to continue to the airport of ultimate destination, including any intermediate stops, or to proceed to a place where suitable repairs can be made. Within Class D airspace there are no transponder requirements, therefore if the aircraft's transponder fails while in Class D airspace, no deviation will be necessary.

ALL
4376 [C] — (J01) — AIM ¶1-4(b)
To use a VOT service, tune in the VOT frequency with the VOR receiver. With the Course Deviation Indicator (CDI) centered, the omni-bearing selector (OBS) should read 0° with a FROM flag, or 180° with a TO flag. The question asks for radial, so you are "0° or 360° FROM."

ALL
4377 [B] — (J01) — AIM ¶1-4
A VOR receiver ground checkpoint requires setting the OBS on a designated radial (found in the A/FD) from a nearby VOR. It requires ±4° accuracy with a FROM indication.

ALL
4378 [B] — (J01) — AIM ¶1-4
If no check signal or point is available while in flight, select a VOR radial that lies along the centerline of an established VOR airway and maneuver the aircraft directly over a prominent ground point. Note the VOR bearing indicated by the receiver. The maximum permissible variation between the published radial and the indicated bearing is ±6°.

ALL
4379 [C] — (J19) — AIM ¶5-84
"Minimum Fuel" indicates that an aircraft's fuel supply has reached a state where, upon reaching the destination, it can accept little or no delay. This is not an emergency situation, but merely indicates an emergency situation is possible should any undue delay occur.

ALL
4380 [B] — (J14) — AIM ¶4-88(d)
When ATC has not used the term "at pilot's discretion" nor imposed any climb or descent restrictions, pilots should initiate climb or descent promptly on acknowledgment of the clearance. Descend or climb at an optimum rate consistent with the operating characteristics of the aircraft to 1,000 feet above or below the assigned altitude, and then attempt to descend or climb at a rate of between 500 and 1,500 fpm until the assigned altitude is reached.

ALL
4381 [A] — (J14) — AIM ¶6-2
Pilots should not hesitate to declare an emergency when faced with a distress condition, such as fire, mechanical failure, or structural damage. A distress condition is defined as being threatened by serious and/or imminent danger and requiring immediate assistance.

ALL
4382 [A] — (J01) — AIM ¶1-4
The maximum permissible indicated bearing error is ±4°, when using a VOT to check the equipment. A valid VOT signal should be indicated as the 360° radial FROM the station. Since RMI arrows point TO the station, an RMI should indicate 180° ±4° on any OBS setting when using a VOT facility, which is the case in RMI 1.

ALL
4383 [B] — (J01) — AIM ¶1-4
If dual system VOR (units independent of each other except for the antenna) is installed in the aircraft, the pilot may check one system against the other. Both systems shall be tuned to the same VOR ground facility and note the indicated bearings to that station. The maximum permissible variation between the two indicated bearings is 4° on the ground or in the air.

ALL
4384 [B] — (J01) — AIM ¶1-4(f)(1)
If a test signal is not available at the airport of intended departure, use a point on the airport surface designated as a VOR system checkpoint. The maximum permissible bearing error is ±4° with a FROM indication.

ALL
4385 [C] — (J01) — AIM ¶1-4
Using an RMI to perform a dual VOR check, the bearing pointers must be within 4° of each other, which is the case in RMI 4.

ALL
4386 [C] — (J01) — AIM ¶1-4(f)(2)
Locations of airborne checkpoints, ground checkpoints and VOTs are published in the Airport/Facility Directory and on the A/G Voice Communication Panel of the Enroute Low Altitude Chart.

ALL
4387 [C] — (J01) — AIM ¶1-4
To test for VOR receiver accuracy center the CDI needle. The omnibearing selector should read 0° with the TO/FROM flag showing FROM and 180° with the flag showing TO. The allowable bearing error is ±4°. Therefore, a 176° TO indication for one of the VOR receivers and a 003° FROM for the other VOR is an acceptable combination.

ALL
4388 [C] — (J01) — AIM ¶1-4(f)(2)
A list of VOR receiver check points and VOR Test Facilities (VOT) can be found in the back of the Airport/Facility Directory.

ALL
4389 [A] — (J01) — AIM ¶1-4
The indicated VOR bearing when over the airborne (or over the ground point) checkpoint cannot be more than ±6° from the designated radial. The flag will have a FROM indication.

ALL
4390 [B] — (J11) — AIM ¶4-189c)(2)
Adjust transponder to reply on the Mode A/3 code specified by ATC and if equipped, to reply on Mode C with altitude reporting capability activated; that is, unless deactivation is directed by ATC, or unless the installed aircraft equipment has not been tested and calibrated as required by FAR §91.413.

ALL
4391 [A] — (J01) — AIM ¶1-4
If dual system VOR (units independent of each other except for the antenna) is installed in the aircraft, the pilot may check one system against the other. Both systems shall be tuned to the same VOR ground facility and note the indicated bearings to that station. The maximum permissible variation between the two indicated bearings is 4° on the ground or in the air.

ALL
4392 [B] — (J14) — AIM ¶4-82
A cruise clearance is used in an ATC clearance to authorize a pilot to conduct flight at any altitude from the minimum IFR altitude up to, and including, the altitude specified in the clearance. The pilot may level-off at any intermediate altitude within this block of airspace. Climb/descent within the block is to be made at the discretion of the pilot. However, once the pilot starts descent, and verbally reports leaving an altitude in the block, he/she may not return to that altitude without additional ATC clearance.

ALL
4393 [C] — (J14) — AIM ¶4-88(d)
When ATC has not used the term "At Pilot's Discretion" nor imposed any climb or descent restrictions, pilots should initiate climb or descent promptly and acknowledgment of the clearance. Descend or climb at an optimum rate consistent with the operating characteristics of the aircraft to 1,000 feet above or below the assigned altitude, and then attempt to descend or climb at a rate of between 500 and 1,500 fpm until the assigned altitude is reached.

ALL
4394 [B] — (J16) — AIM ¶5-24(a)
If operating from an airport not served by a control tower, the pilot may receive a clearance containing a provision that if the flight has not departed by a specific time, the clearance is void. In this situation, the pilot who does not depart prior to the void time must advise ATC as soon as possible, but no later than 30 minutes, of their intentions.

ALL
4395 [B] — (J14) — AIM ¶4-85(b)
Pilots of airborne aircraft should read back those parts of ATC clearances and instructions containing altitude assignments or vectors as a means of mutual verification. The readback of the "numbers" serves as a double check between pilots and controllers and reduces the kinds of communications errors that occur when a number is either "misheard" or is incorrect.

ALL
4396 [B] — (J14) — AIM ¶5-22
The clearance as issued will include the destination airport filed in the flight plan. An enroute altitude will be stated in the clearance or the pilot will be advised to expect an assigned or filed altitude within a given time frame or at a certain point after departure. ATC procedures now require the controller to state the SID name, the current number and the SID Transition name.

ALL
4397 [A] — (I08) — AC 61-27C, page 119
The DME indicator displays slant range distance in nautical miles.

ALL
4398 [C] — (J14) — AIM ¶5-22
An abbreviated clearance will also include an altitude phrase either directly stated or indirectly stated by assigning a SID.

ALL
4399 [B] — (I08) — AC 61-27C, page 119
The greatest slant-range error occurs when flying directly over the DME facility (VORTAC, VOR/DME), when the displayed distance is the height above the facility in nautical miles. Slant-range error is negligible if the aircraft is 1 mile or more from the ground facility for each 1,000 feet of altitude above the elevation of the facility.

ALL
4400 [C] — (J01) — AIM ¶1-8(d)
Normal usable radius distance for high-altitude VOR/VORTAC navaids (H) within the contiguous United States between 14,500 feet and 17,999 feet is 100 NM. A direct-route flight should have a distance of no more than 200 NM between two H-class navaids.

ALL
4401 [B] — (J25) — AIM ¶7-13
Runway Visual Range (RVR) is the maximum horizontal distance down a specified instrument runway at which a pilot can see and identify standard high-intensity runway lights. It is always determined using a transmissiometer and is reported in hundreds of feet.

ALL
4402 [C] — (J28) — AC 61-27C, page 31
With the altimeter set to the current reported altimeter setting, note any variation between the known field elevation and the altimeter indication. If the variation exceeds ±75 feet, the accuracy of the altimeter is questionable and the problem should be referred to an appropriately-rated repair station for evaluation and possible correction.

ALL
4403 [B] — (J11) — AIM ¶4-13(b)
ATIS broadcasts shall be updated upon the receipt of any official weather, regardless of content change and reported values. A new recording will also be made when there is a change in other pertinent data such as runway change, instrument approach in use, etc.

ALL
4404 [A] — (J11) — AIM ¶4-13(b)
If the weather is above a ceiling/sky condition of 5,000 feet and the visibility is 5 miles or more, inclusion of the ceiling/sky condition, visibility, and obstructions to vision in the ATIS message is optional.

ALL
4405 [B] — (J15) — AIM ¶5-1(a)
Every pilot is urged to receive a preflight briefing and to file a flight plan. This briefing should consist of the latest or most current weather, airport, and enroute NAVAID information. Briefing service may be obtained from an FSS either by telephone or interphone, by radio when airborne, or by a personal visit to the station.

ALL
4406 [B] — (J15) — AIM ¶5-3(b)
FDC NOTAMs are transmitted via Service A and are kept on file at the FSS until published or canceled.

ALL
4407 [A] — (B07) — FAR §91.123(d)
Each pilot-in-command who, though not deviating from a rule, is given priority by ATC in an emergency, shall submit a detailed report of that emergency within 48 hours to the manager of that ATC facility, if requested by ATC.

ALL
4408 [B] — (J03) — AIM ¶2-8(d)
In Class B, C, D, or E airspace, operation of the airport beacon during the hours of daylight often indicates that the ground visibility is less than 3 miles and/or the ceiling is less than 1,000 feet. Pilots should not rely solely on the operation of the airport beacon to indicate if weather conditions are IFR or VFR.

ALL
4409 [C] — (J11) — AIM ¶4-17(c)(1)
Class C Service provides approved separation between IFR and VFR aircraft, and sequencing of VFR arrivals to the primary airport.

ALL
4410 [A] — (J10) — AIM ¶1-3(c)
The only positive method of identifying a VOR is by its Morse Code identification or by the recorded automatic voice identification which is always indicated by use of "VOR" following the range's name. During periods of maintenance, the coded facility identification is removed. If a navigational signal is being received without the corresponding Morse Code identification, the VOR is out for maintenance.

ALL
4411 [C] — (J01) — AIM ¶1-3(c)
The only positive method of identifying a VOR is by its Morse Code identification or by the recorded automatic voice identification which is always indicated by use of "VOR" following the range's name. During periods of maintenance, the coded facility identification is removed. If a navigational signal is being received without the corresponding Morse Code identification, the VOR is out for maintenance.

ALL
4412 [C] — (J01) — AIM ¶1-7(f)
VOR/DME, VORTAC, ILS/DME, and LOC/DME facilities are identified by synchronized identifications which are transmitted on a time-share basis. When either the VOR or the DME is inoperative, it is important to recognize which identifier is retained for the operative facility. A single-coded identification with a repetition interval of approximately 30 seconds indicates that the DME is operative.

ALL
4413 [B] — (I08) — AIM ¶1-7(b)
Distance information received from DME equipment is actual distance and not horizontal distance. It is measured in nautical miles, and since 6,000 feet equals 1 NM, the DME will show 1.0 NM above the VORTAC.

ALL
4414 [B] — (J16) — AIM ¶5-22
An abbreviated clearance always includes the name of the destination airport or clearance limit, altitude, and if a SID is to be flown, the SID name, current SID number, and the SID transition name.

ALL
4415 [C] — (J11) — AIM ¶4-9(d)
Airport Advisory Service (AAS) is a service provided by an FSS physically located on an airport which does not have a control tower, or where the tower is temporarily closed or operated on a part-time basis.

ALL
4416 [B] — (J11) — AIM ¶4-9(d)
Airport Advisory Service (AAS) is a service provided by an FSS physically located on an airport which does not have a control tower, or where the tower is temporarily closed or operated on a part-time basis.

ALL
4417 [C] — (J40) — AIM ¶5-25(a)(2)
If, for any reason, the pilot does not wish to use a SID, he/she is expected to advise ATC. Notification may be accomplished by writing "NO SID" in the remarks section of the flight plan, or by the less desirable method of verbally advising ATC.

ALL
4418 [C] — (J40) — AIM ¶5-25(b)
The question can be solved using the Rate-of-Climb Table (FAA Legend 18). At a ground speed of 140 knots, a climb rate of 210 feet per NM must be interpolated between 200 and 250: 583 – 467 = 116 x 20% = 23.2, added to 467 fpm equals 490.2 fpm.

ALL
4419 [C] — (J40) — AIM ¶5-25(a)
The use of a SID requires pilot possession of at least the textual description of the approved effective SID.

ALL
4420 [B] — (J16) — AIM ¶5-24
The pilot should contact departure control on the assigned frequency upon release from the control tower.

ALL
4421 [A] — (J11) — AIM ¶4-14(c)
Radar traffic information is given by the controller which includes: The azimuth from the aircraft's ground track (not the heading) in terms of a 12-hour clock, and each clock position is 30°. 2 o'clock is about 60° from the aircraft's course, but since the pilot is already holding a 20° right cross-wind, the pilot will only have to look 40° to the right of the airplane's nose.

ALL
4422 [A] — (J33) — Pilot/Controller Glossary
"Resume own navigation" is used by ATC to advise a pilot to resume his/her own navigational responsibility. It is issued after completion of a radar vector or when radar contact is lost while the aircraft is being radar vectored.

ALL
4423 [B] — (J33) — Pilot/Controller Glossary
Radar contact is used by ATC to inform an aircraft that it is identified on the radar display and radar flight following will be provided until radar identification is terminated. Radar service may also be provided within the limits of necessity and capability. When a pilot is informed of "radar contact," he/she automatically discontinues reporting over compulsory reporting points.

ALL
4424 [C] — (J33) — Pilot/Controller Glossary
"Resume own navigation" is used by ATC to advise a pilot to resume his/her own navigational responsibility. It is issued after completion of a radar vector or when radar contact is lost while the aircraft is being radar vectored.

RTC

4425 [A] — (B07) — FAR §91.177(a)(2)(ii)

Except when necessary for takeoff or landing, or unless otherwise authorized by the Administrator, no person may operate an aircraft in nonmountainous terrain under IFR below an altitude 1,000 feet above the highest obstacle within a horizontal distance of 4 NM from the course to be flown (4.6 SM).

ALL

4426 [C] — (B08) — FAR §91.131(c)

No person may operate an aircraft within Class B airspace unless that aircraft is equipped with—

1. *For IFR operation.* An operable VOR or TACAN receiver; and
2. *For all operations.* An operable two-way radio capable of communications with ATC on appropriate frequencies for that Class B airspace area and a transponder having Mode C capability.

ALL

4427 [C] — (B08) — FAR §91.173

No person may operate an aircraft in controlled airspace under IFR unless that person has:

1. Filed an IFR flight plan; and
2. Received an appropriate ATC clearance.

RTC

4428 [C] — (B08) — FAR §91.177

In the case of operations over an area designated as mountainous terrain, no person may operate an aircraft under IFR (except when necessary for takeoff or landing) below an altitude of 2,000 feet above the highest obstacle, within a horizontal distance of 4 nautical miles from the course to be flown.

ALL

4429 [A] — (J33) — Pilot/Controller Glossary

Minimum Enroute IFR Altitude (MEA) is the lowest published altitude between radio fixes which assures acceptable navigational signal coverage and meets obstacle clearance requirements between those fixes. The MEA prescribed for a Federal airway or segment thereof, area navigation low or high route or other direct route applies to the entire width of the airway, segment or route between the radio fixes defining the airway, segment or route.

ALL

4430 [B] — (B11) — FAR §91.159

The VFR-On-Top clearance is the ATC authorization for an IFR aircraft to operate in VFR conditions at any appropriate VFR altitude (as specified in FAR and as restricted by ATC). A pilot receiving this authorization must comply with VFR visibility, distance from cloud criteria, and the minimum IFR altitudes.

ALL

4431 [C] — (B11) — FAR §91.179

A pilot operating on an IFR flight plan, in VFR conditions, with an ATC authorization to "Maintain VFR-On-Top, maintain VFR conditions," must fly at appropriate VFR altitudes which are VFR cardinal altitudes plus 500 feet.

ALL

4432 [C] — (J33) — Pilot/Controller Glossary

The Pilot/Controller Glossary states that the MEA is the lowest published altitude between radio fixes which assures acceptable navigational signal coverage and meets obstacle clearance requirements between those fixes.

ALL

4433 [C] — (B11) — FAR §91.179

A pilot operating on an IFR flight plan, in VFR conditions, with a VFR-On-Top clearance, must fly at appropriate VFR altitudes, which are VFR cardinal altitudes plus 500 feet. On a magnetic course of 180° through 359°, any even thousand-foot MSL altitude plus 500 feet must be flown.

ALL

4434 [B] — (J09) — AIM ¶3-34

Military Operations Areas (MOAs) consist of airspace of defined vertical and lateral limits established for the purpose of separating certain military training activities from IFR traffic.

ALL

4435 [A] — (J33) — Pilot/Controller Glossary

Minimum Reception Altitude (MRA) is the lowest altitude required to receive adequate signals to determine specific fixes. Reception of signals from a radio facility located off the airway being flown may be inadequate at the designated MEA, in which case, a MRA is designated for the fix.

ALL

4436 [C] — (B06) — FAR §91.177

Obstruction clearance is normally at least 1,000 feet in non-mountainous areas above the highest terrain or obstruction 4 nautical miles either side of centerline of the airway or route.

ALL

4437 [A] — (J33) — Pilot/Controller Glossary

If a normal climb, commenced immediately after passing a fix beyond which a higher MEA applies, would not assure adequate obstruction clearance, then a MCA is specified. Otherwise, cross the radio fix at the MEA at which it is approached, and then climb to the next higher MEA.

ALL
4438 [A] — (B08) — FAR § 91.215(d)(3)
ATC may authorize deviations on a continuing or individual basis for operations of aircraft without a transponder, in which case the request for a deviation must be submitted to the ATC facility having jurisdiction over the airspace concerned at least 1 hour before the proposed operation.

ALL
4439 [C] — (B08) — FAR §91.215(d)(3)
ATC may authorize deviations on a continuing or individual basis for operation of an aircraft that is not equipped with a transponder. The request must be made at least one hour before the proposed operation.

ALL
4440 [A] — (B08) — FAR §91.131(c)(d)
No person may operate an aircraft in Class B airspace unless the aircraft is equipped with—

1. *For IFR operation.* An operable VOR or TACAN receiver; and
2. *For all operations.* An operable two-way radio capable of communications with ATC on appropriate frequencies for that Class B airspace area, and a transponder with Mode C capability.

ALL
4441 [A] — (B08) — FAR §91.181
Unless otherwise authorized by ATC, no person may operate an aircraft within controlled airspace under IFR except as follows:

1. On a Federal airway, along the centerline of that airway.
2. On any other route, along the direct course between the navigational aids or fixes defining that route.

However, this section does not prohibit maneuvering the aircraft to pass well clear of other air traffic or maneuvering the aircraft, in VFR conditions, to clear the intended flight path both before and during climb or descent.

ALL
4442 [C] — (J14) — AIM ¶5-25, ¶5-40, ¶5-59
Contact approaches will only be authorized when requested by the pilot, and the reported ground visibility at the destination airport is at least 1 statute mile. ATC may initiate a visual approach when it will be operationally beneficial and certain conditions can be met. SIDs and STARs are assigned as determined necessary by ATC.

ALL
4443 [B] — (J14) — AIM ¶4-82(d)(3)
A cruise clearance is used in an ATC clearance to authorize a pilot to conduct flight at any altitude from the minimum IFR altitude up to and including the altitude specified in the clearance. The pilot may level-off at any intermediate altitude within this block of airspace. Climb/descent within the block is to be made at the discretion of the pilot. However, once the pilot starts descent and verbally reports leaving an altitude in the block, he/she may not return to that altitude without additional ATC clearance.

ALL
4444 [C] — (J14) — AIM ¶7-41(b)
The cruising altitude or flight level of aircraft shall be maintained by reference to an altimeter which shall be set, when operating at or above 18,000 feet MSL, to 29.92" Hg (standard setting).

ALL
4445 [C] — (J26) — AC 61-27C, page 29
En route at FL290, the pilot would have the altimeter set to 29.92" Hg, since the altitude is at or above FL180. If the pilot failed to reset it to the local altimeter setting of 30.57 during descent, the altimeter will not indicate field elevation upon landing. Since:

$$\begin{array}{r} 30.57 \\ -\ 29.92 \\ \hline 0.65 \end{array}$$

and since 1 inch per 1,000 feet equals 0.1 inch per 100 feet and 0.01 inch per 10 feet, the altimeter will, in this case, indicate an error of 650 feet. Because the flight went from a low-pressure area (29.92) to a higher-pressure area (30.57), the altimeter error is subtracted from the field elevation of 650 feet.

650 – 650 = 0 feet, or the altimeter will indicate sea level.

ALL
4446 [B] — (J26) — AC 61-27C, page 31
Pressure altitude is the altitude indicated when the altimeter setting window is adjusted to 29.92. Because the pilot is flying at FL250 the altimeter will be set at 29.92" Hg (since the pilot is above FL180). Consequently the pressure altitude will be 25,000 feet.

ALL
4447 [A] — (J08) — AIM ¶4-86(h)
VFR-On-Top is not permitted in Class A airspace. Consequently, IFR flights operating VFR-On-Top are not allowed to operate above FL180.

ALL
4448 [C] — (B08) — FAR §91.205(e)
For flight at and above 24,000 feet MSL, if VOR navigational equipment is required, no person may operate a U.S. registered civil aircraft within the 50 states, and the District of Columbia, unless that aircraft is equipped with approved distance measuring equipment (DME). When DME fails at and above FL240, the pilot-in-command of the aircraft shall notify ATC immediately, and may then continue operations at and above FL240 to the next airport of intended landing at which repairs or replacement of the equipment can be made.

ALL
4449 [B] — (J14) — AIM ¶4-86(e)
When operating in VFR conditions with an ATC authorization to "Maintain VFR-On-Top/Maintain VFR Conditions," pilots on IFR flight plans must:

1. Fly at the appropriate VFR altitude as prescribed in FAR §91.159.
2. Comply with the VFR visibility and distance from cloud criteria in FAR §91.155 (Basic VFR Weather Minimums).
3. Comply with instrument flight rules that are applicable to this flight (i.e., minimum IFR altitudes, position reporting, radio communications, course to be flown, adherence to ATC clearance, etc.).

ALL
4450 [A] — (J14) — AIM ¶4-86(d)
ATC may not authorize VFR-On-Top operations unless the pilot requests the VFR operation while operating in VMC weather.

ALL
4451 [A] — (J14) — AIM ¶4-86(b)
Pilots desiring to climb through a cloud, haze, smoke, or other meteorological formation, and then either cancel their IFR flight plan or operate VFR-On-Top, may request a climb to VFR-On-Top.

ALL
4452 [B] — (J14) — AIM ¶4-86(e)
When operating on a VFR-On-Top clearance, the pilot must fly VFR cruising altitudes which are based on Magnetic Course.

ALL
4453 [C] — (J14) — AIM ¶4-86(h)
ATC will not authorize VFR or VFR-On-Top operations in Class A airspace.

ALL
4454 [B] — (J14) — AIM ¶4-86(e)
When operating on a VFR-On-Top clearance, the pilot must fly appropriate VFR cruising altitudes. On a magnetic course of 180° through 359°, the pilot must fly any even thousand-foot MSL altitude plus 500 feet.

ALL
4455 [A] — (J14) — AIM ¶4-86(e)
When operating in VFR conditions with an ATC authorization to "Maintain VFR-On-Top" pilots on IFR flight plans must comply with instrument flight rules that are applicable to this flight (i.e., minimum IFR altitudes, position reporting, radio communications, course to be flown, adherence to ATC clearance, etc.).

ALL
4456 [B] — (B07) — AIM ¶5-32(a)(2)
The following reports should be made to ATC or FSS facilities without a specific ATC request when not in radar contact:

1. When leaving final approach fix inbound on final approach.
2. A corrected estimate at anytime it becomes apparent that an estimate previously submitted is in error in excess of 3 minutes.

ALL
4457 [C] — (J14) — AIM ¶4-86
VFR-On-Top flight must be conducted at or above the minimum IFR altitude and in basic VFR weather conditions.

ALL
4458 [A] — (J14) — AIM ¶4-82(d)(3)
A cruise clearance is used in an ATC clearance to authorize a pilot to conduct flight at any altitude from the minimum IFR altitude up to and including the altitude specified in the clearance. The pilot may level-off at any intermediate altitude within this block of airspace. Climb/descent within the block is to be made at the discretion of the pilot. However, once the pilot starts descent and verbally reports leaving an altitude in the block, he/she may not return to that altitude without additional ATC clearance.

ALL
4459 [C] — (B08) — FAR §91.205(e)
For flight at and above 24,000 feet MSL, if VOR navigational equipment is required, no person may operate a U.S.-registered civil aircraft within the 50 states, and the District of Columbia, unless that aircraft is equipped with approved distance measuring equipment (DME). When DME fails at and above FL240, the pilot-in-command of the aircraft shall notify ATC immediately, and may then continue operations at and above FL240 to the next airport of intended landing at which repairs or replacement of the equipment can be made.

ALL
4460 [A] — (B08) — FAR §91.187(a)
The pilot-in-command of each aircraft operated in controlled airspace under IFR, shall report as soon as practical to ATC any malfunctions of navigational, approach, or communication equipment occurring in flight.

ALL
4461 [A] — (B08) — FAR §91.123(c)
Each pilot-in-command who, in an emergency, deviates from an ATC clearance or instruction shall notify ATC of that deviation as soon as possible.

ALL
4462 [B] — (B08) — FAR §91.185(c)
If the clearance limit is not a fix from which an approach begins, leave the clearance limit at the expected-further-clearance time if one has been received. Then proceed to a fix from which an approach begins, and commence descent, or descent and approach, as close as possible to the estimated time of arrival as calculated from the filed, or amended (with ATC), estimated time en route.

ALL
4463 [B] — (B08) — FAR §91.185(c)
If the clearance limit is not a fix from which an approach begins, leave the clearance limit at the expected-further-clearance time if one has been received. Then proceed to a fix from which an approach begins, and commence descent, or descent and approach, as close as possible to the estimated time of arrival as calculated from the filed, or amended (with ATC), estimated time en route.

ALL
4464 [A] — (J24) — AIM ¶6-31(a)
If an aircraft with a coded radar beacon transponder experiences a loss of two-way radio capability, the pilot should set transponder to code 7600. During a flight in IMC, the pilot should continue flight on the assigned route and fly at the last assigned altitude or the MEA, whichever is higher.

ALL
4465 [A] — (J24) — AIM ¶6-30(c)(2)
If the failure occurs in VFR conditions, or if VFR conditions are encountered after the failure, the pilot shall continue the flight under VFR and land as soon as practicable.

ALL
4466 [A] — (J24) — AIM ¶6-30
If two-way radio communications are lost in IMC, the flight should be continued via the last assigned route by ATC, the route ATC advised or may be expected to advise, or by the flight plan. The altitude to be used is the highest of the altitude assigned in the last ATC clearance, the MEA, or the flight level ATC may be expected to assign.

ALL
4467 [C] — (J25) — AIM ¶7-5
AIRMETs are broadcast at 15 minutes past the hour and 45 minutes past the hour during the first hour after issuance.

ALL
4468 [A] — (J25) — AIM ¶7-12(b)(4)
The controller's primary function is to provide safe separation between aircraft. Any additional service, such as weather avoidance assistance, can only be provided to the extent that it does not degrade the primary function. The separation workload is generally greater than normal when weather disrupts the usual flow of traffic. ATC radar limitations and frequency congestion may also be a factor in limiting the controller's capability to provide additional service.

ALL
4469 [C] — (J10) — FAR §91.129
When flying in Class D airspace, communications must be established with the tower. Class D airspace has a radius of at least 4.4 nautical miles, which equals 5 statute miles.

ALL
4470 [C] — (J16) — AC 61-27C, page 186
A "T" within a black triangle in the minimums section of the instrument approach plate indicates that takeoff minimums are not standard and/or departure procedures are published.

ALL
4471 [C] — (J19) — AIM ¶5-77
When meteorological conditions permit, regardless of type of flight plan or whether or not under control of a radar facility, the pilot is responsible to see and avoid other traffic, terrain, or obstacles.

ALL
4472 [B] — (I07) — AC 61-27C, page 119
Slant-range error is negligible if the aircraft is 1 mile or more from the ground facility for each 1,000 feet of altitude above the elevation of the facility.

ALL
4473 [A] — (J08) — AIM ¶3-15(e)
A Transition Area is Class E airspace starting from 700 feet or more above the surface when designated in conjunction with an airport for which an instrument approach procedure has been prescribed, and extends up to the overlying controlled airspace.

ALL
4474 [B] — (J08) — AIM ¶3-11(a)
Class A airspace includes airspace within the continental U.S. from 18,000 feet MSL up to and including FL600.

ALL
4475 [A] — (J07) — AIM ¶3-20
Class G airspace is that portion of the airspace that has not been designated as Class A, B, C, D, or E airspace. If flight is conducted outside of these areas, ATC does not exercise control over that flight.

ALL
4476 [C] — (J08) — AIM ¶3-15(e)
A Transition Area is Class E airspace starting from 700 feet or more above the surface when designated in conjunction with an airport for which an instrument approach procedure has been prescribed, and extends up to the overlying controlled airspace.

ALL
4477 [A] — (I04) — AC 61-27C, page 28
Pressure altitude can always be determined from the altimeter whether in flight or on the ground. Set the altimeter at the standard altimeter setting of 29.92 inches, and the altimeter will indicate pressure altitude.

ALL
4478 [A] — (I04) — AC 61-27C, page 28
Pressure altitude can always be determined from the altimeter whether in flight or on the ground. Set the altimeter at the standard altimeter setting of 29.92 inches, and the altimeter will indicate pressure altitude.

ALL
4479 [B] — (I04) — AC 61-27C, page 28
With the altimeter set at the standard altimeter setting of 29.92 inches, the altimeter will indicate pressure altitude.

ALL
4480 [C] — (J26) — FAR §91.121
In the case of an aircraft not equipped with a radio, set the altimeter to the elevation of the departure airport or use an appropriate altimeter setting available prior to departure.

ALL
4481 [C] — (J26) — AC 61-27C, page 29
En route at FL290, the pilot would have the altimeter set to 29.92 inches, since the altitude is above FL180. Since the pilot failed to reset it to the local altimeter setting of 30.26 during descent, the altimeter will not indicate field elevation upon landing. Since:

$$\begin{array}{r} 30.26 \\ -\ 29.92 \\ \hline 0.34 \end{array}$$

and one inch per 1,000 feet equals 0.1 inch per 100 feet and 0.01 inch per 10 feet, the altimeter will, in this case, indicate an error of 340 feet.

Because the flight went from a low-pressure area (29.92) to a higher-pressure area (30.26), the altimeter error is subtracted from the field elevation of 134 feet.

134	airport elevation
– 340	error
-206	feet below sea level

ALL
4482 [C] — (J26) — AIM ¶7-40(a)(1)
For flight below 18,000 feet MSL, the altimeter should be set to the current reported altimeter setting of a station along the route, and within 100 NM of the aircraft, or if there is no station within this area, the current reported altimeter setting of an appropriate available station. When an aircraft is en route on an instrument flight plan, ATC will furnish this information to the pilot at least once while the aircraft is in their area of jurisdiction.

ALL
4483 [C] — (I04) — AC 61-27C, page 29
The small hand indicates thousands of feet. The long, thin hand indicates hundreds of feet. The very thin hand with the triangular tip indicates tens of thousands of feet. An altimeter indicating 12,000 feet has the small hand on the 2, the long, thin hand on the 0, and the very thin hand with the triangular tip past the 1.

ALL
4484 [B] — (I04) — AC 61-27C, page 29
The small hand indicates thousands of feet. The long, thin hand indicates hundreds of feet. The very thin hand with the triangular tip indicates tens of thousands of feet. An altimeter indicating 8,000 feet has the small hand on the 8, the long, thin hand on the 0, and the very thin hand with the triangular tip before the 1.

ALL
4485 [A] — (J17) — AIM ¶5-33(a)(2)
Jet Route is a route designed to serve aircraft operations from 18,000 feet MSL up to, and including flight level 450.

ALL
4486 [B] — (J16) — AIM ¶5-22(c)
The abbreviated clearance will include the destination airport filed in the flight plan.

ALL
4487 [B] — (I07) — AC 61-27C, page 119
Slant-range error is negligible if the aircraft is 1 mile or more from the ground facility for each 1,000 feet of altitude above the elevation of the facility.

ALL
4488 [C] — (J07) — AC 61-27C, page 138
When flying out on the back course of an ILS, you receive front course indications, so indicator #2 in FAA Figure 86 shows you are to the right of course. VOR indicators #3 and #4 show the 062 radial selected, which is the only way to determine WAGGE intersection. Visualize yourself flying out on Squaw Valley Radial 062. Indicator #3 shows you to the left of the 062 radial; therefore, you are to the north of and approaching WAGGE intersection.

ALL
4489 [A] — (J16) — AIM ¶5-25(a)
The departure route description for the Washoe Two Departure is to climb via I-RNO localizer south course to WAGGE intersection then via radar vectors to assigned route.

ALL
4490 [B] — (J16) — AIM ¶5-25(a)
The lost communication procedures state: if not in contact with departure control within one minute after takeoff, or if communications are lost before reaching 9,000 feet, continue climb via I-RNO localizer south course to WAGGE intersection. Turn left, proceed direct FMG VOR. Cross FMG VOR at or above MCA, thence via assigned route or climb in holding pattern northeast on FMG R-041. Left turns to cross FMG VOR at or above MCA for assigned route.

ALL
4491 [A] — (J16) — AIM ¶5-25
See "note" on the plan view of the approach plate. It states: minimum climb rate at 400 feet per NM to 9,000 feet required.

ALL
4492 [C] — (J16) — AIM ¶5-26(b)(3)
See "note" on the plan view of the approach plate. It states the minimum climb rate of 400 feet per nautical mile to 9,000 feet required. Use the rate-of-climb table located in FAA Legend 18 to convert to feet per minute. Find the GS of 150 knots and the required climb rate of 400 feet. The required rate of 1,000 feet per minute can be read at the intersection of the two.

ALL
4493 [C] — (J17) — AIM ¶ 5-35(b)
VOR changeover points (COP) are the points at which the pilot should change navigation receiver frequency from the station behind the aircraft to the station ahead and are:

1. halfway between stations, if no other designation is given;
2. at a course change in the route;
3. at the COP symbol.

In this case, the COP symbol indicates the change should be made 30 NM from Daisetta (DAS).

ALL
4494 [A] — (J17) — AIM ¶ 5-36
Determine the holding pattern by placing your pencil on the holding fix and dragging it on the holding radial given by ATC, then returning back to the fix. Then draw the pattern from the fix with turns in the direction specified. In this question, ATC has specified standard turns (right turns).

The entry procedure is based on the aircraft's heading. To determine which entry procedure to use, draw a line at a 70 angle from the holding fix, and cutting the outbound leg at about one-third its length. With the aircraft at STRUT intersection heading eastbound, we are in the largest piece of the pie, so a direct entry would be used. *See* the figure to the right.

ALL
4495 [A] — (I08) — AC 61-27C, page 138
The #1 VOR is tuned to Beaumont VORTAC (BPT). With the OBS set to 264, FROM, and the CDI needle deflected right, the aircraft is south of V222. The #2 VOR is tuned to Daisetta VORTAC (DAS). With the OBS set to 139, FROM, and the CDI needle deflected right, the aircraft is right (east) of the DAS 139 radial. Since the 139 radial is east of the 142 radial (which defines FALSE intersection), the aircraft is east of FALSE intersection.

ALL
4496 [C] — (J35) — Enroute Low Altitude Chart Legend
The localizer course symbol at Jefferson Co Field indicates an ILC Localizer Course with ATC function is available.

ALL
4497 [C] — (J35) — Enroute Low Altitude Chart Legend
Frequencies 122.2 and 121.5 are normally available at all FSS's. In addition, 122.3 is listed on the top of the Lake Charles VOR identifier box, which indicates there is a transceiver at that site remoted to the DE RIDDER flight service station.

ALL
4498 [C] — (J35) — Enroute Low Altitude Chart Legend
An ILS Localizer Course with ATC function is available. The feathered side indicates Blue Sector. The "back course" indicates that it is a back course ILS Localizer Course with ATC function.

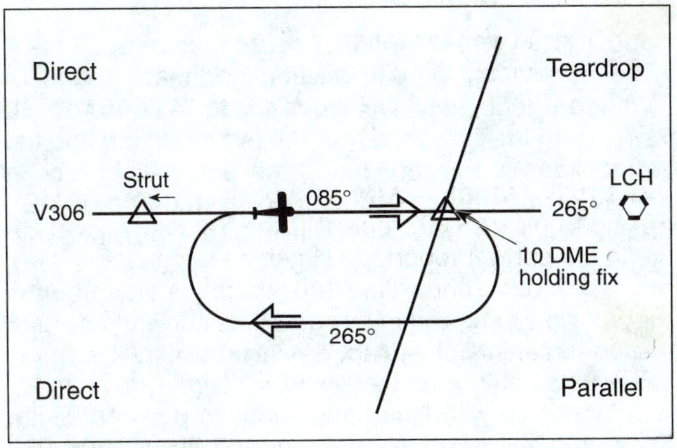

Question 4494

ALL
4499 [A] — (J35) — Enroute Low Altitude Chart Legend
The changeover point (COP) is located midway between the navigation facilities for straight route segments. In the case of dogleg route segments, the COP is located at the intersection of radials or courses forming the dogleg. When the COP is not located at the midway point, the location will be identified with a COP symbol, giving the mileage to the radio aids.

There is no COP symbol on this route segment; therefore, the COP is at the halfway point.

ALL
4500 [C] — (J24) — AIM ¶6-30 and FAR §91.185(c)(3)
If a pilot experiences a loss of two-way radio capability while on an IFR flight, the pilot should:

1. Adjust the transponder to code 7600.
2. Leave the clearance limit to begin the approach as close as possible to the expect-further-clearance time.

ALL
4501 [B] — (J35) — Enroute Low Altitude Chart Legend
The Minimum Crossing Altitude (MCA) is the lowest altitude at which an aircraft must cross when proceeding in the direction of a higher Minimum Enroute Altitude (MEA). The MCA southbound on V235 is 11,400 feet and 12,000 feet eastbound on V293. The 12,000 feet eastbound must be reached by the Cedar City VOR.

ALL
4502 [B] — (J35) — Enroute Low Altitude Chart Legend
There is an FSS located on the field at Cedar City (area D) that is in operation part-time. The frequencies available at Cedar City are 122.6 (published), 121.5 and 122.2 (the box is bold). The pilot can also use 122.1 to transmit to the FSS, and they will respond through the MLF VORTAC on 112.1 (area C).

ALL
4503 [C] — (B11) — FAR §91.211
The lowest appropriate altitude northeast bound on V382 is 16,000 feet. At altitudes above 14,000 feet MSL, the minimum flight crew must be provided with and use supplemental oxygen during the entire flight time at those altitudes. At altitudes above 15,000 feet MSL, each occupant of the aircraft must be provided with supplemental oxygen.

ALL
4504 [C] — (J35) — AIM ¶7-4(a)
EFAS (Enroute Flight Advisory Service) provides enroute aircraft with weather advisories pertinent to the type of flight, route to be flown, and the altitude. The service is normally available throughout the conterminous United States from 6 a.m. to 10 p.m. local time. Aircraft flying between 5,000 feet AGL and 18,000 feet AGL can obtain the service by calling "Flight Watch" on radio frequency 122.0 MHz.

ALL
4505 [B] — (B10) — FAR §91.185
If a two-way radio communication failure occurs in VFR conditions, or if VFR conditions are encountered after the failure, each pilot shall continue the flight under VFR and land as soon as practicable.

ALL
4506 [C] — (J35) — Enroute Low Altitude Chart Legend
Discrete frequencies are normally designated for each control sector in enroute terminal facilities. Since the COP for southwest-bound traffic between Hanksville and Page on V208 is still in Salt Lake City ARTCC's boundary, the frequency should be the one contained in the Hanksville Remote Site scalloped box by the Hanksville VORTAC: 133.6.

ALL
4507 [A] — (I08) — AC 61-27C, page 135
The OBS #1 indicates you are to the right of V382 (033° radial FROM BCE) which coincides with OBS #2's indication that you to the left of V8 (046° radial TO HVE).

ALL
4508 [C] — (J35) — Enroute Low Altitude Chart Legend
Class G airspace is that portion of the airspace that has not been designated as Class A through E. On the Enroute Low Altitude Chart, open areas (white) indicate controlled airspace, while shaded areas (brown) indicate uncontrolled airspace up to 14,500. The Federal Airways are in Class E airspace areas, and unless otherwise specified, extend upward from 1,200 feet to but not including 18,000 feet MSL. Since Bryce Canyon is an uncontrolled field in a white area, the airspace is Class G from the surface to 1,200 feet AGL.

ALL
4509 [B] — (J35) — Enroute Low Altitude Chart Legend
DBS VORTAC can be found in the bottom left-hand quadrant of the FAA figure. The minimum crossing altitude (MCA) is identified in the middle of the DBS VORTAC compass rose, beneath the flag with the "X" inside. The minimum crossing altitude at DBS VORTAC for an IFR flight northbound on V257 is 8,600 feet.

ALL
4510 [C] — (J35) — Enroute Low Altitude Chart Legend
The eastbound course on V343 from DBS VORTAC to RANEY intersection requires VFR-On-Top altitudes of odd plus 500 feet above the MEA (15,000 feet MSL), and below Class A (18,000 feet MSL).

ALL
4511 [C] — (I13) — AC 61-27C

1. Convert wind to magnetic using the 17° variation:

   260° – 17°E = 243°

2. Calculate the ground speed using your E6-B or CX-1a:

Wind direction	243°
Wind speed	24 knots
Course	186°
TAS	185 knots
GS	170.8 knots

3. Calculate your ETE (leg time) using your E6-B or CX-1a:

Distance	111 NM
GS	170.8 knots
Time	00:38:59

ALL
4512 [C] — (J03) — Enroute Low Altitude Chart Legend
The Jackson Hole Airport is located in the bottom right quadrant of FAA Figure 91. The "L" in the circle indicates pilot controlled lighting.

ALL
4513 [C] — (B11) — FAR §91.211(a)
Eastbound V520 is in the bottom, center of FAA Figure 91. The MEA for V520 eastbound from DBS VORTAC is 15,000 feet; therefore the required crew must be provided with, and use, supplemental oxygen.

ALL
4514 [B] — (I13) — AC 61-27C

1. Calculate your ground speed between CPN and DIVID intersection (upper left quadrant of FAA Figure 91) using your E6-B or CX-1a:

Distance	9 NM
Time	4 minutes
GS	135.0 knots

2. Calculate your leg time between DIVID and DBS VORTAC:

Distance	110 NM
GS	135 knots
Time	00:48:53 or 49 minutes

3. Calculate your ETA at DBS VORTAC:

   0854 + 49 minutes = 0943

ALL
4515 [B] — (S32) — Enroute Low Altitude Chart Legend
A Remote Communications Outlet (RCO) is an unmanned communications facilities remotely controlled by air traffic personnel. RCOs serve FSS's. An RCO may be UHF or VHF and will extend the communications range of the air traffic facility.

ALL
4516 [B] — (J35) — Enroute Low Altitude Chart Legend
The changeover point (COP) symbol indicates the point where the pilot should change navigation receiver frequency from the station behind the aircraft to the station ahead. The distance from each VORTAC can be read above and below the symbol, pointing to the VORTAC which it is referring to. The COP from DBS to JAC on V520 is 60 NM from DBS VORTAC.

ALL
4517 [B] — (J35) — Enroute Low Altitude Chart Legend
SABAT Intersection has a flag with an "X" in it, indicating a Minimum Crossing Altitude (MCA). The MCA at SABAT when eastbound from DBS VORTAC on V298 is 11,100 feet.

ALL
4518 [B] — (B09) — FAR §91.155
In Class E airspace, at or above 10,000 feet MSL, the VFR weather minimums are flight visibility 5 statute miles, and cloud clearance of 1,000 feet below, 1,000 feet above, and 1 statute mile horizontal.

ALL
4519 [A] — (B09) — FAR §91.155
In Class E airspace, less than 10,000 feet MSL, the VFR weather minimums are flight visibility 3 statute miles, and cloud clearance of 500 feet below, 1,000 feet above, and 2,000 feet horizontal.

ALL
4520 [C] — (B09) — FAR §91.155
In Class E airspace, less than 10,000 feet MSL, the VFR weather minimums are flight visibility 3 statute miles, and cloud clearance of 500 feet below, 1,000 feet above, and 2,000 feet horizontal (regardless of whether it is day or night).

ALL
4521 [B] — (B09) — FAR §91.155
In Class E airspace, at or above 10,000 feet MSL, the VFR weather minimums are flight visibility 5 statute miles, and cloud clearance of 1,000 feet below, 1,000 feet above, and 1 statute mile horizontal (regardless of whether it is day or night).

ALL
4522 [C] — (B09)— FAR §91.155
In Class G airspace, more than 1,200 feet AGL and at or above 10,000 feet MSL, during daylight hours, the VFR weather minimums are flight visibility 5 miles, and cloud clearance of 1,000 feet above, 1,000 feet below, and 1 statute mile horizontal.

ALL
4523 [A] — (B09) — FAR §91.155
In Class G airspace, more than 1,200 feet AGL but less than 10,000 feet MSL, during daylight hours, the VFR weather minimums are flight visibility 1 statute mile, and cloud clearance of 1,000 feet above, 500 feet below, and 2,000 feet horizontal.

ALL
4524 [B] — (B09) — FAR §91.155
In Class G airspace, less than 1,200 feet AGL, during daylight hours, the VFR weather minimums are flight visibility 1 statute mile, and the pilot must stay clear of clouds.

ALL
4525 [C] — (B09) — FAR §91.157
Special VFR operations below 10,000 feet MSL at night require the pilot to remain clear of clouds, and ground visibility must be at least 1 statute mile, or if ground visibility is not reported, flight visibility must be at least 1 statute mile.

ALL
4526 [B] — (J06) — AIM ¶3-10
The floor of Class E airspace when designated in conjunction with an airway is 1,200 feet AGL.

ALL
4527 [B] — (J06) — AIM ¶3-14
Class D airspace is that airspace from the surface to 2,500 feet above the airport elevation (charted in MSL) surrounding those airports that have an operational control tower.

ALL
4528 [B] — (J06) — AIM ¶3-10
Assuming Class E airspace has not been designated to the surface, the floor of Class E airspace is 700 feet AGL when designated in conjunction with an airport which has an approved instrument approach procedure.

ALL
4529 [C] — (J06) — AIM ¶3-11
Class A airspace is that airspace from 18,000 feet MSL up to and including FL600, including that airspace overlying the waters within 12 nautical miles of the coast of the 48 contiguous states and Alaska.

ALL
4530 [B] — (J06) — AIM ¶3-10
The maximum altitude Class G airspace exists is 14,500 feet MSL.

ALL
4531 [B] — (J06) — AIM ¶3-12
The shape of Class B airspace is individually tailored to fit the specific requirements of the terrain and traffic flows. However, the vertical dimension is generally from the surface to 10,000 feet MSL.

ALL
4532 [C] — (J06) — AIM ¶3-10
The normal lateral limits for Class D airspace is 4 NM.

ALL
4533 [C] — (J06) — AIM ¶3-10 and 3-11
Class A airspace exists from 18,000 feet MSL up to and including FL600.

ALL
4534 [B] — (J05) — AIM ¶2-37
Mandatory instruction signs have a red background with a white inscription and are used to denote:

1. An entrance to a runway or critical area; and
2. Areas where an aircraft is prohibited from entering.

ALL
4535 [B] — (J05) — AIM ¶2-37
The holding position sign is a mandatory instruction sign, used to hold an aircraft on a taxiway located in the approach or departure area for a runway so the aircraft does not interfere with operations on that runway.

ALL
4536 [A] — (J05) — AIM ¶2-37
Holding position signs are mandatory instruction signs, and have a red background with a white inscription.

ALL
4537 [B] — (J05) — AIM ¶2-34
Holding position markings indicate where an aircraft is supposed to stop. They consist of four yellow lines: two solid and two dashed, spaced six inches apart and extending across the width of the taxiway or runway. The solid lines are always on the side where the aircraft is to hold.

ALL
4538 [C] — (J13) — AIM ¶4-59(c)
Pilots should state their position on the airport when calling the tower for takeoff from a runway intersection.

ALL
4539 [A] — (J06) — AIM ¶3-13(d)(2)
A Mode C transponder and two-way communications are required in order to operate within Class C airspace.

ALL
4540 [B] — (J18) — AIM ¶5-44
Minimum Sector Altitudes provide at least 1,000 feet of clearance above the highest obstacle in the defined sector to a distance of 25 NM from the facility. Navigational course guidance is not assured at the MSA.

ALL
4541 [C] — (B08) — FAR §91.177(a)(2)(i)
Except when necessary for takeoff or landing, or unless otherwise authorized by the Administrator, no person may operate an aircraft in an area designated as a mountainous area under IFR below an altitude of 2,000 feet above the highest obstacle within a horizontal distance of 4 NM from the course to be flown.

ALL
4542 [C] — (B08) — Pilot/Controller Glossary
Minimum Enroute IFR Altitude (MEA) is the lowest published altitude which assures acceptable navigational signal coverage and meets obstacle clearance requirements between those fixes.

ALL
4543 [C] — (B08) — FAR §91.159
Each person operating an aircraft under the ATC clearance "VFR Conditions On-Top," shall maintain an altitude or flight level appropriate to the magnetic course.

ALL
4544 [B] — (J19) — Pilot/Controller Glossary
MRA (Minimum Reception Altitude) is the lowest altitude required to receive adequate signals to determine specific fixes. Reception of signals from a radio facility located off the airway being flown may be inadequate at the designated MEA, in which case, an MRA is designated for the fix.

ALL
4545 [A] — (J19) — Pilot/Controller Glossary
Minimum Obstruction Clearance Altitude (MOCA) is the lowest published altitude in effect between radio fixes on VOR airways, off-airway routes, or route segments which meets obstacle clearance requirements for the entire route segment and which assures acceptable navigational signal coverage only within 25 statute (22 nautical) miles of a VOR.

ALL
4546 [B] — (J10) — AIM ¶3-41
Military Training Routes (MTRs) are charted on IFR Low Altitude Enroute charts, VFR Sectional charts, and Area Planning charts.

ALL
4547 [B] — (J19) — Pilot/Controller Glossary
Minimum Obstruction Clearance Altitude (MOCA) is the lowest published altitude in effect between radio fixes on VOR airways, off-airway routes, or route segments which meets obstacle clearance requirements for the entire route segment and which assures acceptable navigational signal coverage only within 25 statute (22 nautical) miles of a VOR.

ALL
4548 [C] — (I07) — AC 61-27C, page 118
Full-needle deflection from the center position to either side of the dial indicates the aircraft is 10° or more off course, assuming normal needle sensitivity.

ALL
4549 [C] — (I08) — AC 61-27C, page 137
Approach to the station is indicated by flickering of the TO/FROM indicator and CDI as the aircraft flies into the "cone of confusion" (no-signal area). Station passage is shown by the first positive, complete reversal of the TO/FROM indicator.

ALL
4550 [C] — (I08) — AC 61-27C, page 137
Station passage is shown by the first positive, complete reversal of the TO/FROM indicator.

ALL
4551 [C] — (I08) — AC 61-27C, page 134
Course sensitivity may be checked by noting the number of degrees of change in the course selected as you rotate the OBS to move the CDI from center to the last dot on either side. This should be between 10° and 12°.

ALL
4552 [B] — (I08) — AC 61-27C, page 137
Aircraft displacement from course is approximately 200 feet per dot per nautical mile. For example, at 30 NM from the station, one dot deflection indicates approximately 1 NM displacement of the aircraft from the course centerline. Therefore, a three-dot deflection would mean the aircraft is approximately 3 NM from the course centerline.

ALL
4553 [B] — (I08) — AC 61-27C, page 137
Aircraft displacement from course is approximately 200 feet per dot per nautical mile. For example, at 30 NM from the station, a 1-dot deflection indicates approximately 1 NM displacement from the course centerline. Assuming a receiver with normal course sensitivity and full-scale deflection being five dots, a half-scale deflection would be two and one-half dots. Therefore, at 30 NM, the aircraft would be approximately 2.5 NM from course centerline.

ALL
4554 [C] — (I07) — AC 61-27C, page 137
Full-needle deflection from the center position to either side of the dial indicates the aircraft is 10° or more off course, assuming normal needle sensitivity. Therefore, a half-scale deflection would indicate an angular deviation of one-half of 10°, or 5°.

ALL
4555 [C] — (J14) — AIM ¶4-88(d)
When ATC has not used the term "At Pilot's Discretion," nor imposed any climb or descent restrictions, pilots should initiate climb or descent promptly on acknowledgment of the clearance. Descend or climb at an optimum rate consistent with the operating characteristics of the aircraft to 1,000 feet above or below the assigned altitude, and then attempt to descend or climb at a rate of 500 to 1,500 feet per minute until the assigned altitude is reached.

ALL
4556 [C] — (I08) — AC 61-27C, page 137
Aircraft displacement from course is approximately 200 feet per dot per nautical mile. Therefore, in this instance, the aircraft is two and one-half dots, or 500 feet, from the radial per nautical mile from the station (at 10 miles, 5,000 feet; at 20 miles, 10,000 feet, and so on). The constant deflection indicates the airplane is flying away from the radial.

ALL
4557 [A] — (I08) — AC 61-27C, page 137
Aircraft displacement from course is approximately 200 feet per dot per nautical mile. For example, at 30 NM from the station, a one dot deflection indicates approximately a 1 NM displacement of the aircraft from the course centerline. In this case, the CDI is displaced 2.5 dots at 60 NM, which equals 2 NM times 2.5 dots, or approximately a 5 NM displacement from course centerline.

ALL
4558 [C] — (I08) — AC 61-27C, page 135
The TO/FROM indicator is the triangular shaped pointer(s) (▲). When the indicator points to the head of the course arrow, it indicates that the course selected, if properly intercepted and flown, will take the aircraft TO the selected facility, and vice versa. In this case, the indicator is pointing to the TAIL of the course arrow, so if flown, the course selected (350) would take the aircraft away FROM the selected facility. Each dot represents 2°. In this case, the aircraft is two and one-half dots (5°) to the LEFT of the selected course, therefore the aircraft is on R-345.

ALL
4559 [B] — (I08) — AC 61-27C, page 135
Currently, the course selector is set to 350 (the arrow head of the needle is pointing to 350), with a FROM indication (the triangular shaped pointers are pointing to the tail of the needle), and a left 2.5 dot needle deflection (5° left of course). This means the aircraft is on R-345 (350 – 5). To center the CDI and change the ambiguity indication to a TO, the OBS should be turned to the reciprocal of 345° (345 – 180), or 165°.

ALL
4560 [C] — (I08) — AC 61-27C, page 137
Full-scale deflection is 10°, so each dot represents 2°. The CDI is displaced two dots, therefore the lateral displacement is 4°.

ALL
4561 [A] — (I08) — AC 61-27C, page 137
Full-scale deflection is 10°, so each dot represents 2°. The CDI is displaced two dots (4°). The OBS is set at 170° with a FROM indication, and the aircraft is 4° to the RIGHT of course (or on the 174° radial). Simply rotating the OBS to 174° would center the CDI.

ALL
4562 [C] — (I08) — AC 61-27C, page 137
Rotating the OBS until the reciprocal of 170° (350°) is shown under the course arrow will cause the ambiguity indication to change to TO. The CDI now indicates that the aircraft is 4° to the right of course (or on the 354° course TO the selected facility). Simply rotating the OBS to 354° will center the CDI.

ALL
4563 [A] — (I08) — AC 61-27C, page 53
HSI Indicator "A" is set up with the head of the arrow pointing to 270°. The Course Deviation Indicator is centered; therefore, the aircraft is on the extended centerline of RW #9 and #27. With a heading of 360°, Indicator "A" represents an airplane at position #6 or #9. *See* the figure below.

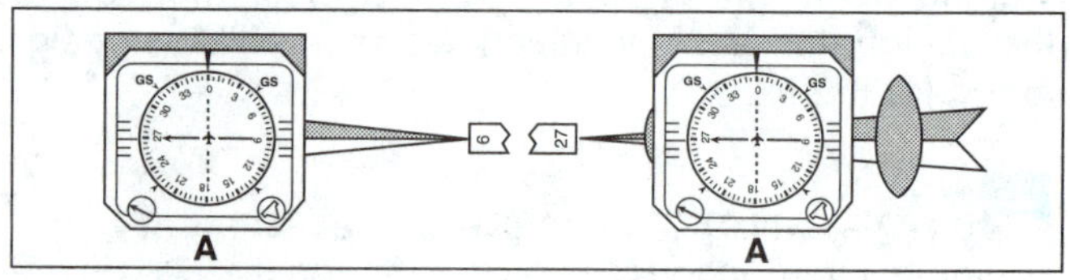

Question 4563

ALL
4564 [B] — (I08) — AC 61-27C, page 53
HSI Indicator "B" is set up with the head of the arrow pointing to 090°. The CDI indication is reversed and the aircraft is actually to the south of the extended centerline. Indicator "B" then, with the aircraft flying on a heading of 090°, could be at position #13 and #5. Remember that the localizer receiver does not know where you are in relation to the antenna site. *See* the figure below.

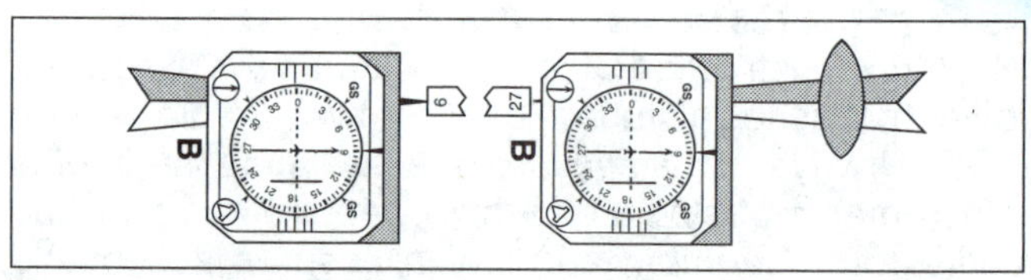

Question 4564

ALL
4565 [C] — (I08) — AC 61-27C, page 53
HSI Indicator "C" is set up with the head of the arrow pointing to 090°. With the CDI centered, the aircraft is obviously on the extended centerline. With a heading of 090°, position #12 is the only one which would have that indication. *See* the figure below.

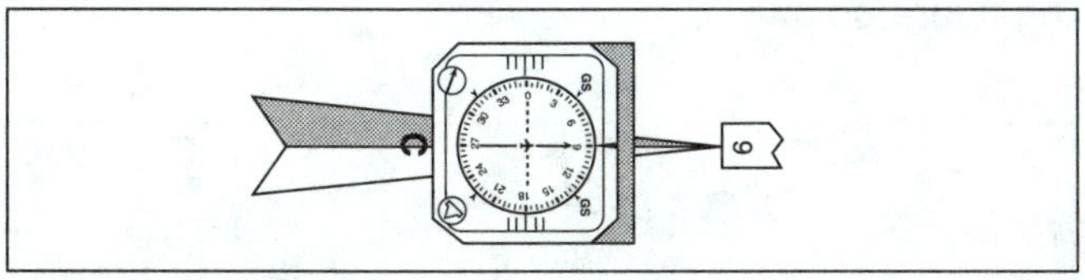

Question 4565

ALL
4566 [C] — (I08) — AC 61-27C, page 53
HSI Indicator "D" is set up with the head of the arrow pointing to 090°. The aircraft is to the south of course, on a heading of 310° with position #2 being the only possible choice. *See* the figure below.

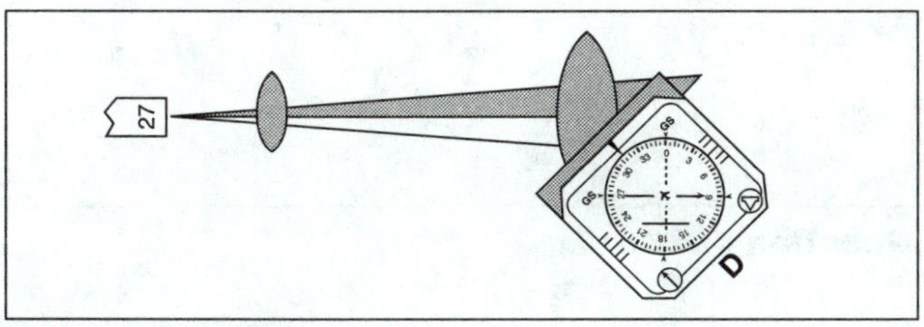

Question 4566

ALL
4567 [C] — (I08) — AC 61-27C, page 53
HSI Indicator "E" is set up with the head of the arrow pointing to 090°. The aircraft is to the south of the extended centerline, with position #8 or #3 being the only possible answer for an aircraft on the heading of 045°. *See* the figure below.

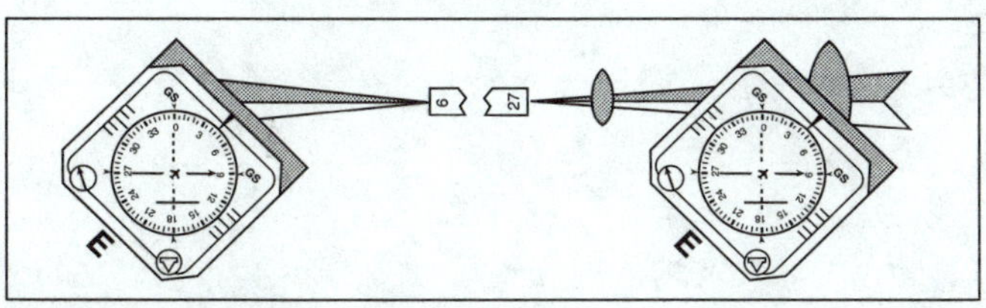

Question 4567

ALL
4568 [A] — (I08) — AC 61-27C, page 53
HSI Indicator "F" is set up with the head of the arrow pointing to 270°, and the CDI centered so the aircraft is on the extended centerline. Position #4 is the only one that would be correct for an aircraft on a heading of 270°. *See* the figure below.

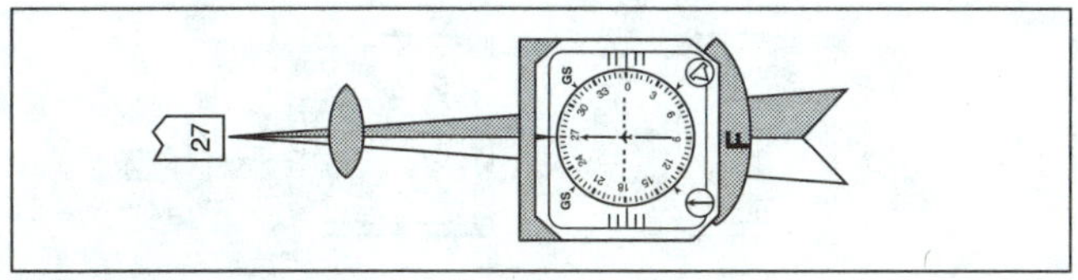

Question 4568

ALL
4569 [B] — (I08) — AC 61-27C, page 53
HSI Indicator "G" is set up with the head of the arrow pointing to 270°. The CDI is deflected to the left, so the aircraft is to the north on a heading of 270°. Again, the indicator does not know which side of the antenna you are on, so position #7 or #11 would be appropriate selections. *See* the figure below.

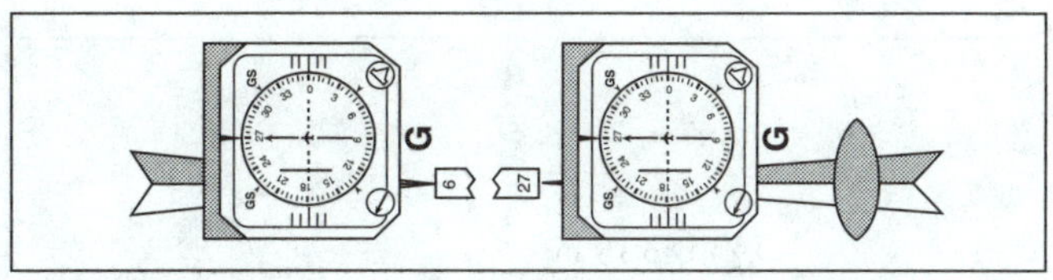

Question 4569

ALL
4570 [B] — (I08) — AC 61-27C, page 53
HSI Indicator "H" is set up with the head of the arrow pointing to 270°. The CDI is deflected to the left, so the aircraft is to the north on a heading of 215°. Position #1 is the only position corresponding to the CDI presentation. *See* the figure below.

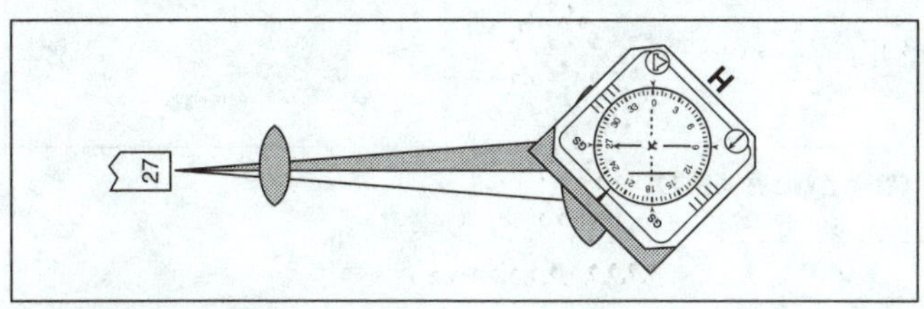

Question 4570

ALL

4571 [C] — (I08) — AC 61-27C, page 53

HSI Indicator "I" is set up with the head of the arrow pointing to 090°. The aircraft is to the north of the extended centerline on a heading of 270°. The CDI is using reverse sensing, so position #7 or #11 would be appropriate. *See* the figure below.

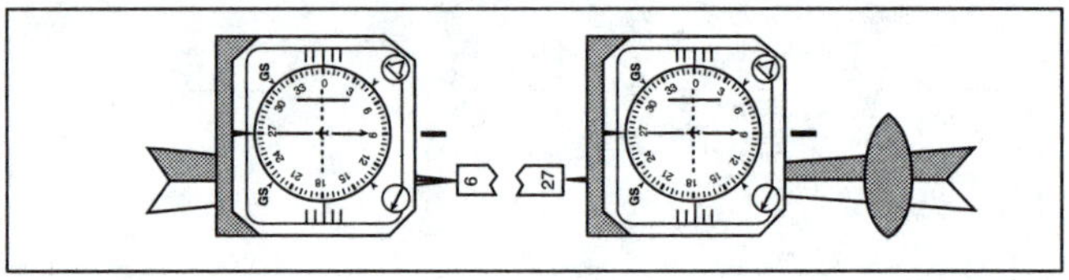

Question 4571

ALL

4572 [C] — (I08) — AC 61-27C, page 53

HSI Indicator "D" has a course selection of 180°, and the TO/FROM indicator is pointing to the tail of the course arrow. So the aircraft is flying away FROM the station, and is south of R-270 and R-090. The CDI bar is to the left, which means the aircraft is west of R-180. The aircraft heading is 180°, which describes aircraft 17. *See* the figure below.

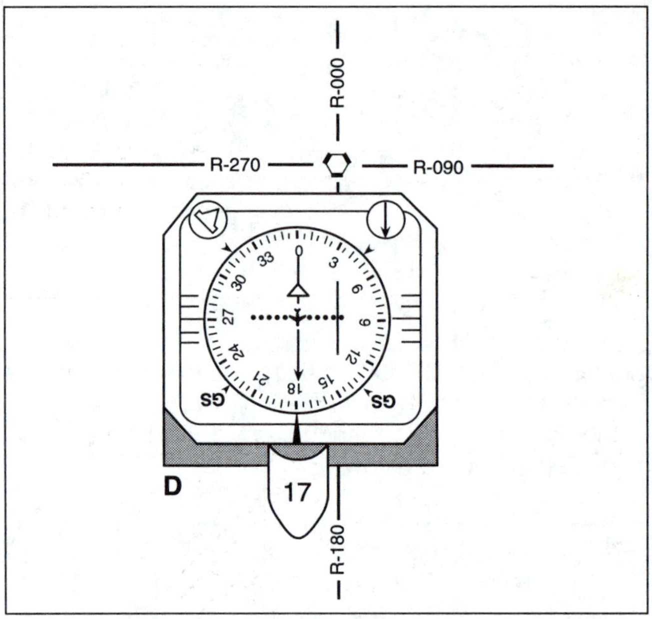

Question 4572

ALL

4573 [B] — (I08) — AC 61-27C, page 53

HSI Indicator "E" has a course selection of 360°, and the TO/FROM indicator is pointing to the tail of the course arrow. So the aircraft is flying away FROM the station, and is north of R-270 and R-090. The CDI bar is to the left, which means the aircraft is east of R-000. The aircraft heading is 360°, which describes aircraft 6. *See* the figure below.

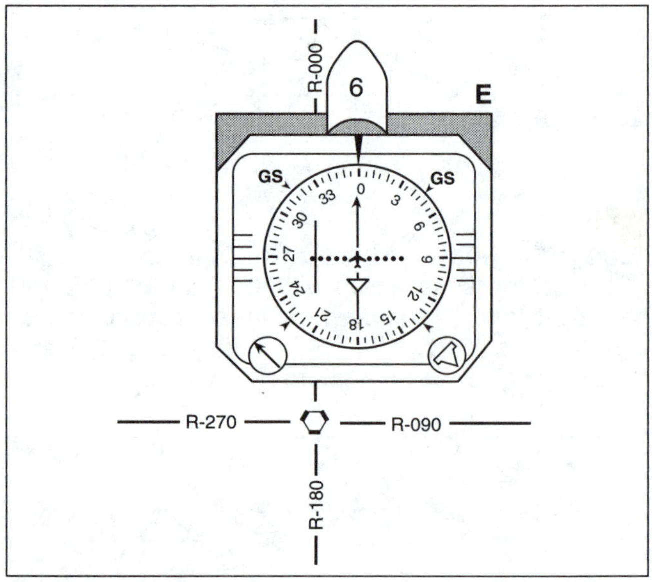

Question 4573

ALL

4574 [C] — (I08) — AC 61-27C, page 53

HSI Indicator "F" has a course selection of 180°, and the TO/FROM indicator is pointing to the tail of the course arrow. So the aircraft is flying away FROM the station, and is south of R-270 and R-090. The CDI bar is centered, which means the aircraft is on the extended centerline. The aircraft heading is 045°, which describes aircraft 16. *See* the figure on the next page.

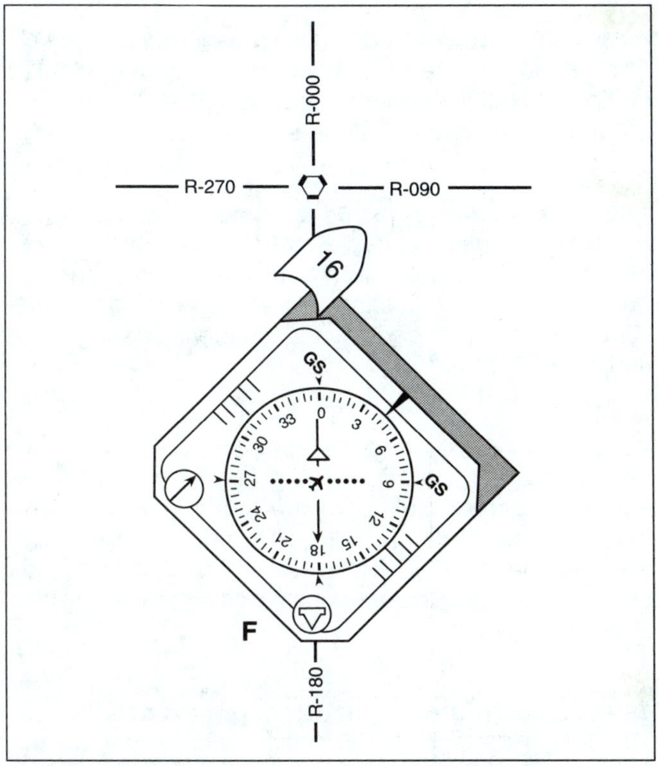

Question 4574

ALL
4575 [A] — (I08) — AC 61-27C, page 53
HSI Indicator "A" has a course selection of 090°, and the TO/FROM indicator is pointing to the head of the course arrow. So the aircraft is flying TO the station, and is west R-000 and R-180. The CDI is deflected to the right, which means the aircraft is north of R-270. The aircraft heading is 205°, which describes aircraft 1. *See* the figure below.

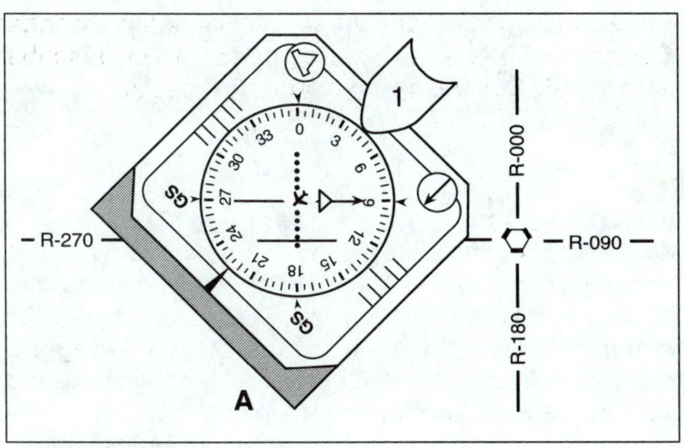

Question 4575

ALL
4576 [C] — (I08) — AC 61-27C, page 53
HSI Indicator "B" has a course selection of 270°, and the TO/FROM indicator is pointing to the tail of the course arrow. So the aircraft is flying FROM the station, and is west of R-000 and R-180. The CDI is deflected right, which means the aircraft is south of R-270. The aircraft heading is 135°, which describes aircraft 19. *See* the figure below.

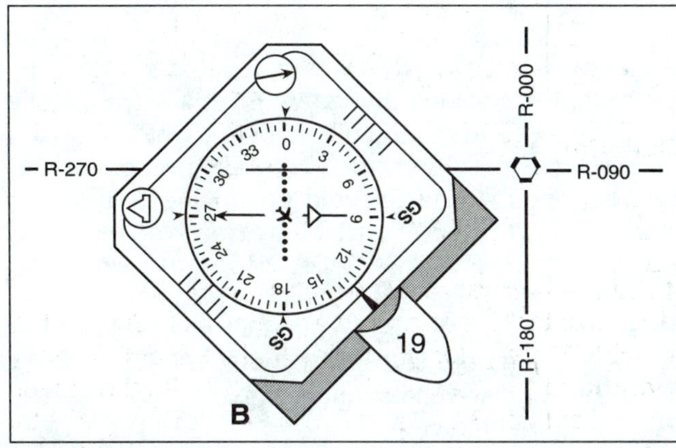

Question 4576

ALL
4577 [C] — (I08) — AC 61-27C, page 53
HSI Indicator "C" has a course selection of 360°, and the TO/FROM indicator is pointing to the head of the course arrow. So the aircraft is flying TO the station, and is south of R-270 and R-090. The CDI is deflected left, which means the aircraft is east of R-180. The aircraft heading is 310°, which describes aircraft 12. *See* the figure below.

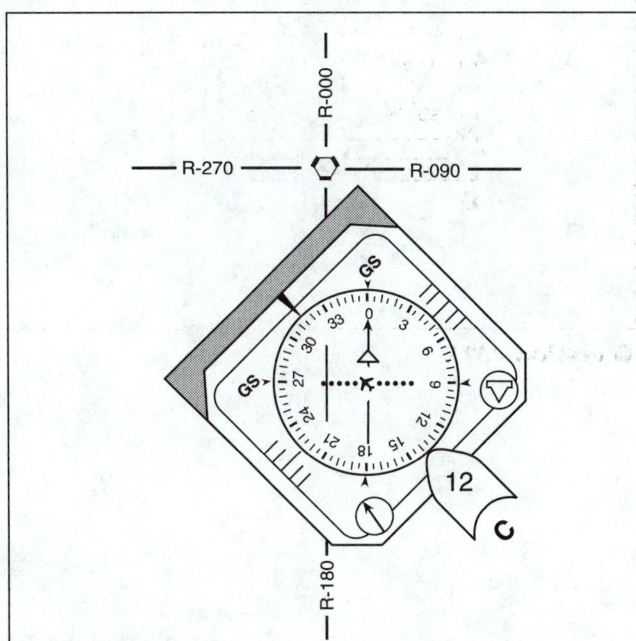

Question 4577

ALL
4578 [B] — (I08) — AC 61-27C, page 146
The magnetic bearing to the station is the sum of the Magnetic (compass) Heading and the Relative Bearing, or the formula: MH + RB = MB TO the station.

In this case, MH (350) + RB (270) = 620.

Since there are only 360° in a complete circle, subtract 360 from 620, which equals 260°, which is the MB TO the station.

ALL
4579 [B] — (I08) — AC 61-27C, page 52
The magnetic heading of the aircraft is always directly under the index at the top of the instrument (assuming no compass deviation error). The bearing pointer displays bearings TO the selected station, the tail displays bearings FROM the station. RMI 3 depicts the magnetic bearing of 235° FROM the station since that is where the tail is pointing. The aircraft heading is also 235° indicating it is tracking outbound on the 235° radial. Since the wind is blowing from 050° at 20 knots (which is a direct tailwind) a wind correction angle would not be required. *See* the figure below.

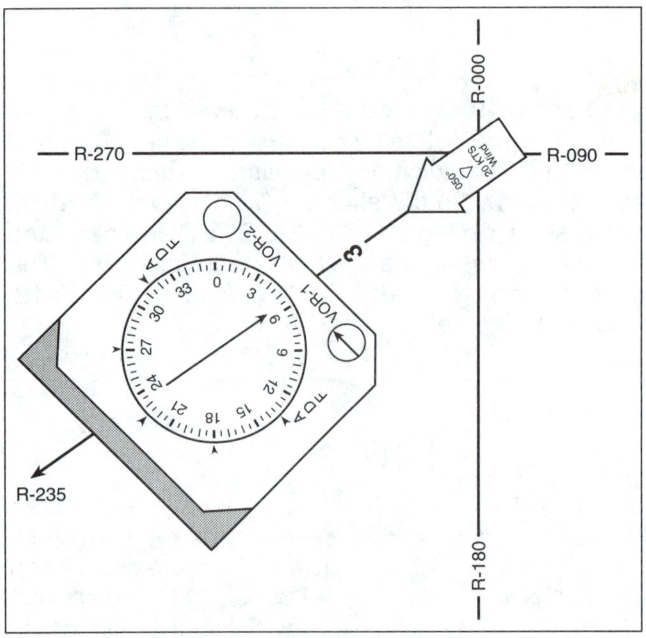

Question 4579

ALL
4580 [B] — (I08) — AC 61-27C, page 52
The bearing pointer displays bearing TO the selected station. In RMI 4, the needle is pointing to 055°, which is the magnetic bearing TO the station.

ALL
4581 [A] — (I08) — AC 61-27C, page 52
An aircraft southwest of the station and moving closer TO the station would have both the heading and the bearing pointer indicating northeast. This describes RMI 1 which has a heading, and magnetic bearing to the station, of 055°.

ALL
4582 [B] — (I08) — AC 61-27C, page 52
The radial, or magnetic bearing FROM the station, is determined from the tail of the bearing pointer. The aircraft is located on the radial when the aircraft heading matches the radial. RMI 2 is indicating 055° with the tail of the needle, and the aircraft located on that radial with a 055° heading.

ALL
4583 [C] — (I08) — AC 61-27C, page 147
Magnetic Heading plus Relative Bearing equals the Magnetic Bearing TO the Station. Using the information from the FAA Figure, the Magnetic Heading is 215°, the Relative Bearing is 140°, therefore:

215° + 140° = 355°	Magnetic Bearing TO the Station

ALL
4584 [A] — (I08) — AC 61-27C, page 147
Magnetic Heading plus Relative Bearing equals the Magnetic Bearing TO the Station. To find Magnetic Bearing FROM the Station, you add or subtract 180° from the Magnetic Bearing TO the Station. Using the information from the FAA Figure, the Magnetic Heading is 215°, the Relative Bearing is 140°, therefore:

215° + 140° = 355°	Magnetic Bearing TO the Station
355° – 180° = 175°	Magnetic Bearing FROM the Station

ALL
4585 [B] — (I08) — AC 61-27C, page 147
Magnetic Heading plus Relative Bearing equals the Magnetic Bearing TO the Station. To find Magnetic Bearing FROM the Station, you add or subtract 180° from the Magnetic Bearing TO the Station. Using the information from the FAA Figure, the Magnetic Heading is 330°, the Relative Bearing is 270°, therefore:

330° + 270° = 600°	Magnetic Bearing TO the Station
600° – 360° = 240°	Magnetic Bearing TO the Station
240° – 180° = 060°	Magnetic Bearing FROM the Station

ALL
4586 [B] — (I08) — AC 61-27C, page 147
Magnetic Heading plus Relative Bearing equals the Magnetic Bearing TO the Station. Using the information from the FAA Figure, the Magnetic Heading is 330°, the Relative Bearing is 270°, therefore:

330° + 270° = 600° Magnetic Bearing TO the Station

600° – 360° = 240° Magnetic Bearing TO the Station

ALL
4587 [A] — (I08) — AC 61-27C, page 52
VOR radials are FROM the station, and are indicated on a radio magnetic indicator (RMI) by the tail of the bearing indicator. Illustration 1 shows the tail of the indicator on 115°, which indicates the aircraft is on the 115° radial.

ALL
4588 [C] — (I08) — AC 61-27C, page 52
VOR radials are FROM the station, and are indicated on a radio magnetic indicator (RMI) by the tail of the bearing indicator. Illustration 4 shows the tail of the indicator on 335°, which indicates the aircraft is on the 335° radial.

ALL
4589 [A] — (I08) — AC 61-27C, page 52
VOR radials are FROM the station, and are indicated on a radio magnetic indicator (RMI) by the tail of the bearing indicator. Illustration 2 shows the tail of the indicator on 315°, which indicates the aircraft is on the 315° radial.

ALL
4590 [C] — (I08) — AC 61-27C, page 52
VOR radials are FROM the station, and are indicated on a radio magnetic indicator (RMI) by the tail of the bearing indicator. Illustration 3 shows the tail of the indicator on 010°, which indicates the aircraft is on the 010° radial.

ALL
4591 [C] — (I08) — AC 61-27C, page 147
Airplane 7 is on a magnetic heading of 270°. To determine which ADF would indicate the airplane is on the 120° magnetic bearing FROM the station, we must calculate the relative bearing, and find the ADF which illustrates this relative bearing. For this problem, we must first convert the magnetic bearing FROM the station to TO the station by adding 180°: 120° + 180° = 300° magnetic bearing TO the station. The following formula is used to find the relative bearing:

Magnetic Heading + Relative Bearing = Bearing TO the Station

270° + ? = 300° Magnetic Bearing TO the Station

? = 30° Relative Bearing

ADF 5 indicates a 30° relative bearing, which with the magnetic heading of 270° found in airplane 7, would indicate the airplane is on the 120° magnetic bearing FROM the station.

ALL
4592 [C] — (I08) — AC 61-27C, page 147
Airplane 5 is on a magnetic heading of 180°. To determine which ADF would indicate the airplane is on the 210° magnetic bearing FROM the station, we must calculate the relative bearing, and find the ADF which illustrates this relative bearing. For this problem, we must first convert the magnetic bearing FROM the station to TO the station by subtracting 180°: 210° – 180° = 300° magnetic bearing TO the station. The following formula is used to find the relative bearing:

Magnetic Heading + Relative Bearing = Bearing TO the Station

180° + ? = 300° Magnetic Bearing TO the Station

? = 210° Relative Bearing

ADF 4 indicates a 210° relative bearing, which with the magnetic heading of 180° found in airplane 5, would indicate the airplane is on the 210° magnetic bearing FROM the station.

ALL
4593 [B] — (I08) — AC 61-27C, page 147
Airplane 3 is on a magnetic heading of 090°. To determine which ADF would indicate the airplane is on the 120° magnetic bearing TO the station, we must calculate the relative bearing, and find the ADF which illustrates this relative bearing. The following formula is used to find the relative bearing:

Magnetic Heading + Relative Bearing = Bearing TO the Station

090° + ? = 120° Magnetic Bearing TO the Station

? = 030° Relative Bearing

ADF 5 indicates a 030° relative bearing, which with the magnetic heading of 090° found in airplane 3, would indicate the airplane is on the 120° magnetic bearing TO the station.

ALL
4594 [A] — (I08) — AC 61-27C, page 147
Airplane 1 is on a magnetic heading of 360°. To determine which ADF would indicate the airplane is on the 060° magnetic bearing TO the station, we must calculate the relative bearing, and find the ADF which illustrates this relative bearing. The following formula is used to find the relative bearing:

Magnetic Heading + Relative Bearing = Bearing TO the Station

360° + ? = 060° Magnetic Bearing TO the Station

? = 060° Relative Bearing

ADF 2 indicates a 060° relative bearing, which with the magnetic heading of 360° found in airplane 1, would indicate the airplane is on the 060° magnetic bearing TO the station.

ALL
4595 [B] — (I08) — AC 61-27C, page 147
Airplane 2 is on a magnetic heading of 045°. To determine which ADF would indicate the airplane is on the 255° magnetic bearing TO the station, we must calculate the relative bearing, and find the ADF which illustrates this relative bearing. The following formula is used to find the relative bearing:

Magnetic Heading + Relative Bearing = Bearing TO the Station

045° + ? = 255° Magnetic Bearing TO the Station
? = 210° Relative Bearing

ADF 4 indicates a 210° relative bearing, which with the magnetic heading of 045° found in airplane 2, would indicate the airplane is on the 255° magnetic bearing TO the station.

ALL
4596 [A] — (I08) — AC 61-27C, page 147
Airplane 4 is on a magnetic heading of 135°. To determine which ADF would indicate the airplane is on the 135° magnetic bearing TO the station, we must calculate the relative bearing, and find the ADF which illustrates this relative bearing. The following formula is used to find the relative bearing:

Magnetic Heading + Relative Bearing = Bearing TO the Station

135° + ? = 135° Magnetic Bearing TO the Station
? = 000° (or 360°) Relative Bearing

ADF 1 indicates a 360° relative bearing, which with the magnetic heading of 135° found in airplane 4, would indicate the airplane is on the 135° magnetic bearing TO the station.

ALL
4597 [B] — (I08) — AC 61-27C, page 147
Airplane 6 is on a magnetic heading of 225°. To determine which ADF would indicate the airplane is on the 255° magnetic bearing FROM the station, we must calculate the relative bearing, and find the ADF which illustrates this relative bearing. For this problem, we must first convert the magnetic bearing FROM the station to TO the station by subtracting 180°: 255° – 180° = 075° magnetic bearing TO the station. The following formula is used to find the relative bearing:

Magnetic Heading + Relative Bearing = Bearing TO the Station

225° + ? = 075° Magnetic Bearing TO the Station
? = 210° Relative Bearing

ADF 4 indicates a 210° relative bearing, which with the magnetic heading of 225° found in airplane 6, would indicate the airplane is on the 255° magnetic bearing FROM the station.

ALL
4598 [C] — (I08) — AC 61-27C, page 147
Airplane 8 is on a magnetic heading of 315°. To determine which ADF would indicate the airplane is on the 090° magnetic bearing FROM the station, we must calculate the relative bearing, and find the ADF which illustrates this relative bearing. For this problem, we must first convert the magnetic bearing FROM the station to TO the station by adding 180°: 090° + 180° = 270° magnetic bearing TO the station. The following formula is used to find the relative bearing:

Magnetic Heading + Relative Bearing = Bearing TO the Station

315° + ? = 270° Magnetic Bearing TO the Station
? = 315° Relative Bearing

ADF 6 indicates a 315° relative bearing, which with the magnetic heading of 315° found in airplane 8, would indicate the airplane is on the 090° magnetic bearing FROM the station.

ALL
4599 [A] — (I08) — AC 61-27C, page 147
Airplane 5 is on a magnetic heading of 180°. To determine which ADF would indicate the airplane is on the 240° magnetic bearing TO the station, we must calculate the relative bearing, and find the ADF which illustrates this relative bearing. The following formula is used to find the relative bearing:

Magnetic Heading + Relative Bearing = Bearing TO the Station

180° + ? = 240° Magnetic Bearing TO the Station
? = 060° Relative Bearing

ADF 2 indicates a 060° relative bearing, which with the magnetic heading of 180° found in airplane 5, would indicate the airplane is on the 240° magnetic bearing TO the station.

ALL
4600 [C] — (I08) — AC 61-27C, page 147
Airplane 8 is on a magnetic heading of 315°. To determine which ADF would indicate the airplane is on the 315° magnetic bearing TO the station, we must calculate the relative bearing, and find the ADF which illustrates this relative bearing. The following formula is used to find the relative bearing:

Magnetic Heading + Relative Bearing = Bearing TO the Station

315° + ? = 315° Magnetic Bearing TO the Station
? = 000° (or 360°) Relative Bearing

ADF 1 indicates a 360° relative bearing, which with the magnetic heading of 315° found in airplane 8, would indicate the airplane is on the 315° magnetic bearing TO the station.

ALL

4601 [B] — (I08) — AC 61-27C, page 135

A FROM indication, with a 360° course selection, means the aircraft is in the northern hemisphere. A left CDI needle means the aircraft is in the right quadrant. Airplane 2 fits this description. Remember that the VOR works without regard to the aircraft heading.

ALL

4602 [B] — (I08) — AC 61-27C, page 144

The #2 (double-barred) needle is pointing toward the right wing-tip reference, so it is evidently tuned to the facility being used for a right-hand arc. The general rule is if a crosswind is blowing you away from the facility (as it is in this case), turn until the bearing pointer is ahead of the wing-tip reference.

ALL

4603 [A] — (I08) — AC 61-27C, page 144

The #2 (double-barred) needle is pointing toward the left wing-tip reference, so it is evidently tuned to the facility being used for a left-hand arc. The general rule is if a crosswind is blowing you away from the facility (as it is in this case), turn until the bearing pointer is ahead of the wing-tip reference.

ALL

4604 [B] — (I08) — AC 61-27C, page 138

Calculate as follows:

1. $\text{Time to Station} = \dfrac{60 \times \text{Minutes Flown}}{\text{Degrees of Bearing Change}}$

or:

$\text{Time} = \dfrac{60 \times 1.5}{5} = 18 \text{ minutes}$

2. $\text{Distance to Station} = \dfrac{\text{TAS} \times \text{Minutes Flown}}{\text{Degrees of Bearing Change}}$

or:

$\text{Distance} = \dfrac{95 \times 1.5}{60} = 28.5 \text{ NM}$

ALL

4605 [B] — (I12) — AIM ¶5-31(c)3

Pilots should resume normal position reporting when ATC advises, "Radar Contact Lost," or "Radar Service Terminated."

ALL

4606 [A] — (I08) — AC 61-27C, page 135

The course selected is 180° and the TO/FROM indicator is pointing to the course, which means the aircraft is north of the course. The CDI needle is deflected to the right, which means the aircraft is left (or east) of course. Therefore, the aircraft is northeast of the VORTAC. *See* the following figure.

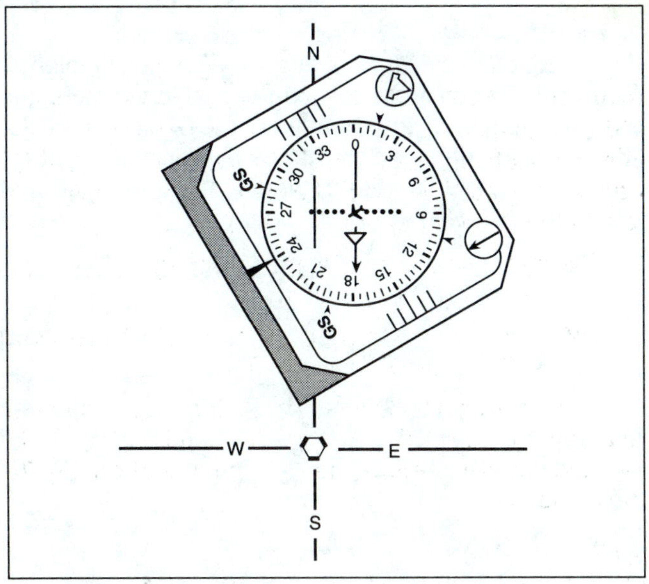

Question 4606

ALL

4607 [C] — (I08) — AC 61-27C, page 135

The course selected is 060° and the TO/FROM indicator is pointing away from the course, which means the aircraft is northeast of the course. The CDI needle is deflected right, which means the aircraft is left (or north) of course. Therefore, the aircraft is northeast of the VORTAC. *See* the figure below.

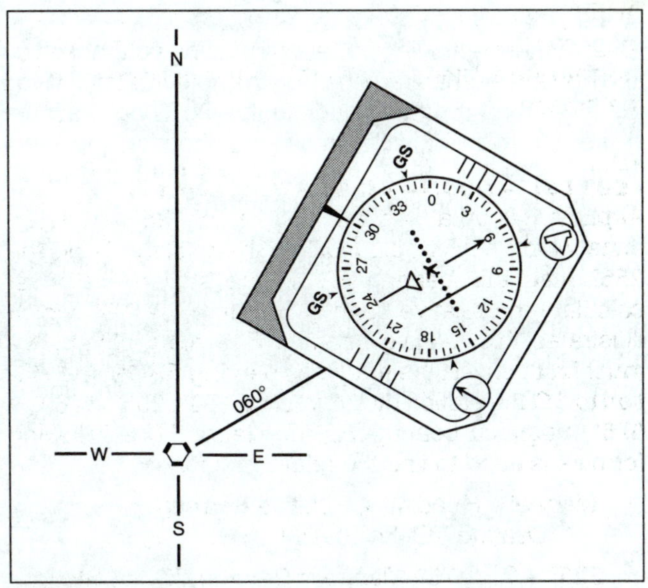

Question 4607

ALL

4608 [C] — (I08) — AC 61-27C, page 135

The course selected is 360° and the TO/FROM indicator is pointing away from the course, which means the aircraft is north of the course. The CDI needle is deflected to the right, which means the aircraft is left (or west) of course. Therefore, the aircraft is northwest of the VORTAC. *See* the figure below.

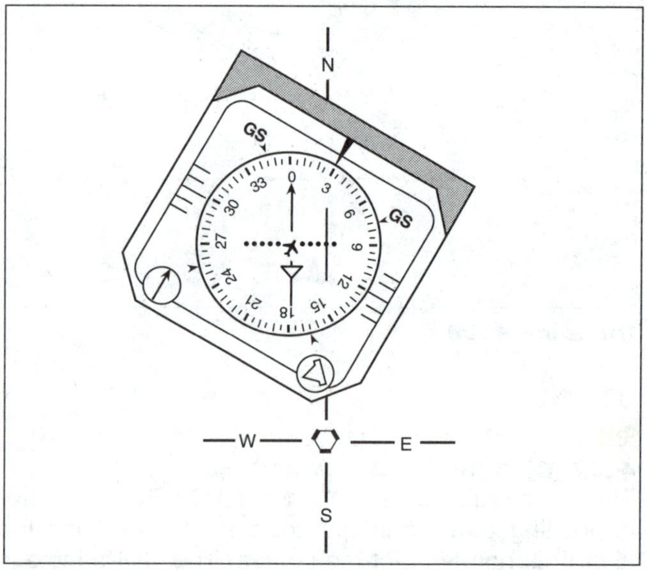

Question 4608

ALL

4609 [B] — (J17) — AIM ¶5-36

Determine the holding pattern by placing your pencil on the holding fix and dragging it on the holding radial given by ATC, then returning back to the fix. Then draw the pattern from the fix with turns in the direction specified. In this question, ATC has specified left-hand turns, which corresponds to holding pattern 1.

The entry procedure is based on the aircraft's heading. To determine which entry procedure to use, draw a line at a 70° angle from the holding fix, and cutting the outbound leg at about one-third its length. With a heading of 350°, we are in the largest piece of the pie, so a direct entry would be used. *See* the following figure.

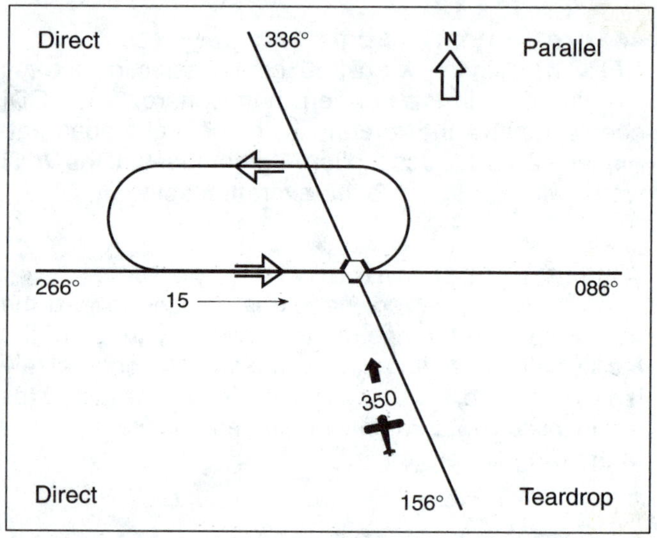

Question 4609

ALL

4610 [A] — (J17) — AIM ¶5-36

Determine the holding pattern by placing your pencil on the holding fix and dragging it on the holding radial given by ATC, then returning back to the fix. Then draw the pattern from the fix with turns in the direction specified. Holding east on the 090° radial with left turns means you will be south of R-090.

The entry procedure is based on the aircraft's heading. To determine which entry procedure to use, draw a line at a 70° angle from the holding fix, and cutting the outbound leg at about one-third its length. With a heading of 055°, we are in the second-largest piece of the pie, so a parallel entry would be used. *See* the figure below.

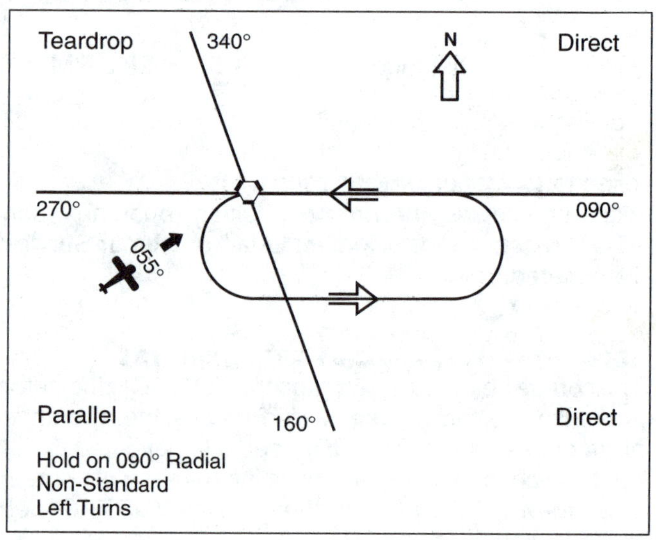

Question 4610

ALL
4611 [B] — (J17) — AIM ¶5-36
Determine the holding pattern by placing your pencil on the holding fix and dragging it on the holding radial given by ATC, then returning back to the fix. Then draw the pattern from the fix with turns in the direction specified. Holding south on the 180° radial with right turns means you will be east of R-180.

The entry procedure is based on the aircraft's heading. To determine which entry procedure to use, draw a line at a 70° angle from the holding fix, and cutting the outbound leg at about one-third its length. With a heading of 055°, we are in the largest piece of the pie, so a direct entry would be used. *See* the figure below.

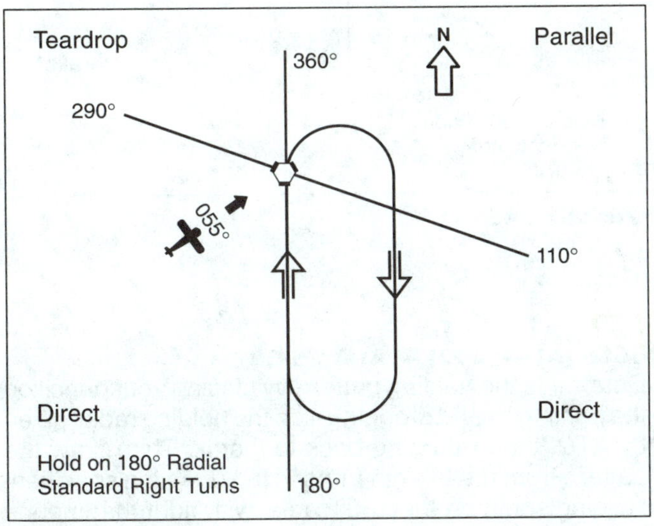

Question 4611

ALL
4612 [C] — (J17) — AIM ¶5-36
Determine the holding pattern is determined by placing your pencil on the holding fix and dragging it on the holding radial given by ATC, then returning back to the fix. Then draw the pattern from the fix with turns in the direction specified. Holding north on the 360° radial with left turns means you will be east of R-360.

The entry procedure is based on the aircraft's heading. To determine which entry procedure to use, draw a line at a 70° angle from the holding fix, and cutting the outbound leg at about one-third its length. With a heading of 055°, we are in the smallest piece of the pie, so a teardrop entry would be used. *See* the following figure.

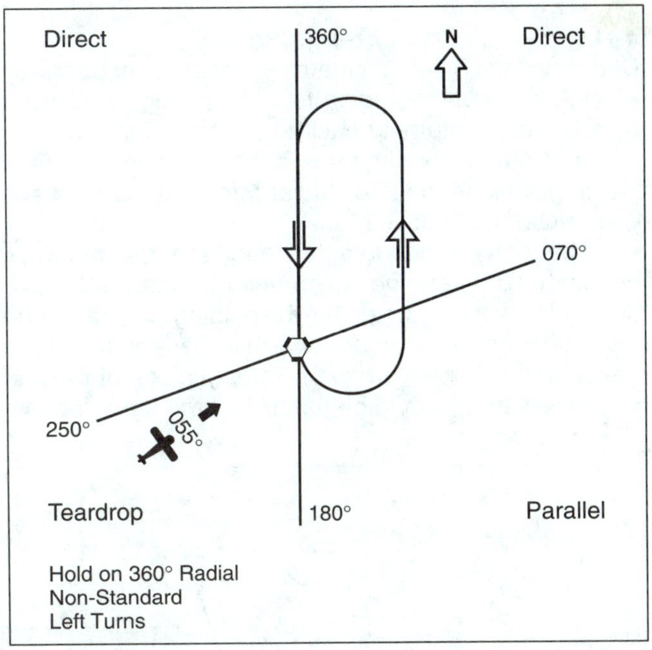

Question 4612

ALL
4613 [B] — (J17) — AIM ¶5-36
Determine the holding pattern by placing your pencil on the holding fix and dragging it on the holding radial given by ATC, then returning back to the fix. Then draw the pattern from the fix with turns in the direction specified. Holding west on the 270° radial with right turns means you will be south of R-270.

The entry procedure is based on the aircraft's heading. To determine which entry procedure to use, draw a line at a 70° angle from the holding fix, and cutting the outbound leg at about one-third its length. With a heading of 055°, we are in the largest piece of the pie, so a direct entry would be used. *See* the figure below.

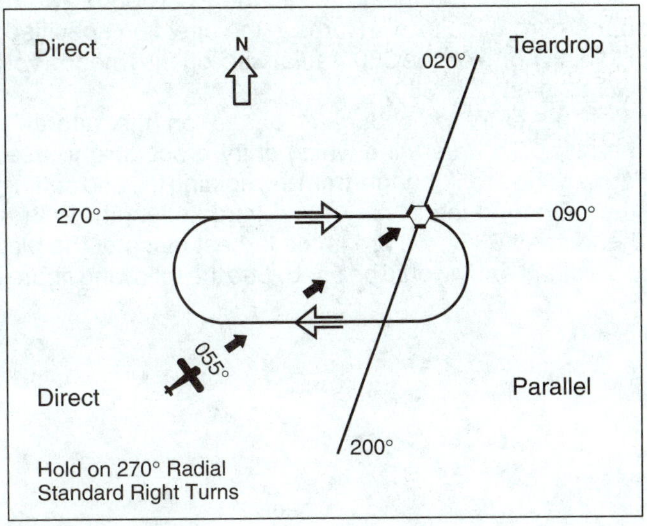

Question 4613

ALL
4614 [C] — (J17) — AIM ¶5-36
Determine the holding pattern by placing your pencil on the holding fix and dragging it on the holding radial given by ATC, then returning back to the fix. Then draw the pattern from the fix with turns in the direction specified. Holding west on the 270° radial with right turns means you will be south of R-270.

The entry procedure is based on the aircraft's heading. To determine which entry procedure to use, draw a line at a 70° angle from the holding fix, and cutting the outbound leg at about one-third its length. With a heading of 055°, we are in the largest piece of the pie, so a direct entry would be used. *See* the figure below.

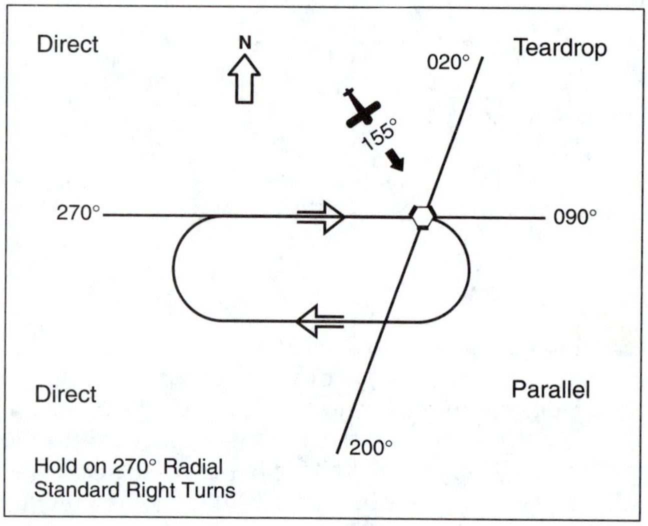

Question 4614

AIR
4615 [C] — (J17) — AIM ¶5-36
Determine the holding pattern by placing your pencil on the holding fix and dragging it on the holding radial given by ATC, then returning back to the fix. Then draw the pattern from the fix with turns in the direction specified. Holding north on the 360° radial with left turn means you will be east of R-360.

The entry procedure is based on the aircraft's heading. To determine which entry procedure to use, draw a line at a 70° angle from the holding fix, and cutting the outbound leg at about one-third its length. With a heading of 055°, we are in the largest piece of the pie, so a direct entry would be used. *See* the following figure.

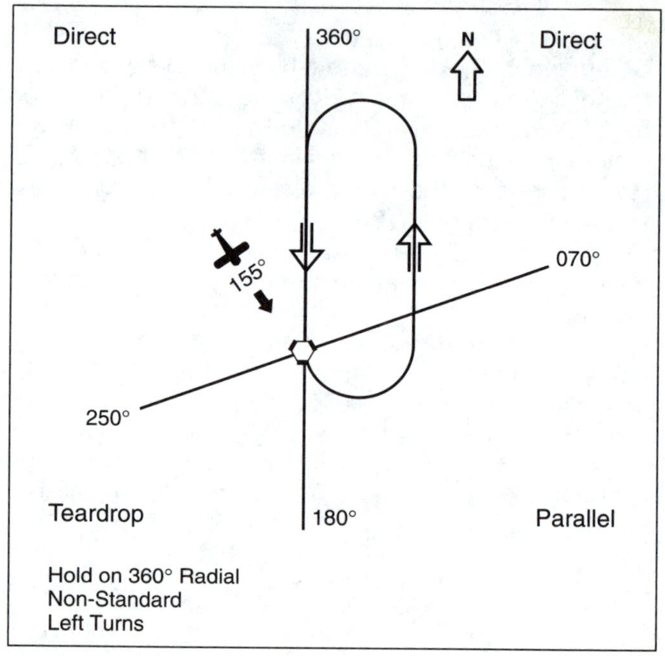

Question 4615

ALL
4616 [A] — (J17) — AIM ¶5-36
Determine the holding pattern by placing your pencil on the holding fix and dragging it on the holding radial given by ATC, then returning back to the fix. Then draw the pattern from the fix with turns in the direction specified. Holding south on the 180° radial with right turn means you will be east of R-180.

The entry procedure is based on the aircraft's heading. To determine which entry procedure to use, draw a line at a 70° angle from the holding fix, and cutting the outbound leg at about one-third its length. With a heading of 055°, we are in the smallest piece of the pie, so a teardrop entry would be used. *See* the figure below.

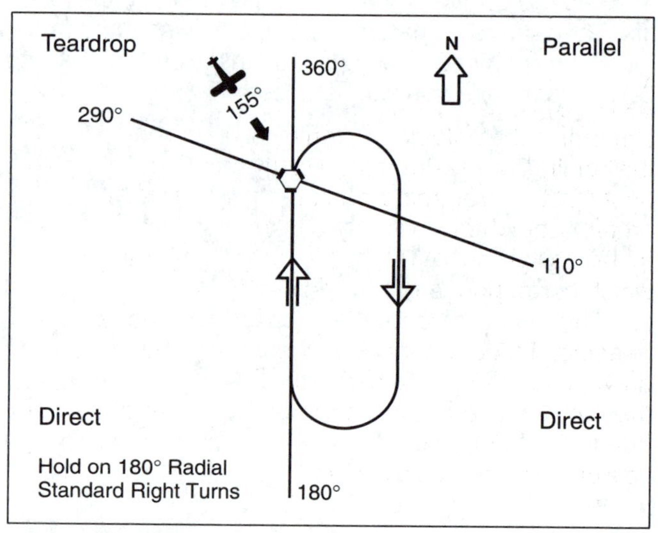

Question 4616

ALL
4617 [B] — (J17) — AIM ¶5-36
The maximum airspeed above 14,000 feet is 175 KIAS for all propeller-driven aircraft, and 265 KIAS for civil turbojet aircraft.

ALL
4618 [C] — (J17) — AIM ¶5-36
Determine the holding pattern by placing your pencil on the holding fix and dragging it on the holding radial given by ATC, then returning back to the fix. Then draw the pattern from the fix with turns in the direction specified. In this question, ATC has specified left-hand turns, which corresponds to holding pattern 1.

The entry procedure is based on the aircraft's heading. To determine which entry procedure to use, draw a line at a 70° angle from the holding fix, and cutting the outbound leg at about one-third its length. With a heading of 350°, we are in the largest piece of the pie, so a direct entry would be used. *See* the figure below.

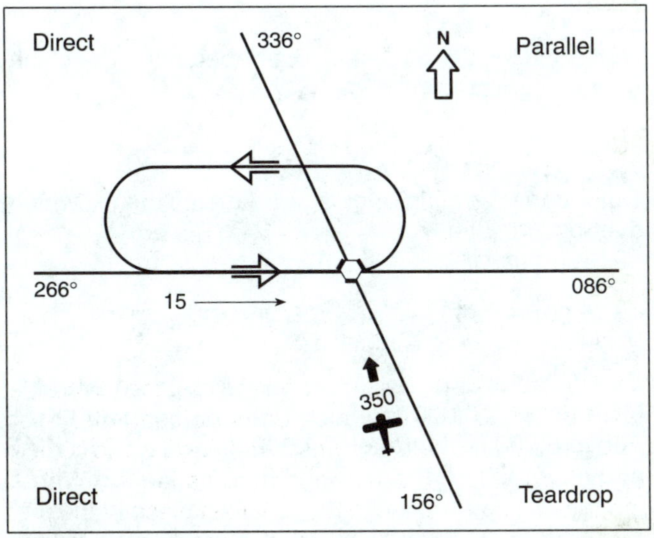

Question 4618

ALL
4619 [B] — (J17) — AIM ¶5-36
Determine the holding pattern by placing your pencil on the holding fix and dragging it on the holding radial given by ATC, then returning back to the fix. Then draw the pattern from the fix with turns in the direction specified. In this question, ATC has specified left-hand turns, which corresponds to holding pattern 2.

The entry procedure is based on the aircraft's heading. To determine which entry procedure to use, draw a line at a 70° angle from the holding fix, and cutting the outbound leg at about one-third its length. With a heading of 350°, we are in the largest piece of the pie, so a direct entry would be used. *See* the following figure.

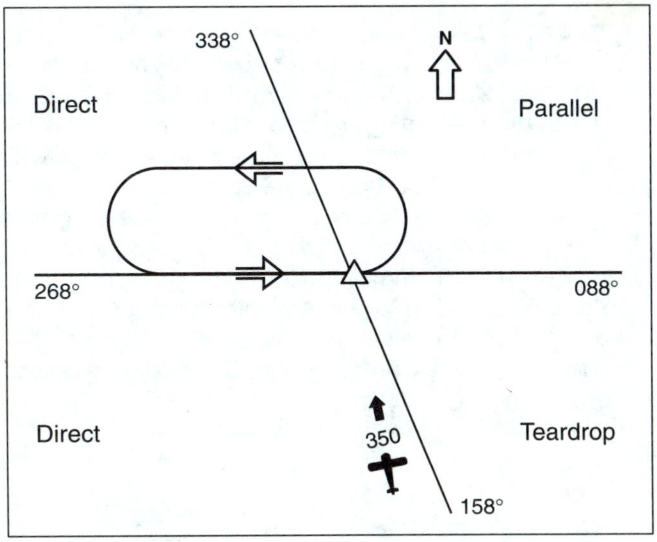

Question 4619

ALL
4620 [C] — (J17) — AIM ¶5-36
Outbound leg timing begins over/abeam the fix, whichever occurs later. If the abeam position cannot be determined, start timing when the turn to the outbound is completed.

ALL
4621 [C] — (J17) — AIM ¶5-36
Determine the holding pattern by placing your pencil on the holding fix and dragging it on the holding radial given by ATC, then returning back to the fix. Then draw the pattern from the fix with turns in the direction specified. In this question, ATC has specified left-hand turns.

The entry procedure is based on the aircraft's heading. To determine which entry procedure to use, draw a line at a 70° angle from the holding fix, and cutting the outbound leg at about one-third its length. With a heading of 055°, we are in the second-largest piece of the pie, so a parallel entry would be used. *See* the figure below.

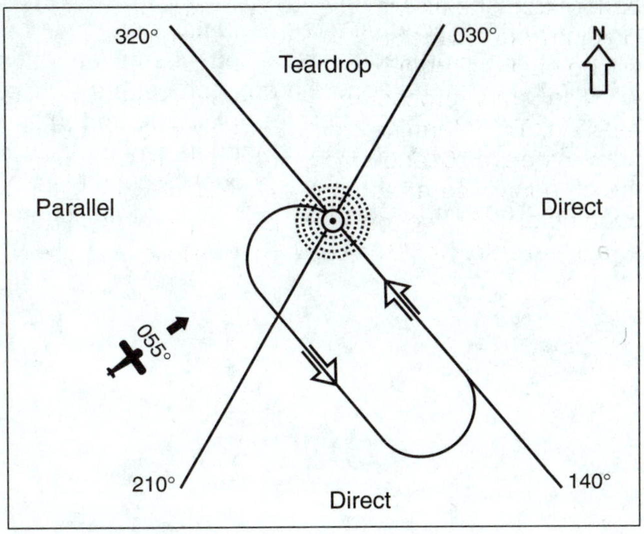

Question 4621

ALL
4622 [B] — (J17) — AIM ¶5-36
Determine the holding pattern by placing your pencil on the holding fix and dragging it on the holding radial given by ATC, then returning back to the fix. Then draw the pattern from the fix with turns in the direction specified. In this question, ATC has specified left-hand turns.

The entry procedure is based on the aircraft's heading. To determine which entry procedure to use, draw a line at a 70° angle from the holding fix, and cutting the outbound leg at about one-third its length. With a heading of 055°, we are in the smallest piece of the pie, so a teardrop entry would be used. *See* the figure below.

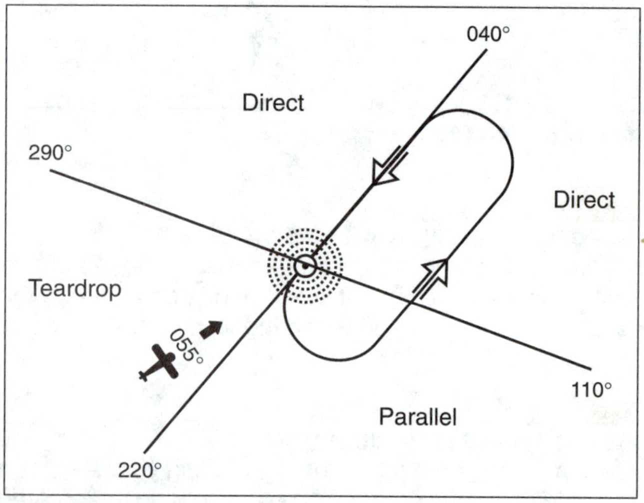

Question 4622

ALL
4623 [A] — (J17) — AIM ¶5-36
Determine the holding pattern by placing your pencil on the holding fix and dragging it on the holding radial given by ATC, then returning back to the fix. Then draw the pattern from the fix with turns in the direction specified. In this question, ATC has not specified which way to turn, so the standard direction is right turns.

The entry procedure is based on the aircraft's heading. To determine which entry procedure to use, draw a line at a 70° angle from the holding fix, and cutting the outbound leg at about one-third its length. With a heading of 055°, we are in the largest piece of the pie, so a direct entry would be used. *See* the following figure.

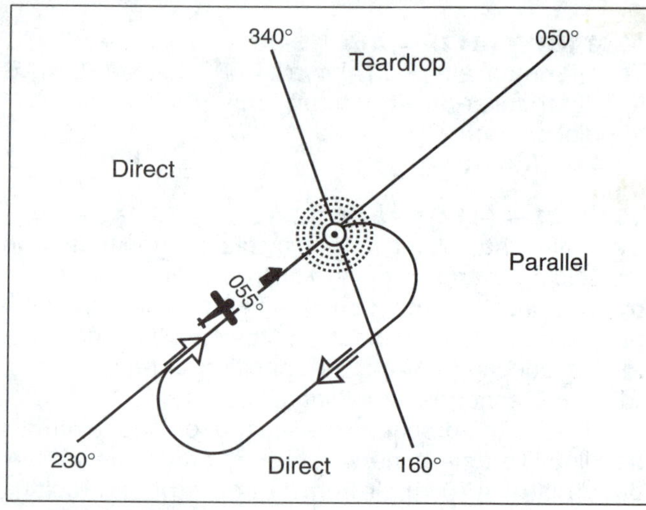

Question 4623

ALL
4624 [A] — (J17) — AIM ¶5-36
Outbound leg timing begins over/abeam the fix, whichever occurs later.

ALL
4625 [C] — (I12) — AIM ¶5-36
Outbound leg timing begins over/abeam the fix, whichever occurs later.

AIR
4626 [B] — (J17) — AIM ¶5-36(j)(2)
Maximum Holding Airspeeds:

Civil Turbojet	Maximum Airspeed
MHA through 6,000 feet	200 KIAS
Above 6,000 feet through 14,000	210 KIAS
Above 14,000 feet	265 KIAS

ALL
4627 [B] — (J18) — AIM ¶5-48(a)
If only one missed approach procedure is available, the reported ceiling and visibility minimums must be equal to or greater than the highest prescribed circling minimums for the instrument approach procedure.

ALL
4628 [B] — (J18) — AIM ¶5-48(a)
Prior to conducting timed approaches from a holding fix, a control tower must be in operation at the airport where the approaches are to be conducted.

ALL
4629 [C] — (J18) — AIM ¶5-48
Each pilot in an approach sequence will be given advance notice as to the time they should leave the holding point on approach to the airport. When a time to leave the holding point has been received, the pilot should adjust the flight path to leave the fix as closely as possible to the designated time.

ALL
4630 [A] — (B08) — FAR §91.169(c)(1)
If the pilot elects to proceed to the selected alternate airport, the alternate ceiling and visibility minimums are disregarded, and the published landing minimums are applicable for the new destination, utilizing facilities as appropriate to the procedure. In other words, the alternate airport becomes a new destination, and the pilot uses the landing minimums appropriate to the type of procedure selected.

ALL
4631 [C] — (J18) — AIM ¶5-58(c)
If visual reference is lost while circling to land from an instrument approach, the missed approach specified for that particular procedure must be followed (unless an alternate missed approach procedure is specified by ATC). To become established on the prescribed missed approach course, the pilot should make an initial climbing turn toward the landing runway and continue the turn until he/she is established on the missed approach course.

ALL
4632 [C] — (J18) — AIM ¶5-47(a)(2)
When the approach procedure involves a procedure turn, a maximum speed no more than 250 KIAS should be observed, and the turn should be executed within the distance specified in the profile view.

ALL
4633 [C] — (J19) — AIM ¶4-86(d)
ATC may not authorize VFR-On-Top/VFR conditions operations unless the pilot requests the VFR operation, or a clearance to operate in VFR conditions will result in noise abatement benefits where part of the IFR departure route does not conform to an FAA-approved noise abatement route or altitude.

ALL
4634 [C] — (B08) — AIM ¶4-93(b)
During climbs and descents in flight conditions which permit visual detection of other traffic, pilots should execute gentle banks, left and right at a frequency which permits continuous visual scanning of the airspace about them.

ALL
4635 [B] — (I10) — AC 61-27C, page 172
When the glide slope becomes inoperative on a precision approach, the approach, if continued, becomes a localizer approach. Looking at the minimums section of the chart for S-LOC, 12L indicates an MDA of 1360 would be applicable for all categories assuming a straight-in landing.

ALL
4636 [A] — (I10) — AC 61-27C, page 184
The absence of the procedure turn barb in the Plan View indicates that a procedure turn is not authorized for that procedure.

ALL
4637 [C] — (I10) — AC 61-27C, page 189
If the pilot elects to proceed to the selected alternate airport, the alternate ceiling and visibility minimums are disregarded, and the published landing minimums are applicable for the new destination, utilizing facilities as appropriate to the procedure. In other words, the alternate airport becomes a new destination, and the pilot uses the landing minimums appropriate to the type of procedure selected.

ALL
4638 [B] — (J16) — AIM ¶5-25
The use of a SID requires pilot possession of at least the textual description of the approved effective SID.

LTA
4639 [C] — (J15) — AIM ¶5-7(c)(2)(c)
For operation off established airways below 18,000 feet MSL, pilots should use aids not more than 80 NM apart. These aids are depicted on Enroute Low Altitude Charts.

ALL
4640 [B] — (J17) — AIM ¶5-40
Standard Terminal Arrival Routes have been established to simplify clearance delivery procedures for arriving aircraft at certain areas having high density traffic.

ALL
4641 [C] — (J18) — AIM ¶5-73(a)(3)(b)
Upon receipt of an approach clearance while on an unpublished route or being radar vectored, the pilots must maintain the last assigned altitude until established on a segment of a published route or IAP, at which time published altitudes apply.

ALL
4642 [B] — (J33) — Pilot/Controller Glossary
The Final Approach Fix (FAF) is the fix from which the final approach (IFR) to an airport is executed and which identifies the beginning of the final approach segment. On government charts it is designated by the Maltese Cross for non-precision approaches and by the lightning bolt symbol or glide slope intercept for precision approaches. The profile view of FAA Figure 119 depicts the lightning bolt symbol pointing from the glide slope intercept altitude of 2200 to Roman LOM which is also the non-precision FAP.

LTA
4643 [C] — (J33) — AIM ¶4-86(e)
If the ATC clearance assigns "VFR conditions on top" the pilot shall maintain an appropriate VFR altitude or flight level as prescribed by FAR §91.159 (which describes the VFR hemispheric rule).

LTA
4644 [C] — (J14) — AIM ¶4-82(d)(3)
The term "cruise" may be used instead of "maintain" to assign a block of airspace, to a pilot, from the minimum IFR altitude up to and including the altitude specified in the cruise clearance. The pilot may level off at any intermediate altitude within this block of airspace. Climb/descent within the block is to be made at the discretion of the pilot. However, once the pilot starts descent and verbally reports leaving an altitude in the block, the pilot may not return to that altitude without additional ATC clearance.

ALL
4645 [C] — (J35) — Enroute Low Altitude Chart Legend
In the absence of a minimum crossing altitude, you would cross at the MEA. Looking at GYMME intersection, the minimum enroute altitude eastbound is 7,000 feet and westbound is 6,500 feet. Since the question asks the altitude while traveling west to east, 7,000 is the minimum altitude to be flown crossing GYMME.

ALL
4646 [A] — (J35) — Enroute Low Altitude Chart Legend
Magnetic bearings are read directly from the VOR indicator. You can easily identify Angoo intersection while on the radial of YKM with one VOR. Identify the intersecting radial of DLS simply by alternately tuning the single VOR to each station.

ALL
4647 [C] — (J35) — Enroute Low Altitude Chart Legend
TROTS intersection has a flag with an "x" in it, which means that there is a Minimum Crossing Altitude (MCA). Under the intersection name the MCA is shown as 11,500 on V468 when proceeding to the northeast.

ALL
4648 [B] — (J42) — U.S. Terminal Procedures, IAP Legend
The FAF identifies the beginning of the final approach segment. On precision approaches a lightning bolt symbol indicates the FAF. For DEN ILS RWY 35R, the FAF intercept altitude is 7,500 feet.

ALL
4649 [A] — (J42) — U.S. Terminal Procedures, IAP Legend
Minimum Safe Altitude (MSA) is defined as altitudes depicted on approach charts which provide at least 1,000 feet of obstacle clearance within a 25-mile radius upon which the procedure is predicated. In this case, it is the Denver VORTAC. MSA DEN 25 NM.

ALL
4650 [B] — (J42) — U.S. Terminal Procedures, IAP Legend
Glide slope intercept altitude is defined as the minimum altitude to intercept the glide slope/path on a precision approach. The intersection of the published intercept altitude with the glide slope/path is designated on government charts by the lightning bolt symbol which is 2,500 feet MSL for the ILS RWY 30R procedure at DSM.

ALL
4651 [B] — (J42) — AC 61-27C, page 172
When the ILS glide slope is inoperative or not utilized, the published straight-in localizer minimums apply. In this case, the minimum is 1,320 feet for the S-LOC 30R.

ALL
4652 [C] — (J42) — U.S. Terminal Procedures, IAP Legend
The bottom right-hand portion of the IAP, above the time and speed table, states "FAF to MAP 5.2 NM."

ALL
4653 [B] — (B11) — AC 61-27C, page 180
VOR/DME procedure number means that both operative VOR and DME receivers and ground equipment in normal operation are required to use the procedure. In the VOR/DME procedure, when either the VOR or DME is inoperative, the procedure is not authorized.

ALL
4654 [B] — (J42) — Pilot/Controller Glossary
The Minimum Safe Altitude (MSA) is identified in the bottom right quadrant of the VOR/DME-A procedure at 7D3. The symbol in the middle of the MSA circle indicates the type of navigation aid the altitude is based. The statement above the circle, "MSA HIC 25 NM," indicates White Cloud VORTAC (HIC) as the facility identifier for the MSA.

Editorial Note: The FAA question currently refers to Figure 13. This question should use FAA Figure 123, which is the figure required to derive the correct answer.

ALL
4655 [A] — (J18) — AC 61-27C, page 184
When a teardrop procedure turn is depicted and a course reversal is required, this type turn must be executed.

ALL
4656 [A] — (B97) — AC 61-27C, page 184
A procedure turn is the maneuver prescribed when it is necessary to reverse direction to establish the aircraft inbound on an intermediate or final approach course. The maneuver must be completed within the distance specified in the profile view. DME and timing are the only practical methods of determining that you remain within the 10-mile boundary.

ALL
4657 [B] — (B97) — AIM ¶5-47(a)(3)
Once the pilot is cleared for the approach and established on part of the published route or procedure, the pilot may descent to the published altitude. LNK is the IAF, and upon crossing LNK the pilot may commence the approach.

ALL
4658 [C] — (B97) — AIM ¶5-46(e)
When cleared for an approach, the pilot has authorization to execute the specific approach cleared for. When cleared for the S-LOC 17R approach from over Touhy Intersection, the approach specifies an interception of the localizer with no procedure turn allowed (NoPT).

ALL
4659 [B] — (B97) — U.S. Terminal Procedures, IAP Legend
Under the DME minimums for a circling approach for RWY 31, Category C aircraft are required to use the MDA of 860 feet MSL and 1-1/2 statute mile visibility.

ALL
4660 [C] — (B97) — Pilot/Controller Glossary
A straight-in approach is one begun without first having executed a procedure turn, not necessarily completed with a straight-in landing or made to straight-in landing minimums.

ALL
4661 [A] — (B97) — U.S. Terminal Procedures, IAP Legend
The minimum for a circling approach in CAT C aircraft without DME is 880 and 1-1/2. With DME it is 860 and 1-1/2. The only change is the 20-foot decrease in the MDA.

ALL
4662 [A] — (B97) — AC 61-27C, page 183
Follow the published feeder route descending to 2,700 outbound on ZZV R-246 "NoPT." From CRISY proceed to CASER LOM via the 277° bearing TO at 2,700. Passing CASER, begin timing and descend to the MDA of 1,620 on the 280° bearing FROM.

ALL
4663 [B] — (J01) — AIM ¶1-7(f)
VOR/DME, VORTAC, ILS/DME and LOC/DME facilities are identified by synchronized identifications which are transmitted on a time share basis when either the VOR or the DME is inoperative. It is important to recognize which identifier is retained for the operative facility. A single coded identification with a repetition interval of approximately 30 seconds indicates that the DME is operative. DME is identified by a coded tone modulated at 1350 Hz.

ALL
4664 [A] — (J01) — AIM ¶1-10(e)
When installed with the ILS and specified in the approach procedure, DME may be used in lieu of the OM.

ALL
4665 [C] — (J01) — AIM ¶1-17
Some installed aircraft Loran-C are approved for IFR enroute operations. This can be confirmed by checking the supplement section of the Airplane Flight Manual or an FAA Form 337.

ALL
4666 [B] — (I07) — AC 61-27C, page 137
Full-scale deflection of the CDI represents 10° or more or a five-dot deflection to one side or the other from center. This will be indicated when the needle moves from the center of the scale to either far side of the scale.

ALL
4667 [A] — (J18) — AIM ¶5-58(b)
When an early missed approach is executed, pilots should fly the IAP as specified on the approach plate to the missed approach point at or above the MDA or DH before executing a turning maneuver to ensure terrain/obstacle clearance.

ALL
4668 [C] — (J17) — AIM ¶5-48
If timed approaches are in use, aircraft are expected to proceed inbound on final approach when clearance is received. Pilots are responsible for advising ATC if they elect to make additional circuits to lose altitude or become established on course. There is no AIM language regarding ATC approval, but it is implied.

ALL
4669 [A] — (J01) — AIM ¶1-7
DME/TACAN channels are associated with LOC frequencies and are on the plan view of the approach chart.

ALL
4670 [A] — (J18) — AIM ¶5-53(b)
Pilots will monitor tower frequency for advisories and instructions when simultaneous approaches are in progress.

ALL
4671 [A] — (J18) — AIM ¶5-47(b)(3)
Holding pattern course reversal is not required with radar vectoring or when "NoPT" is shown on the approach course.

ALL
4672 [A] — (J18) — AIM ¶5-44(a)(3)
A pilot adhering to the altitudes, flight paths, and weather minimums depicted on the IAP chart is assured terrain and obstruction clearance and runway or airport alignment during approach for landing.

RTC
4673 [C] — (I10) — FAR §97.3(d)(1)
Choppers use Category A MDA or DH with half the published visibility (but equal to or not less than 1,200 RVR or 1/4 mile). The visibility minimums for the VOR RWY 36 approach, Category A are one mile, so the helicopter minimum is one-half mile visibility.

ALL
4674 [C] — (J01) — AIM ¶1-7(f)
A DME tone is heard once for each three or four VOR/LOC tones. A single code each 30 seconds indicates DME is operative. No ident indicates the NAVAID is unreliable even though signals are received.

ALL
4675 [A] — (I10) — AIM ¶5-36
The missed approach takes you outbound on the PUC 127° radial then right turn direct to PUC on a no-wind course of 310 or more, which will be a direct entry.

RTC
4676 [C] — (I10) — FAR §97.3(d)(1)
Choppers use category A MDA or DH with half the published visibility (but equal to or not less than 1,200 RVR or 1/4 mile). The MDA for a straight-in VOR RWY 36 approach, Category A, is 6,400 feet MSL.

ALL
4677 [B] — (I10) — AC 61-27C, page 168
The overhead view shows 8,000 while on the arc. Once established on R-186, the pilot can proceed down to 6,400. After the 4 DME fix the pilot can proceed down to 6,180, which is the decision height. With the runway environment in sight, the pilot may continue to descend for the landing.

ALL
4678 [A] — (I10) — Pilot/Controller Glossary
Minimum Safe Altitudes (MSA) are published for emergency use on Approach Procedure Charts, utilizing NDB or VOR type facilities. The altitude shown provides at least 1,000 feet of clearance above the highest obstacle in the defined sector to a distance of 25 NM from the facility upon which the procedure is predicated. As many as four sectors may be depicted with different altitudes for each sector displayed in rectangular boxes in the plan view of the chart.

RTC
4679 [A] — (I10) — FAR §97.3(d)(1)
Choppers use Category A MDA or DH with half the published visibility (but equal to or not less than 1,200 RVR or 1/4 mile). The visibility minimum for RNAV RWY 36, Category A, is 1 mile. So for helicopters, the minimum is one-half mile. The weather minimums are below this, so the helicopter is not permitted to continue the approach and land.

ALL
4680 [A] — (I10) — AIM ¶5-36
DME holding patterns and procedure turn entries follow the usual rules except that distances (nautical miles) are used in lieu of time values. The DME will indicate 4 DME (see the holding pattern in the plan view) from LABER when it is time to turn inbound.

ALL
4681 [C] — (J17) — AIM ¶5-36
The missed approach procedure is to fly direct to Bendy WPT. The holding pattern depicted is left-hand turns, so the pilot should execute a teardrop turn to enter the holding pattern.

ALL
4682 [A] — (I10) — AC 61-27C, page 137
The MAP is a waypoint and the TO/FROM indicator will show passage of the "phantom station."

ALL
4683 [C] — (I10) — U.S. Terminal Procedures, IAP Legend
The bottom line "113.9 LIT 198.0°-8" of the information box extending from LABER waypoint gives the frequency, the identifier (Little Rock VOR), and the radial and distance from the reference facility (198°, 8 NM).

ALL
4684 [C] — (I10) — AC 61-27C, page 121
An approved RNAV receiver is required for RNAV approaches.

ALL
4685 [B] — (I07) — AIM ¶1-10(c)
The LDA is not aligned with the runway. Straight-in minimums may be published where alignment does not exceed 30° between the course and runway. Circling minimums only are published where this alignment exceeds 30°.

ALL
4686 [A] — (B97) — AC 61-27C, page 184
The profile view for LDA RWY 6 approach has a note in the upper left-hand corner which states "Remain within 10 NM." The 10 NM is in reference to the Clamm INT. The procedure turn is depicted north of the approach with a right turn, so the pilot must remain on the north side of the approach course.

ALL
4687 [B] — (B97) — U.S. Terminal Procedures, IAP Legend
See the note under in the remarks section of the instrument approach. Circling not authorized northwest of runway 6/24.

ALL
4688 [A] — (B97) — U.S. Terminal Procedures, IAP Legend
The single asterisk alongside the glide slope intercept altitude in the profile view refers you to the note below specifying 4,200 feet MSL when glide slope is not in use.

ALL
4689 [A] — (B97) — U.S. Terminal Procedures, IAP Legend
The LDA Rwy GS 6 approach is considered a precision approach; therefore, the missed approach point is the DH which is 1,540 feet.

ALL
4690 [A] — (B97) — U.S. Terminal Procedures, IAP Legend
The landing minima format for LOC Rwy 5-18 provides the following information: 5560 is the MDA, 50 is the RVR (in 100's of feet), 319 is the HAT, and (300-1) are minimums applicable to military pilots.

ALL
4691 [C] — (B97) — U.S. Terminal Procedures, IAP Legend
The HAT is the height of the Decision Height or Minimum Descent Altitude above the highest runway elevation in the touchdown zone (first 3,000 feet of the runway). HAT is published on instrument approach charts in conjunction with all straight-in minimums. It can be located in the landing minima section of the approach chart. For LOC RWY S-18 approach, the HAT is 319 AGL for all categories of aircraft.

ALL
4692 [B] — (J18) — AIM ¶5-52
Parallel ILS approaches provide aircraft a minimum of two miles separation between successive aircraft on the adjacent localizer course.

ALL
4693 [B] — (B97) — U.S. Terminal Procedures, IAP Legend
Looking in the upper left-hand corner of the approach plate, 127.4 and 288.1 are the approach frequencies when coming from the north. 118.3, 119.5, and 257.8 are the Denver tower frequencies.

ALL
4694 [A] — (B97) — FAR §91.169(c)(1)
If an instrument approach procedure has been published for that airport, the alternate airport minimums are specified in that procedure (by a black triangle with an A inside), or if none are so specified, the following minimums apply: (1) precision approach procedure—ceiling 600 feet and visibility 2 SM, and (2) non-precision approach procedure—ceiling 800 feet and visibility 2 SM.

ALL
4695 [A] — (B97) — U.S. Terminal Procedures, IAP Legend
A VOR/LOC is required for a localizer approach. Stapleton International is in Class B airspace so a transponder with altitude encoding is also required.

ALL
4696 [B] — (B97) — AIM ¶5-47(a)(4)
A procedure turn need not be established when an approach can be made from a properly aligned holding pattern. In such cases, the holding pattern is established over an intermediate fix or a final approach fix. The holding pattern maneuver is completed when the aircraft is established on the inbound course after executing the appropriate entry.

RTC
4697 [A] — (B97) — U.S. Terminal Procedures, IAP Legend
Helicopters may use airplane approach procedures using the Category A Minimum Descent Altitude (MDA) or Decision Height (DH). The required visibility minimum may be reduced to one-half the published visibility minimum for Category A aircraft, but it must not in any case be reduced to less than one-quarter mile or 1,200 feet RVR.

1,160	MDA
+ 40	from note #2 in plan view
1,200	New MDA

ALL
4698 [A] — (B97) — AIM ¶5-36
The pilot will use a direct entry for the holding pattern at Riverside Municipal. The missed approach procedures route the pilot directly to R-078 where they can join the depicted holding pattern.

ALL
4699 [C] — (B10) — FAR §91.175(k)
A compass locator or precision radar may be substituted for the outer or middle marker. DME, VOR, or nondirectional beacon fixes authorized in the standard instrument approach procedure, or surveillance radar may be substituted for the outer marker. With the marker beacon receiver inoperative, the Swan Lake INT (Localizer and R-327 Paradise) can be substituted for the outer marker.

ALL
4700 [C] — (B97) — AC 61-27C, page 186
A step-down fix may be provided on the final, i.e., between the final approach fix and the airport for the purpose of authorizing a lower MDA after passing an obstruction. This step-down fix may be made by an NDB bearing, fan marker, radar fix, radial from another VOR, or by a DME. The Agnes INT needs to be identified for the step-down fix, in order to proceed to the DH 1160.

ALL
4701 [C] — (B97) — AC 61-27C, page 186
A pilot shall maintain the last altitude assigned to that pilot until the aircraft is established on a segment of a published route or instrument approach procedure, unless a different altitude is assigned by ATC. Once cleared from Seal Beach VORTAC, the pilot shall descend and maintain 3,000 feet to Jaser INT, then descend to 2,500 feet once established on the localizer course inbound, then intercept and maintain the glide slope to the decision height (991 feet).

ALL
4702 [A] — (I07) — AIM ¶1-10 and ¶1-11
An LDA is of comparable utility and accuracy to a localizer but is not part of a complete ILS. The LDA usually provides a more precise approach course than the SDF, which may have a course width of 6° or 12°. The LDA course widths are from 3° to 6°.

ALL
4703 [A] — (I07) — AIM ¶1-10
An LDA is of comparable utility and accuracy to a localizer but is not part of a complete ILS. The LDA usually provides a more precise approach course than the SDF, which may have a course width of 6° or 12°. The LDA is not aligned with the runway.

ALL
4704 [B] — (I07) — AIM ¶1-11
The SDF signal emitted from the transmitter is either 6° or 12° wide as needed to provide maximum flyability and optimum course quality.

ALL
4705 [B] — (I07) — AIM ¶1-11
The SDF course may not be aligned with the runway and the course may be wider than an ILS localizer course. The ILS localizer course width varies between 3° and 6° and is tailored to provide 700 feet at the threshold (full-scale limits).

ALL
4706 [B] — (B10) — FAR §91.175(c)(3)(vi)
As long as the flight visibility in not less than the visibility prescribed for the approach, the pilot should continue the approach and use the VASI.

ALL
4707 [C] — (J27) — AIM ¶7-53
A tailwind condition can move the vortices of the preceding aircraft forward into the touchdown zone. The light quartering tailwind requires maximum caution. Pilots should be alert to large aircraft upwind from their approach and takeoff flight paths.

ALL
4708 [C] — (J27) — AIM ¶7-52
The strength of the VORTEX is governed by the weight, speed, and shape of the wing of the generating aircraft. The greatest VORTEX strength occurs when the generating aircraft is heavy, clean, and slow.

ALL
4709 [B] — (J27) — AIM ¶7-53
A crosswind will decrease the lateral movement of the upwind vortex and increase the movement of the downwind vortex. Thus, a light wind of 3 to 7 knots could result in the upwind vortex remaining in the touchdown zone for a period of time and hasten the drift of the downwind vortex toward another runway.

ALL
4710 [C] — (J27) — AIM ¶7-55
When landing behind a large aircraft on the same runway, stay at or above the large aircraft's final approach flight path. Note the touchdown point and land beyond it.

ALL
4711 [C] — (B10) — AC 61-27C, page 212
An ASR (Airport Surveillance Radar) approach is conducted by surveillance radar and provides navigational guidance in azimuth only. This type of approach may be made to any airport or heliport having an approved surveillance approach.

RTC
4712 [A] — (B09) — AIM ¶5-61
Pilots operating in accordance with an IFR flight plan, provided they are clear of clouds and have at least 1 mile flight visibility and can reasonably expect to continue to the destination airport in those conditions, may request ATC authorization for a contact approach. Controllers may authorize a contact approach provided the contact approach is specifically requested by the pilot. ATC cannot initiate this approach.

RTC
4713 [B] — (B10) — FAR §91.175(c)(i)
Where a DH or MDA is applicable, no pilot may operate an aircraft at any airport below the authorized MDA or continue an approach below the authorized DH unless the aircraft is continuously in a position from which a descent to a landing on the intended runway can be made at a normal rate of descent using normal maneuvers.

ALL
4714 [A] — (J33) — AC 61-27C, page 173
If it is necessary to maneuver at speeds in excess of the upper limit of the speed range for each category, the minimum for the next higher approach category should be used.

ALL
4715 [A] — (B97) — Instrument Approach Procedure Legend
The instrument approach commences at the Initial Approach Fix (IAP). Initial Approach Fixes are specifically identified with "IAF."

ALL
4716 [A] — (B10) — FAR §91.175(h)
If RVR minimums for takeoff or landing are prescribed in an instrument approach procedure, but RVR is not reported for the runway of intended operation, the RVR minimum shall be converted to ground visibility shall be the visibility minimum for takeoff or landing on that runway.

ALL
4717 [B] — (J33) — AC 61-27C, page 173
Aircraft approach category means a grouping of aircraft based on a speed of 1.3 V_{SO} and the maximum certificated landing weight are those values as established for the aircraft by the certificating authority of the country of registry.

ALL
4718 [A] — (J18) — AIM ¶5-59 and ¶5-61
Visual approaches are initiated by ATC to reduce pilot/controller workload and expedite traffic by shortening flight paths to the airport. Contact approaches are initiated by the pilot and are provided if the pilot will remain clear of clouds and have at least 1 mile flight visibility.

ALL
4719 [C] — (B10) — FAR §91.169(c)
If the pilot elects to proceed to the selected alternate airport, the alternate ceiling and visibility minimums are disregarded, and the published landing minimum is applicable for the new destination, utilizing facilities as appropriate to the procedure. In other words, the alternate airport becomes a new destination, and the pilot uses the landing minimum appropriate to the type of procedure selected.

ALL
4720 [B] — (K04) — AC 00-54A
When on an approach in a tailwind condition that shears into a calm wind or headwind, initially the IAS and pitch will increase and the aircraft will balloon above the glide slope. Power should first be reduced to correct this condition. Power should be further reduced as the aircraft proceeds through the wind shear, then increased.

ALL
4721 [B] — (I10) — AC 00-54A
The vertical speed indicator (VSI) shows rate of change of altitude. The faster an aircraft descends, the greater the rate of descent shown on the VSI. The slower an aircraft descends, the slower the rate of descent shown on the VSI. The vertical speed required is dependent on the ground speed when flying on a fixed glide slope. Therefore, when ground speed increases, the rate of descent must increase.

RTC
4722 [B] — (J42) — FAR §97.3(b)
Helicopters use Category A procedures, which requires a speed less than 91 knots.

RTC
4723 [A] — (J42) — FAR §97.3(d)(1)
When helicopters use instrument flight procedures designed for fixed-wing aircraft, approach Category A approach minima shall apply, regardless of helicopter weight.

RTC
4724 [C] — (J42) — FAR §97.3(d)(1)
The required visibility minimum may be reduced to one-half the published visibility minimum for Category A aircraft, but must not in any case be reduced to less than one-quarter mile or 1,200 feet RVR.

ALL
4725 [C] — (J19) — AIM ¶5-78(a)(3)
When complying with speed adjustments assignment, maintain an indicated airspeed within ±10 knots or .02 Mach number of the specified speed.

ALL
4726 [B] — (J18) — AIM ¶5-43(d)
Radar vectors and altitude/flight levels will be issued as required for spacing and separating aircraft; therefore, you must not deviate from the headings as issued by approach control. You will normally be informed when it becomes necessary to vector you across the final approach course for spacing or other reasons. If you determine that approach course crossing is imminent and you have not been informed that you will be vectored across it, you should question the controller. You should not turn inbound on the final approach course unless you have received an approach clearance.

ALL
4727 [C] — (K04) — AC 00-54A
When an aircraft that is flying with a tailwind suddenly encounters calm air, it is the same as encountering a headwind. Initially, the IAS and pitch will increase and the aircraft will balloon above the glide slope.

ALL
4728 [C] — (J18) — AIM ¶5-42
Where radar is approved for approach control service, it is used not only for radar approaches (ASR and PAR), but is also used to provide vectors in conjunction with published instrument approach procedures predicated on radio navaids such as ILS, VOR, and NDB.

ALL
4729 [C] — (J10) — AIM ¶1-10(g)
Compass locators transmit two-letter identification groups. The outer locator transmits the first two letters of the localizer identification group, and the middle locator transmits the last two letters.

ALL
4730 [C] — (J01) — AIM ¶1-10(g)
Compass locators transmit two-letter identification groups. The outer locator transmits the first two letters of the localizer identification group, and the middle locator transmits the last two letters.

ALL
4731 [A] — (I12) — AC 61-27C, page 172
If more than one component is inoperative, each minimum is raised to the highest minimum required by any single component that is inoperative.

ALL
4732 [A] — (I12) — AC 61-27C, page 169
The basic ground components of an ILS are the localizer, glide slope, outer marker, middle marker, and when installed for use with Category II or Category III instrument approach procedures, an inner marker. A compass locator or precision radar may be substituted for the outer or middle marker. DME, VOR, or nondirectional beacon fixes authorized in the standard instrument approach procedure or surveillance radar may be substituted for the outer marker.

ALL
4733 [B] — (I12) — FAR §91.175(k)
Based on the Inoperative Components Table (*See* FAA Legend 20), loss of the middle marker would not affect minimums.

ALL
4734 [A] — (J19) — AIM ¶5-46
When operating on an unpublished route or while being radar vectored, the pilot, when an approach clearance is received, shall maintain the last altitude assigned to that pilot until the aircraft is established on a segment of a published route or instrument approach procedure, unless a different altitude is assigned by ATC.

ALL
4735 [C] — (J19) — AIM ¶5-61
Pilots operating in accordance with an IFR flight plan, provided they are clear of clouds and have at least 1 mile flight visibility and can reasonably expect to continue to the destination airport in those conditions, may request ATC authorization for a contact approach. Controllers may authorize a contact approach if it is specifically requested by the pilot. ATC cannot initiate this approach.

ALL
4736 [A] — (J19) — AIM ¶5-80(a)(6)
On a visual approach, radar service is automatically terminated without advising the pilot, when the aircraft is instructed to contact the tower.

ALL
4737 [C] — (J19) — AIM ¶5-61
Pilots operating in accordance with an IFR flight plan, provided they are clear of clouds and have at least 1 mile flight visibility and can reasonably expect to continue to the destination airport in those conditions, may request ATC authorization for a contact approach.

ALL
4738 [B] — (V14) — AC 91-6A, page 5
Hydroplaning may occur as a result of aircraft speed and water on the runway. As the speed of the aircraft and the depth of the water increase, the water layer builds up an increasing resistance to displacement, resulting in the formation of a wedge of water beneath the tire. Dynamic hydroplaning occurs when there is standing water on the runway surface. Water about one-tenth of an inch deep acts to lift the tire off the runway.

ALL
4739 [B] — (K04) — AC 00-54A
When a headwind shears to a constant tailwind: the nose pitches down, thrust must be increased initially to resume normal approach speed and then reduced as the airspeed stabilizes, vertical speed increases, and indicated airspeed decreases and then increases to approach speed. *See* the figure below.

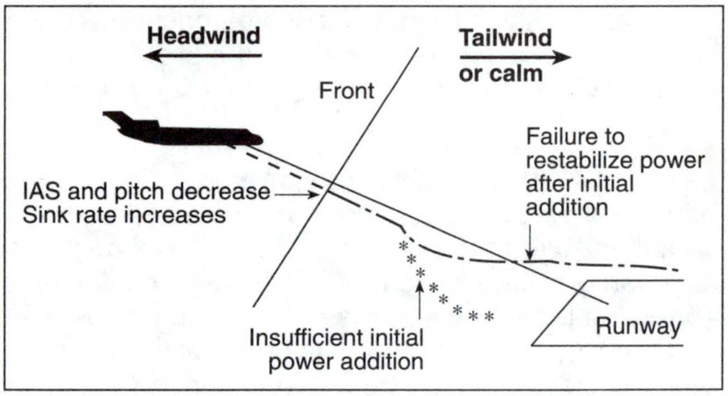

Question 4739

ALL
4740 [B] — (J18) — AIM ¶5-56(b)
Aircraft that execute a side-step maneuver will be cleared for a specified approach and landing on the adjacent parallel runway. Example, "Cleared for ILS runway 07 left approach, side-step to runway 07 right." Pilots are expected to commence the side-step maneuver as soon as possible after the runway or runway environment is in sight.

ALL
4741 [C] — (J18) — AIM ¶5-49(c)2
An ASR approach is one in which a controller provides navigation guidance in azimuth only. The pilot will be advised when to commence descent to the MDA. In addition, the pilot will be advised of the location of the Missed Approach Point (MAP) prescribed for the procedure and the aircraft's position each mile on final from the runway, airport or heliport or MAP, as appropriate.

ALL
4742 [B] — (J01) — AIM ¶1-10
The basic ground components of an ILS are the localizer, glide slope, outer marker, middle marker, and when installed for use with Category II or Category III instrument approach procedures, an inner marker. A compass locator or precision radar may be substituted for the outer or middle marker.

ALL
4743 [C] — (J18) — AIM ¶5-59(a)
When it will be operationally beneficial, ATC may authorize an aircraft to conduct a visual approach to an airport or to follow another aircraft when flight to, and landing at, the airport can be accomplished in VFR weather. The pilot must have the airport or the identified aircraft in sight before the clearance is issued.

ALL
4744 [A] — (J18) — AIM ¶5-58(b)
An ILS approach is a precision approach procedure, because it has an electronic glide slope. The Missed Approach Point (MAP) on a precision approach is always the Decision Height (DH).

ALL
4745 [C] — (K04) — AC 00-54A, ¶6(a)
The vertical speed required is dependent on the ground speed when flying on a fixed glide slope. In order to stay on the glide slope, the vertical speed must be decreased if the ground speed is decreased.

ALL
4746 [C] — (B97) — Pilot/Controller Glossary
Initial Approach Fixes are the fixes depicted on the Instrument Approach Procedure Charts that identify the beginning of the initial approach segment(s). The letters IAF identify Initial Approach Fixes on instrument approach charts.

ALL
4747 [B] — (I07) — AC 61-27C, page 128
The Inner Marker (IM), where installed, is located on the front course between the middle marker and the landing threshold. It indicates the point at which an aircraft is at a designated Decision Height (DH) on the glidepath. The IM is modulated at 3000 Hz and identified with continuous dots at the rate of 6 per second, and a white marker beacon light.

ALL
4748 [C] — (K04) — AC 00-54A, ¶6(a)
The vertical speed required is dependent on the ground speed when flying on a fixed glide slope. In order to stay on the glide slope, the vertical speed must be increased if the ground speed is increased.

ALL
4749 [C] — (J18) — AIM ¶5-57(d)
When either the normal rate of descent or the runway alignment factor of 30° is exceeded, a straight-in-minimum is not published and a circling minimum applies. The fact that a straight-in-minimum is not published does not preclude the pilot from landing straight-in, if the active runway is in sight and there is sufficient time to make a normal approach for landing.

ALL
4750 [A] — (J18) — AIM ¶5-61
A contact approach is an approach wherein an aircraft on an IFR flight plan, having an air traffic control authorization, operating clear of clouds with at least 1 mile flight visibility and a reasonable expectation of continuing to the destination airport on those conditions, may deviate from the standard instrument approach procedure (SIAP) and proceed to the destination airport by visual reference to the surface. This approach will only be authorized when requested by the pilot and the reported ground visibility at the destination airport is at least 1 statute mile.

ALL
4751 [C] — (J18) — AIM ¶5-40(c)
Pilots of IFR civil aircraft destined to locations for which STARs have been published may be issued a clearance containing a STAR whenever ATC deems it appropriate. Use of STARs requires pilot possession of at least the approved textual description. As with any ATC clearance or portion thereof, it is the responsibility of each pilot to accept or refuse an issued STAR. A pilot should notify ATC if he/she does not wish to use a STAR by placing "NO STAR" in the remarks section of the flight plan, or by the less desirable method of verbally stating the same to ATC.

ALL
4752 [C] — (J01) — AC 00-54A, ¶6(a)
The vertical speed required is dependent on the ground speed when flying on a fixed glide slope. For example, in order to stay on the glide slope, the vertical speed must be increased if the ground speed is increased.

ALL
4753 [B] — (J01) — AIM ¶1-10(f)(2)(b)
The MM indicates a position at which an aircraft is approximately 3,500 feet from the landing threshold. This will also be the position at which an aircraft on the glidepath will be at an altitude of approximately 200 feet above the elevation of the touchdown zone.

ALL
4754 [C] — (B10) — FAR §91.175(h)(2)
Use the RVR/Meteorological Visibility Comparable Values in FAA Legend 13 to convert RVR (feet) to visibility (statute miles). 2,400 RVR can be substituted for a ground visibility of 1/2 SM.

ALL
4755 [A] — (K04) — AC 00-54A, ¶6(a)(1)
When a headwind shears to a tailwind the airspeed decreases, the nose pitches down, and the aircraft will begin to drop below the glide slope. *See* the figure below.

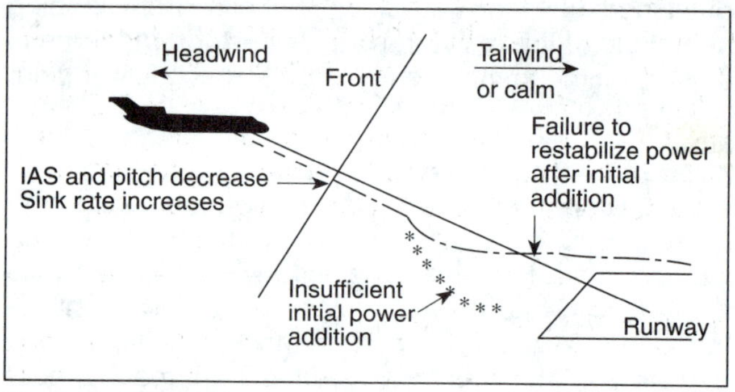

Question 4755

ALL
4756 [B] — (I05) — AC 61-27C, page 141
The heaviest demand on pilot technique occurs during descent from the outer marker to the middle marker, when you maintain the localizer course, adjust pitch attitude to maintain the proper rate of descent, and adjust power to maintain proper airspeed.

ALL
4757 [C] — (J18) — AIM ¶5-42(b)
Radar vectors and altitude/flight levels will be issued as required for spacing and separating aircraft; therefore, pilots must not deviate from the headings as issued by approach control. Aircraft will normally be informed when it becomes necessary to vector across the final approach course for spacing or other reasons. If approach course crossing is imminent and the pilot has not been informed that the aircraft will be vectored across the final approach course, the pilot should query the controller.

ALL
4758 [B] — (J13) — AIM ¶5-75(a)(3)
If operating VFR and compliance with any radar vector or altitude would cause a violation of any FAR, pilots should advise ATC and obtain a revised clearance or instruction.

ALL
4759 [A] — (B10) — FAR §91.175(h)(1)
If RVR minimums for takeoff or landing are prescribed in an instrument approach procedure, but RVR is not reported for the runway of intended operation, the RVR minimum shall be converted to ground visibility and shall be the visibility minimum for takeoff or landing on that runway. *See* FAA Legend 13 for the RVR/Meteorological Visibility Comparable Values.

ALL
4760 [B] — (B10) — FAR §91.169(c)
Unless otherwise authorized by the Administrator, no person may include an alternate airport in an IFR flight plan that has a precision (ILS) approach unless current weather forecasts indicate that, at the estimated time of arrival at the alternate airport, the ceiling will be at least 600 feet and visibility 2 statute miles.

ALL
4761 [C] — (B10) — FAR §91.169(a)
Unless otherwise authorized by ATC, each person filing an IFR or VFR flight plan shall include in it the point of first intended landing and the estimated elapsed time until over that point.

ALL
4762 [C] — (B10) — FAR §91.175(h)(1)
If RVR minimums for takeoff or landing are prescribed in an instrument approach procedure, but RVR is not reported for the runway of intended operation, the RVR minimum shall be converted to ground visibility and shall be the visibility minimum for takeoff or landing on that runway. *See* FAA Legend 13 for the RVR/Meteorological Visibility Comparable Values.

ALL
4763 [A] — (B10) — FAR §91.175
No pilot may operate an aircraft at any airport below the authorized MDA, or continue an approach below the authorized DH unless at least one of the following visual references for the intended runway is distinctly visible and identifiable to the pilot. If the approach lights are not visible upon arrival at the DH, then the pilot must execute the missed approach procedure.

ALL
4764 [B] — (B10) — FAR §91.175
When the glide slope fails, the ILS reverts to a nonprecision localizer approach. With a glide slope failure, the pilot is permitted to switch to a localizer approach and descend to the localizer minimum descent altitude (MDA).

ALL
4765 [C] — (B10) — FAR §91.177(a)(2)(i)
Except when necessary for takeoff or landing, or unless otherwise authorized by the Administrator, no person may operate an aircraft in a designated mountainous area under IFR below an altitude of 2,000 feet above the highest obstacle within a horizontal distance of 4 nautical miles from the course to be flown.

ALL
4766 [B] — (J17) — AIM ¶5-36(j)(2)
While in a holding pattern, the maximum indicated airspeed above 14,000 feet is 175 KIAS for all propeller-driven aircraft, and 265 KIAS for civil turbojet aircraft.

ALL
4767 [A] — (J17) — AIM ¶5-47(b)(3)
Where a holding pattern is specified in lieu of a procedure turn, the holding maneuver must be executed within the 1 minute time limitation or published leg length. The maneuver is completed when the aircraft is established on the inbound course after executing the appropriate holding pattern entry.

ALL
4768 [B] — (J17) — AIM ¶5-48(a)
Timed approaches may be conducted when the following conditions are met:

1. A control tower is in operation at the airport where the approaches are conducted.
2. Direct communications are maintained between the pilot and the center or approach controller until the pilot is instructed to contact the tower.
3. If more than one missed approach procedure is available, none require a course reversal.

ALL
4769 [C] — (B10) — FAR §91.169(c)(2)
If no instrument approach procedure has been published for that airport, the ceiling and visibility minimums are those allowing descent from the MEA, approach, and landing, under basic VFR.

ALL
4770 [B] — (J01) — AIM ¶1-10(j)
Inoperative ILS Components:

Inoperative Localizer—When the localizer fails, an ILS approach is not authorized.

Inoperative Glide Slope—When the glide slope fails, the ILS reverts to a nonprecision localizer approach, and localizer minimums apply.

ALL
4771 [A] — (J18) — AIM ¶5-56(b)
Aircraft that execute a side-step maneuver will be cleared for a specified approach and landing on the adjacent parallel runway. Pilots are expected to commence the side-step maneuver as soon as possible after the runway or runway environment is in sight.

ALL
4772 [B] — (K04) — AC 00-54A
The vertical speed required is dependent on the ground speed when flying on a fixed glide slope. For example, in order to stay on the glide slope, the vertical speed must be increased if the ground speed is increased.

ALL
4773 [A] — (I08) — AC 61-27C, page 141
Drift corrections should be small and reduced proportionately as the course narrows. By the time you reach the outer marker, your drift correction should be established accurately enough on a well-executed approach to permit completion of the approach with heading corrections no greater than 2°.

ALL
4774 [B] — (J03) — AIM ¶2-2(a)(3)
Two-bar VASI installations provide one visual glidepath which is normally set at 3°.

ALL
4775 [B] — (J03) — AIM ¶2-2(a)(3)
When on the glidepath with a two-bar VASI (four light units shown), the pilot will see red-over-white. As the airplane levels off and deviates above the visual glidepath, it flies through the transition zone for the far bars. The far bars will turn from red to white, and the airplane will be above the glidepath. *See* the figure below.

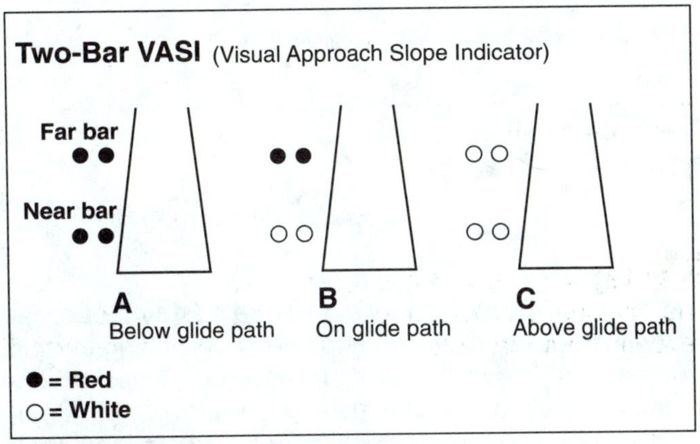

Question 4775

ALL
4776 [C] — (J03) — AIM ¶2-2(a)
Three-bar VASI installations provide two visual glidepaths. The lower path is provided by the near and middle bars and is normally set at 3° higher, while the upper path, provided by the middle and far bars, is normally 1/4 degree higher.

ALL
4777 [A] — (J03) — AIM ¶2-2(c)
Tri-color visual approach slope indicators normally consist of a single light unit projecting a three-color visual approach path into the final approach area of the runway upon which the indicator is installed. The below glidepath indication is red, the above glidepath indication is amber, and the on glidepath indication is green. *See* the figure below.

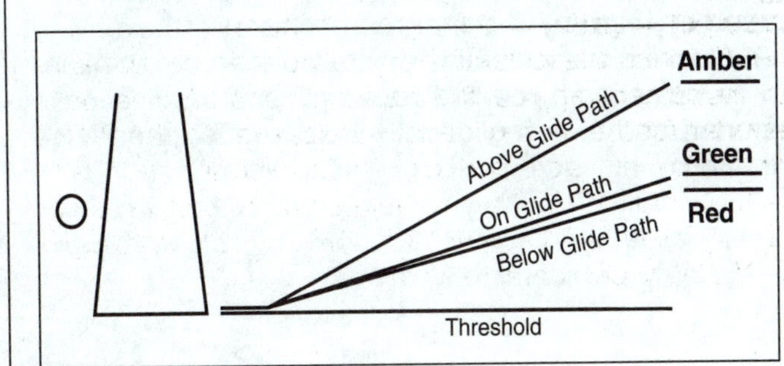

Question 4777

ALL
4778 [A] — (J03) — AIM ¶2-2(a)
When on the glidepath with a two-bar VASI, the pilot will see red-over-white. *See* the figure below.

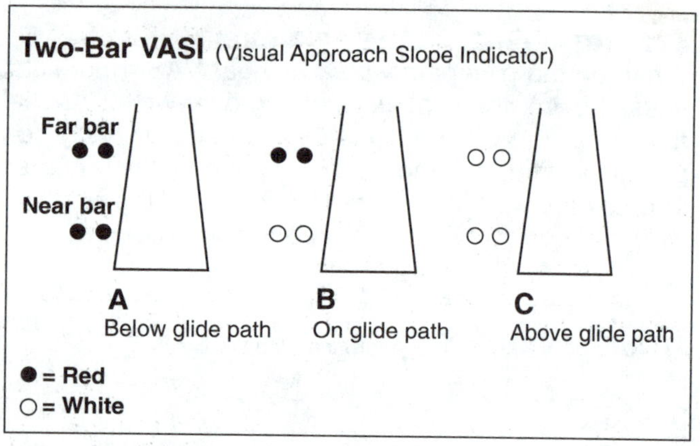

Question 4778

ALL
4779 [C]—(J03)—AIM ¶2-2(a) and FAR §91.129(d)3
See the figure below. An airplane approaching to land on a runway served by a visual approach slope indicator, shall maintain an altitude at or above the glide slope until a lower altitude is necessary for a safe landing. If the VASI is indicating all red lights, the aircraft is below the glidepaths, and the pilot should level off momentarily to intercept the proper approach path.

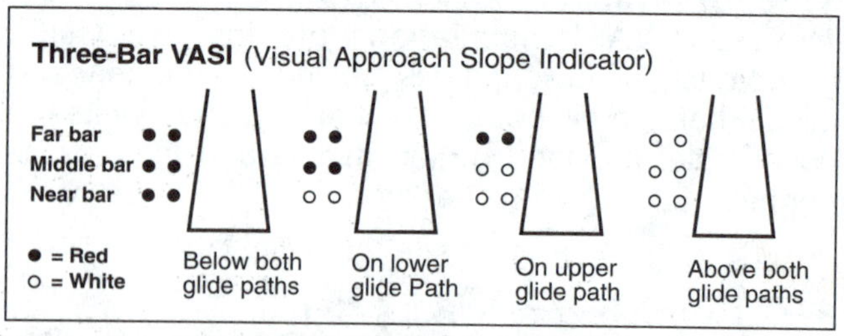

Question 4779, 4783, 4784 and 4785

ALL
4780 [A] — (J03) — AIM ¶2-2(c)
Tri-color visual approach slope indicators normally consist of a single light unit projecting a three-color visual approach path into the final approach area of the runway upon which the indicator is installed. The below-glidepath indication is red, the above-glidepath indication is amber, and the on-glidepath indication is green. *See* the following figure.

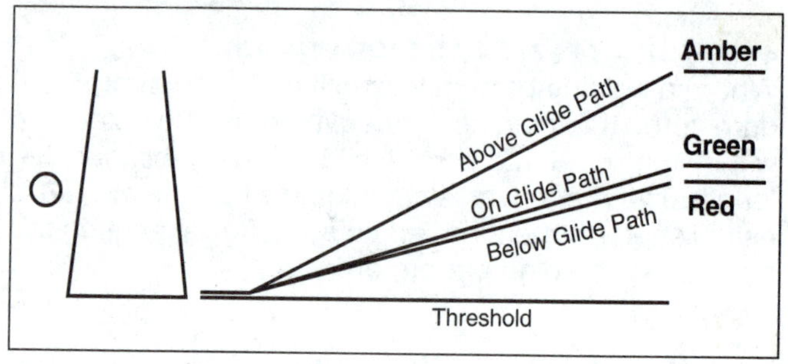

Question 4780

ALL
4781 [B] — (J03) — AIM ¶2-2(a)
The VASI is a system of lights arranged to provide visual descent guidance information during the approach to a runway. These lights are visible from 3 to 5 miles during the day, and up to 20 miles or more at night. The visual glidepath of the VASI provides safe obstruction clearance within plus or minus 10° of the extended runway centerline, and to 4 NM from the runway threshold.

ALL
4782 [C] — (J03) — AIM ¶2-2(a)
Three-bar VASI installations provide two visual glidepaths. The lower glidepath is provided by the near and middle bars and is normally set at 3° while the upper glidepath, provided by the middle and far bars, is normally 1/4-degree higher. This higher glidepath is intended for use only by high cockpit aircraft to provide a sufficient threshold-crossing height. Although normal glidepath angles are 3°, angles at some locations may be as high as 4.5° to give proper obstacle clearance.

ALL
4783 [B] — (J03) — AIM ¶2-2(a)
On a lower glidepath of a 3-bar VASI, the near lights will be white and the middle and far bars will be red, which is illustration 5. *See* the figure at left.

ALL
4784 [C] — (J03) — AIM ¶2-2(a)
Above a glidepath of a 3-bar VASI, all of the lights will be white, which is illustration 7. *See* the figure at left.

ALL
4785 [A] — (J03) — AIM ¶2-2(a)
Below a glidepath of a 3-bar VASI, all of the lights will be red, which is illustration 4. *See* the figure at left.

ALL
4786 [B] — (J03) — AIM ¶2-2(b)
The precision approach path indicator (PAPI) uses light units similar to the VASI, but are installed in a single row of either two- or four-light units. The on-glidepath indication will have 2 red lights and 2 white lights, which is illustration 10.

ALL
4787 [C] — (J03) — AIM ¶2-2(b)
The precision approach path indicator (PAPI) uses light units similar to the VASI, but are installed in a single row of either two- or four-light units. The slightly-low position will be indicated with 3 red lights and 1 white light, which is illustration 11.

ALL
4788 [A] — (J03) — AIM ¶2-2(b)
The precision approach path indicator (PAPI) uses light units similar to the VASI, but are installed in a single row of either two- or four-light units. The high-glidepath indication will have all white lights, which is illustration 8.

ALL
4789 [B] — (J03) — AIM ¶2-2(b)
The precision approach path indicator (PAPI) uses light units similar to the VASI, but are installed in a single row of either two- or four-light units. The slightly-high position will be indicated with 3 white lights, and 1 red light, which is illustration 9.

ALL
4790 [C] — (J03) — AIM ¶2-2(b)
The precision approach path indicator (PAPI) uses light units similar to the VASI, but are installed in a single row of either two- or four-light units. The low-glidepath indication will have all red lights, which is illustration 12.

ALL
4791 [B] — (J05) — AIM ¶2-32(d)
The aiming point marking (or fixed distance marker) serves as a visual aiming point for a landing aircraft. These two rectangular markings consist of a broad white stripe located on each side of the runway centerline and approximately 1,000 feet from the landing threshold.

ALL
4792 [B] — (J05) — AIM ¶2-32(e)
The touchdown zone markings identify the touchdown zone for landing operations and are coded to provide distance information in 500-foot increments. These markings consist of groups of one, two, and three rectangular bars symmetrically arranged in pairs about the runway centerline.

ALL
4793 [B] — (J05) — AIM ¶2-32
The touchdown zone marker is 500 feet from the runway threshold, and the fixed distance marker is 1,000 feet from the runway threshold. Therefore, the distance from the beginning of the touchdown zone marker to the beginning of the fixed distance marker is 500 feet (1,000 – 500).

ALL
4794 [A] — (J05) — AIM ¶2-32(h)
A displaced threshold is a threshold that is not at the beginning of the full strength runway pavement. The paved area behind the displaced runway threshold is available for taxiing, the landing rollout, and the takeoff of aircraft. White arrows are located along the centerline in the area between the beginning of the runway and the displaced threshold.

ALL
4795 [C] — (J03) — AIM ¶2-3
Runway End Identifier Lights (REIL) are installed for rapid and positive identification of the approach end of a runway. The system consists of a pair of synchronized flashing lights, located laterally one on each side of the runway threshold facing the approach area. *See the figure below.*

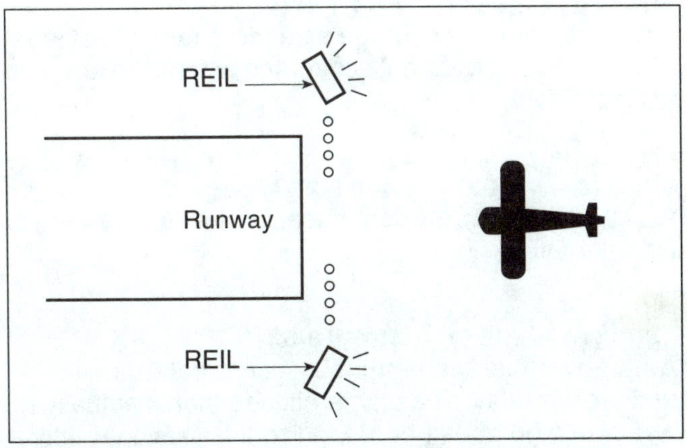

Question 4795

ALL
4796 [A] — (J03) — AIM ¶2-3
Runway End Identifier Lights (REIL) are installed for rapid and positive identification of the approach end of a runway. The system consists of a pair of synchronized flashing lights, located laterally one on each side of the runway threshold facing the approach area.

ALL
4797 [C] — (J05) — AIM ¶2-4(c)
On displaced thresholds, runway edge lights emit red light toward the runway to indicate the end of the runway to a departing aircraft, and green light outward from the runway end to indicate the threshold to landing aircraft. The area behind the displaced threshold is available for taxiing, landing rollout, and takeoff.

ALL
4798 [B] — (J01) — AIM ¶1-13(d)
The identification consists of a three-letter Morse Code identifier preceded by the Morse Code for "M" (— —) (i.e., M-STP). The "M" distinguishes this system from ILS which is preceded by the Morse Code for "I" (••) (i.e., I-STP).

ALL
4799 [A] — (J01) — AIM ¶1-12(b)(3)
The front azimuth coverage extends:

1. Laterally, at least 40° on either side of the runway.
2. In elevation, up to an angle of 15° and to at least 20,000 feet.
3. In range, to a distance of at least 20 NM.

ALL
4800 [C] — (J01) — AIM ¶1-12(b)(3)
The front azimuth coverage extends:

1. Laterally, at least 40° on either side of the runway.
2. In elevation, up to an angle of 15° and to at least 20,000 feet.
3. In range, to a distance of at least 20 NM.

ALL
4801 [C] — (J01) — AIM ¶1-12
The front azimuth coverage extends in range to at least 20 NM. The back azimuth coverage extends in range to at least 7 NM.

ALL
4802 [A] — (I02) — AC 61-27C, page 11
A pilot often interprets centrifugal force as a sensation of rising or falling.

ALL
4803 [A] — (J31) — AIM ¶8-5
A narrower-than-usual runway can create the illusion that the aircraft is at a higher altitude than it actually is. The pilot who does not recognize this illusion will fly a lower approach, with the risk of striking objects along the approach path or landing short.

ALL
4804 [A] — (J31) — AIM ¶8-5
An upsloping runway, upsloping terrain, or both can create the illusion that the aircraft is at a higher altitude than it actually is. The pilot who does not recognize this illusion will fly a lower approach. This is the same effect as a narrower-than-usual runway.

ALL
4805 [A] — (J31) — AIM ¶8-5(b)2
An abrupt head movement in a prolonged constant rate turn that has ceased stimulating the motion-sensing system can create the illusion of rotation or movement on an entirely different axis. The disoriented pilot will maneuver the aircraft into a dangerous attitude in an attempt to stop rotation. This most overwhelming of all illusions (called the Coriolis illusion) may be prevented by not making any sudden, extreme head movements, particularly during prolonged constant-rate turns under IFR conditions.

ALL
4806 [C] — (J31) — AIM ¶8-5(b)
Sloping cloud formations, an obscured horizon, a dark scene spread with ground lights and stars, or certain geometric patterns of ground light can create illusions of not being aligned correctly with the actual horizon. The disoriented pilot will place the aircraft in a dangerous position. This illusion is called "False Horizons."

ALL
4807 [A] — (J31) — AIM ¶8-5(b)
An abrupt change from climb to straight-and-level flight can create the illusion of tumbling backwards. The disoriented pilot will push the aircraft abruptly into a nose-low attitude, possibly intensifying this illusion. This is called "Inversion Illusion."

ALL
4808 [B] — (J31) — AIM ¶8-5(b)
A rapid acceleration during takeoff can create the illusion of being in a nose-up attitude. The disoriented pilot will push the aircraft into a nose-low, or dive attitude. This is called "Somatogravic Illusion."

ALL
4809 [B] — (J31) — AIM ¶8-2(a)(5)
The effects of hypoxia are usually difficult to recognize, especially when they occur gradually.

ALL
4810 [C] — (J31) — AIM ¶8-5
The sensations which lead to illusions during instrument flight conditions are normal perceptions experienced by normal individuals. These undesirable sensations cannot be completely prevented, but they can and must be ignored or sufficiently suppressed by developing absolute reliance upon what the flight instruments are telling about the attitude of the aircraft.

ALL
4811 [C] — (J31) — AIM ¶8-5
To overcome spatial orientation, pilots must become proficient in the use of flight instruments and rely upon them. Sight is the only reliable sense during instrument flight.

ALL
4812 [B] — (J31) — AIM ¶8-6(b)(2)
In darkness, vision becomes more sensitive to light through a process called dark adaptation. Since any degree of dark adaptation is lost within a few seconds of viewing a bright light, the pilot should avoid it, or close one eye when using white light.

ALL
4813 [B] — (J31) — AIM ¶8-5
To overcome spatial orientation, pilots must become proficient in the use of flight instruments and rely upon them. Sight is the only reliable sense during instrument flight.

ALL
4814 [C] — (J31) — AIM ¶8-5
Illusions that lead to spatial disorientation are created by information received from our motion-sensing system located in each inner ear. In flight, the system may be stimulated by motion of the aircraft alone, or in combination with head and body movement. Sight is the only reliable sense when flying solely by reference to flight instruments.

ALL
4815 [C] — (J31) — AIM ¶8-5
Spatial disorientation cannot be completely prevented, but it can and must be ignored or sufficiently suppressed by developing absolute reliance upon what the flight instruments are telling about the attitude of the aircraft.

ALL
4816 [B] — (J31) — AIM ¶8-3(b)
The symptoms of hyperventilation subside within a few minutes after the rate and depth of breathing are consciously brought back under control. The buildup of carbon dioxide in the body can be hastened by controlled breathing in and out with a paper bag held over the nose and mouth.

ALL
4817 [A] — (J31) — AIM ¶8-6(b)
In darkness, vision becomes more sensitive to light through a process called dark adaptation. Since any degree of dark adaptation is lost within a few seconds of viewing a bright light, the pilot should avoid it, or close one eye when using white light.

ALL
4818 [A] — (J31) — AIM ¶8-6(c)
While the eyes can observe an approximate 200° arc of the horizon at one glance, only a very small center area called the fovea, in the rear of the eye, has the ability to send clear, sharply focused messages to the brain. Because the eyes can focus only on this narrow viewing area, effective scanning is accomplished with a series of short, regularly spaced eye movements that bring successive areas of the sky into the central visual field. Each movement should not exceed 10°, and each area should be observed for at least one second to enable detection.

ALL
4819 [C] — (J31) — AIM ¶8-5(b)(3)(e)
Atmospheric haze causes the illusion of being a greater distance from the runway. The pilot who does not recognize this illusion will fly a lower approach.

ALL
4820 [B] — (I05) — AC 61-27C, page 65
As a rule-of-thumb, for errors less than 100 feet, use a half-bar-width correction.

ALL
4821 [A] — (L57) — AC 91-43
If both the ram air input and drainhole are completely blocked by ice, the pressure is trapped in the system and the airspeed indicator may react as an altimeter:

1. During level flight, airspeed indication will not change even when actual airspeed is varied by large power changes.
2. During climb, airspeed indication will increase.
3. During descent, airspeed indication will decrease.

ALL
4822 [C] — (J18) — AIM ¶5-49(c)(3)
During a "no-gyro" approach and prior to being handed off to the final approach controller, the pilot should make all turns at standard rate and should execute the turn immediately upon receipt of instructions.

ALL
4823 [A] — (J18) — AIM ¶5-49(c)(3)
After being handed off to the final approach controller during a "no-gyro" approach, the pilot should make all turns at standard rate and should execute the turn immediately upon receipt of instructions.

ALL
4824 [B] — (I07) — AC 61-27C, page 126
With the CDI two dots to the left, we are to the right of course, and according to FAA Figure 139, at the 1.9 NM location 710 feet to the right of course. With the glide slope indication two dots below, we are above the glide slope and at the 1.9 NM mark on FAA Figure 139, that would put you 140 feet above the glide slope.

ALL
4825 [C] — (I07) — AC 61-27C, page 126
With the CDI a dot and one-half to the right, we are to the left of course, and according to FAA Figure 139, at the 1,300-foot location (you must interpolate between 500 feet and 1,500 feet AGL), we are 322.5 feet to the left of course. The glide slope indication is a dot and a half above the glide slope and at the 1,300-foot location; that would put you 21 feet above the glide slope.

ALL
4826 [A] — (I07) — AC 61-27C, page 126
With the CDI two dots to the right, we are to the left of course, and according to FAA Figure 139, at the 5.6 NM location 1,550 feet to the left of course. The glide slope indication is one dot below the glide slope and at the 5.6 NM mark, that would put you 210 feet below the glide slope.

ALL
4827 [C] — (I04) — AC 61-27C, page 51
The slaving meter indicates the difference between the displayed heading and the magnetic heading. A right deflection indicates a clockwise error of the compass card; a left deflection indicates a counter-clockwise error. In this case, the compass card is indicating 120°, when it should be indicating 110°. It has a left deflection, or a counter-clockwise error. The compass card may be aligned with the magnetic compass by selecting the free-gyro mode and depressing the clockwise heading drive button, which will cause the compass card to rotate in a clockwise direction.

ALL
4828 [A] — (I04) — AC 61-27C, page 51
When the system is in the free-gyro mode, depressing the clockwise manual heading drive button will rotate the compass card to the right to eliminate left compass card error (clockwise adjustment).

ALL
4829 [C] — (I04) — AC 61-27C, page 51
The slaving meter indicates the difference between the displayed heading and the magnetic heading. A right deflection indicates a clockwise error of the compass card; a left deflection indicates a counter-clockwise error. In this case, it has a left deflection, or a counter-clockwise error. The compass card may be aligned with the magnetic compass by selecting the free-gyro mode and depressing the clockwise heading drive button, which will cause the compass card to rotate in a clockwise direction.

ALL
4830 [C] — (L57) — AC 91-43, 3(b)(2)
If both the ram air input and the drain hole are blocked, the pressure is trapped in the system and the airspeed indicator may react as an altimeter:

1. During level flight, the airspeed indication will not change even when actual airspeed is varied by large power changes;
2. During climb, the airspeed indication will increase; and
3. During descent, the airspeed indication will decrease.

ALL
4831 [B] — (I04) — AC 61-27C, page 59
When an aircraft makes a taxiing left turn, the turn coordinator will show the same indications as a level, no-bank left turn in flight. Those indications are: The miniature airplane will show a left turn (the direction of turn and rate) while the ball will show an uncoordinated turn (a skid), by moving to the right.

ALL
4832 [A] — (I05) — AC 61-27C, page 57
The heading indicator provides the most pertinent banking information in unaccelerated straight-and-level flight, since banking means turning, and it is primary for bank. With gyroscopic heading indicator failure, the magnetic compass becomes the only heading indicator.

ALL
4833 [A] — (I05) — AC 61-27C, page 88
The angle of bank necessary for a given rate-of-turn is proportional to the true airspeed. Since the turns are executed at standard rate, the angle of bank must be varied in direct proportion to the airspeed change in order to maintain a constant rate-of-turn. During a reduction of airspeed, you must decrease the angle of bank and increase the pitch attitude to maintain altitude and a standard rate turn.

ALL
4834 [C] — (I05) — AC 61-27C, page 59
Check the magnetic compass card for freedom of movement and be sure that the bowl is full of fluid. Determine compass accuracy by comparing the indicated heading against a known heading, while the airplane is stopped or taxiing straight.

ALL
4835 [A] — (I05) — AC 61-27C, page 59
If the attitude indicator's horizon bar erects to the horizontal position and remains at the correct position for the attitude of the airplane, or if it begins to vibrate after this attitude is reached and then slowly stops vibrating altogether, the instrument is operating properly. If the horizon bar fails to remain in the horizontal position during straight taxiing, or tips in excess of 5° during taxi turns, the instrument is unreliable.

ALL
4836 [A] — (I05) — AC 61-27C, page 77
The attitude indicator is primary only during transitions from one attitude to another. Once stabilized, the attitude indicator becomes a support instrument. The turn coordinator is a supporting instrument in straight flight.

ALL
4837 [A] — (I05) — AC 61-27C, page 77
Primary pitch information during the transition is attitude indicator for pitch, heading indicator for bank, and tachometer or manifold pressure for power. When the airspeed has stabilized, the air speed indicator becomes primary for pitch.

ALL
4838 [B] — (I05) — AC 61-27C, page 81
The turn coordinator displays the movement of the aircraft on the roll axis that is proportional to the roll rate. Therefore, the turn coordinator is the primary bank instrument once a standard rate turn is established.

ALL
4839 [A] — (I05) — AC 61-27C, page 44
The miniature aircraft of the turn coordinator displays rate of roll and rate of turn.

ALL
4840 [C] — (I05) — AC 61-27C, page 57
During attitude instrument training, a pilot must develop three fundamental skills involved in all instrument flight maneuvers:

1. Instrument cross-check
2. Instrument interpretation
3. Aircraft control.

RTC
4841 [C] — (I06) — AC 61-27C, page 99
When making initial pitch attitude corrections to maintain altitude, the changes of attitude should be small and smoothly applied. The initial movement of the horizon bar should not exceed one bar high or low. If further change is required, an additional correction of one-half bar will normally correct any deviation from the desired altitude. This correction (one and one-half bars) is normally the maximum.

ALL
4842 [C] — (I05) — AC 61-27C, page 59
If the horizon bar erects to the horizontal position and remains at the correct position for the attitude of the airplane, or if it begins to vibrate after this attitude is reached and then slowly stops vibrating altogether, the instrument is operating properly. If the horizon bar fails to remain in the horizontal position during straight taxiing, or tips in excess of 5° during taxi turns, the instrument is unreliable.

ALL
4843 [A] — (I03) — AC 61-27C, page 21
The rate-of-turn at any given airspeed depends on the amount of sideward force causing the turn; that is, the horizontal lift component, which varies directly in proportion to bank in a correctly executed turn.

ALL
4844 [B] — (I03) — AC 61-27C, Figure 3-11, page 23
A skidding turn results from excess centrifugal force over the horizontal lift component, pulling the aircraft toward the outside of the turn. As centrifugal force increases, the load factor also increases. *See* the figure below.

ALL
4845 [A] — (I05) — AC 61-27C, page 77
As the power is increased to the approximate setting for the desired rate, simultaneously raise the miniature aircraft to the climbing attitude for the desired airspeed and rate of climb. As the power is increased, the ASI is primary for pitch control until the vertical speed approaches the desired value. The heading indicator is primary for bank and the tachometer, or manifold pressure gauge, is primary for power. As the vertical speed needle stabilizes, it becomes primary for pitch.

RTC
4846 [B] — (I06) — AC 61-27C, page 110
During the initial acceleration, the pitch attitude of the helicopter, as read on the attitude indicator, should be one to two bar widths low.

ALL
4847 [A] — (I03) — AC 61-27C, page 44
The miniature aircraft of the turn coordinator displays only rate-of-roll and rate-of-turn. It does not directly display the bank angle of the aircraft.

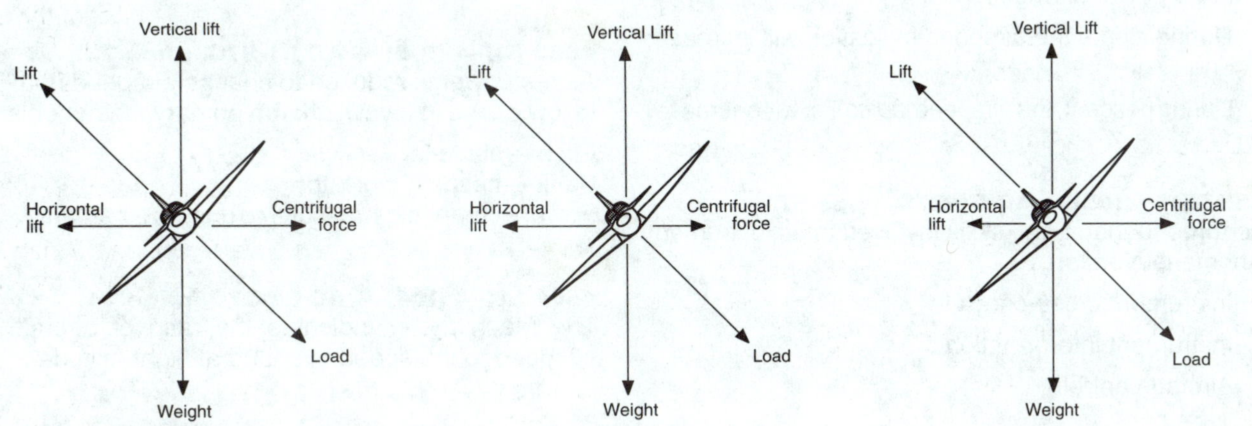

1. Normal Turn
Centrifugal force equals horizontal lift.

2. Slipping Turn
Centrifugal force less than horizontal lift.

3. Skidding Turn
Centrifugal force greater than horizontal lift.

Question 4844

ALL

For a constant airspeed climb (at cruising airspeed), once the airplane stabilizes at a constant airspeed and attitude, the airspeed indicator becomes the primary pitch instrument.

RTC
4849 [B] — (I06) — AC 61-27C, page 110
The airspeed indicator is the primary pitch instrument and should be adjusted to the recommended autorotation speed.

ALL
4850 [A] — (I05) — AC 61-27C, page 83
The altimeter is the primary pitch instrument when establishing a level standard rate turn.

ALL
4851 [C] — (I05) — AC 61-27C, page 83
The attitude indicator is the initial primary bank instrument when establishing a level standard rate turn.

RTC
4852 [C] — (I06) — AC 61-27C, page 110
The pitch attitude of the helicopter should be approximately level as shown by the attitude indicator.

ALL
4853 [B] — (I06) — AC 61-27C, page 77
The attitude indicator and turn coordinator are the supporting bank instruments when entering a constant airspeed climb from straight-and-level flight.

ALL
4854 [A] — (L57) — AC 91-43(3)(b)
If both the ram air input and the drain hole are blocked, the pressure is trapped in the system and the airspeed indicator may react as an altimeter:

1. During level flight, the airspeed indication will not change even when actual airspeed is varied by large power changes;
2. During climb, the airspeed indication will increase; and
3. During descent, the airspeed indication will decrease.

ALL
4855 [B] — (I05) — AC 61-27C, page 57
The three fundamental skills involved in all instrument flight maneuvers are:

1. Instrument cross-check
2. Instrument interpretation
3. Aircraft control.

ALL
4856 [A] — (I05) — AC 61-27C, page 44
The miniature aircraft of the turn coordinator displays only rate-of-roll and rate-of-turn. It only indirectly displays the bank angle of the aircraft.

ALL
4857 [C] — (I04) — AC 61-27C, page 40
After return of the aircraft to straight-and-level, coordinated flight, the miniature aircraft shows a turn in the direction opposite the skid.

ALL
4858 [A] — (I05) — AC 61-27C, page 81
The attitude indicator is the initial primary bank instrument when establishing a level standard rate turn.

ALL
4859 [C] — (I05) — AC 61-27C, page 57
The three fundamental skills involved in all instrument flight maneuvers are:

1. Instrument cross-check
2. Instrument interpretation
3. Aircraft control.

ALL
4860 [B] — (I04) — AC 61-27C, page 40
If a 180° steep turn is made to the right and the aircraft is rolled-out to straight-and-level flight by visual references, the miniature aircraft will show a slight climb and turn to the left.

ALL
4861 [C] — (I04) — AC 61-27C, page 40
During a normal turn, movement of the vanes by centrifugal force causes precession of the gyro toward the inside of the turn.

ALL
4862 [B] — (I05) — AC 61-27C, page 57
The three fundamental skills involved in all instrument flight maneuvers are:

1. Instrument cross-check
2. Instrument interpretation
3. Aircraft control

ALL
4863 [C] — (I05) — AC 61-27C, page 73
As the power is reduced to change airspeed from high to low cruise in level flight the primary instruments are:

Pitch—altimeter
Bank—heading indicator
Power—manifold pressure gauge (momentarily)

ALL
4864 [A] — (I04) — AC 61-27C, page 34
The Mach meter indicates the ratio of aircraft true airspeed to the speed of sound at flight altitude.

ALL
4865 [C] — (I05) — AC 61-27C, page 73
The primary instruments in straight-and-level flight are:

Primary Pitch—Altimeter
Primary Bank—Heading Indicator
Primary Power—Airspeed Indicator

See the figure below.

ALL
4866 [B] — (I05) — AC 61-27C, page 81
On the roll-in to a level standard rate turn, use the attitude indicator to establish the approximate angle of bank (primary), then check the miniature aircraft of the turn coordinator for a standard rate turn indication (secondary).

ALL
4867 [A] — (I05) — AC 61-27C, page 90
When the rate of movement of altimeter and airspeed indicator needles decreases, and the vertical speed indicator reverses its trend, the aircraft is approaching level pitch attitude.

ALL
4868 [B] — (I04) — AC 61-27C, page 45
In a coordinated turn, horizontal lift equals centrifugal force. This is indicated by the ball on the turn-and-slip indicator being centered. *See* the figure below.

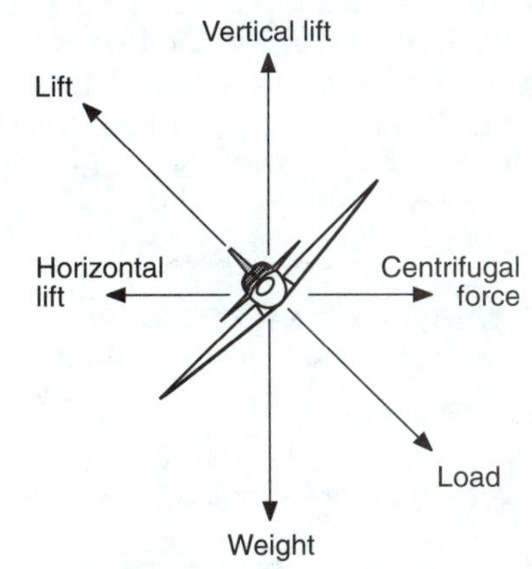

Normal Turn
Centrifugal force equals horizontal lift.

Question 4868

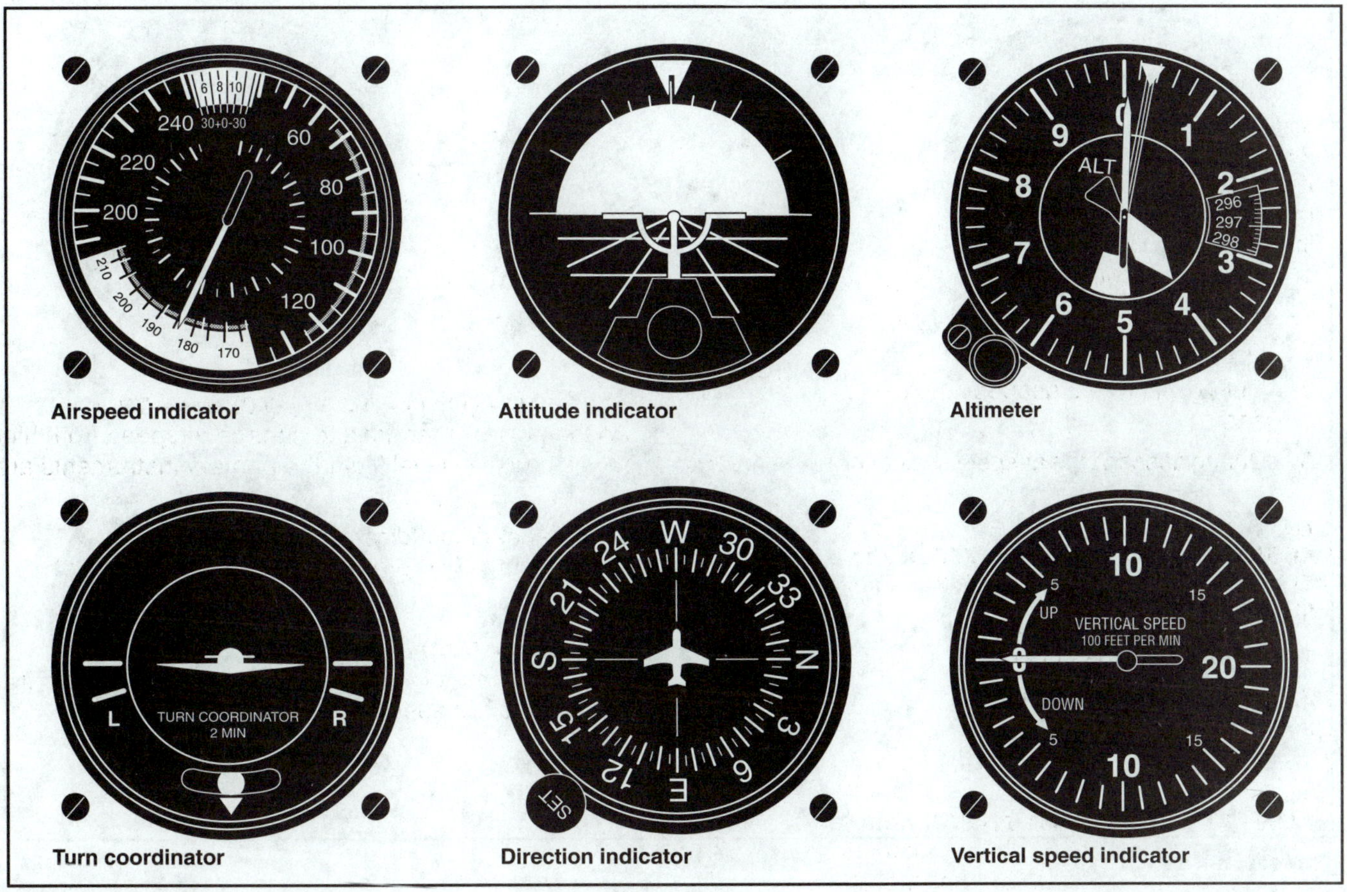

Question 4865

ALL
4869 [C] — (I05) — AC 61-27C, page 56
In addition to the attitude indicator, the altimeter, airspeed indicator, and vertical speed indicator provide pitch information.

ALL
4870 [C] — (I05) — AC 61-27C, page 21
An aircraft requires a sideward force to make it turn. In a normal turn, this force is supplied by banking the aircraft so that lift is exerted inward as well as upward. The horizontal lift component is the sideward force that causes the aircraft to turn.

ALL
4871 [C] — (I05) — AC 61-27C, page 73
The primary instruments in straight-and-level flight are:

Primary Pitch—Altimeter
Primary Bank—Heading Indicator
Primary Power—Airspeed Indicator

ALL
4872 [C] — (I05) — AC 61-27C, page 88
The attitude indicator and the vertical speed indicator are supporting instruments for pitch during change of airspeed in a level turn. *See* the figure below.

ALL
4873 [B] — (I05) — AC 61-27C, page 90
As soon as the unusual attitude is detected, the recovery should be initiated primarily by reference to the airspeed indicator (for pitch attitude), altimeter (for pitch attitude), vertical-speed indicator, and turn coordinator.

ALL
4874 [A] — (I05) — AC 61-27C, page 88
The manifold pressure gauge (or tachometer) is primary for power control while the airspeed is changing in a level turn. As the airspeed approaches the new indication, the airspeed indicator becomes primary for power control.

ALL
4875 [B] — (I05) — AC 61-27C, page 90
Reduce power to prevent excessive airspeed and loss of altitude, correct the bank attitude with coordinated aileron and rudder pressure to straight flight by referring to the turn coordinator. Raise the nose to level-flight attitude by smooth back-elevator pressure.

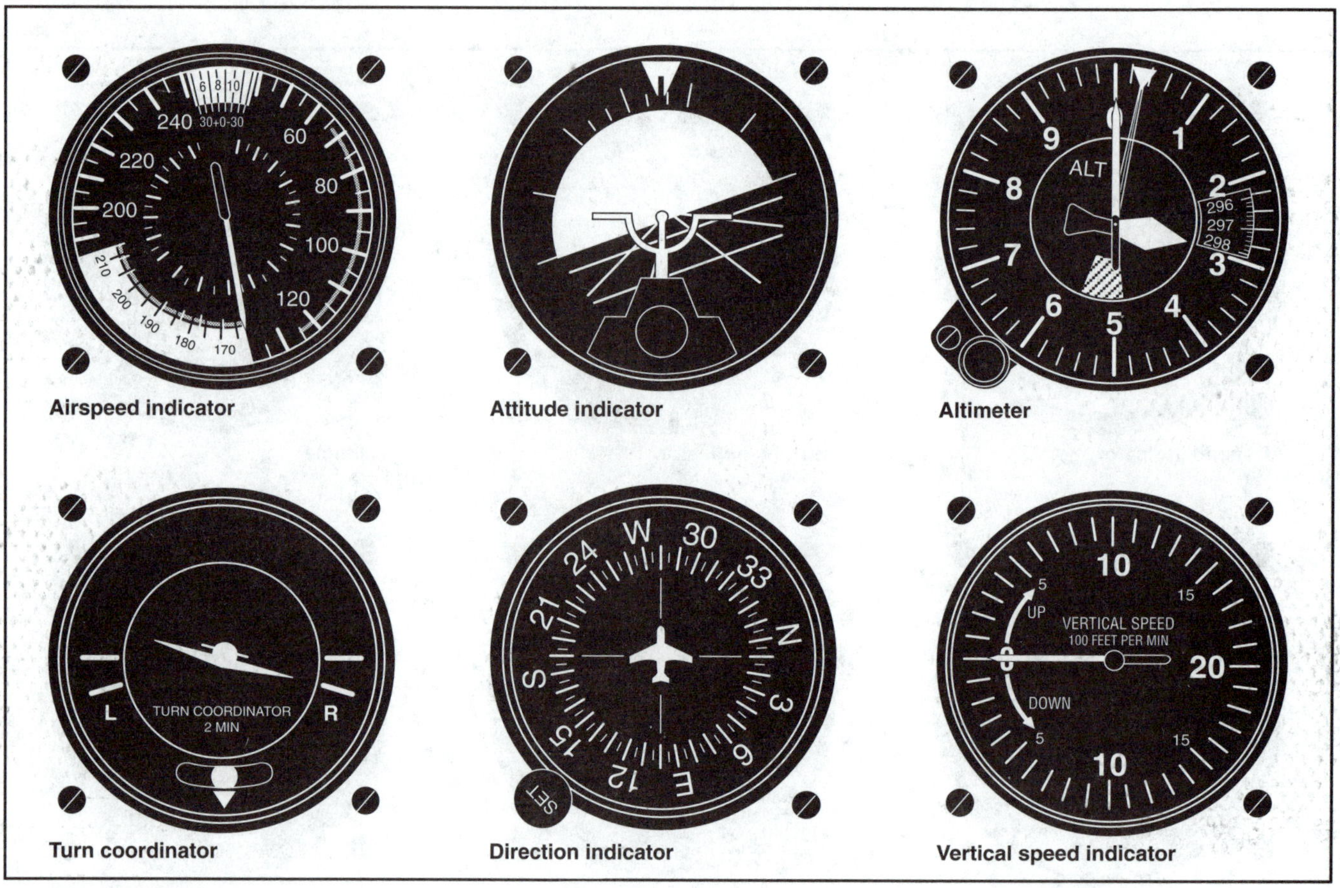

Question 4872

ALL

4876 [C] — (I05) — AC 61-27C, pages 61 and 65

The pitch instruments are the attitude indicator, the altimeter, the vertical-speed indicator and the airspeed indicator. When a pitch error is detected, corrective action should be taken promptly, but with light control pressures.

ALL

4877 [C] — (I05) — AC 61-27C, page 85

When you are on an east or west heading, the compass indicates correctly as you start a turn in either direction.

ALL

4878 [B] — (I05) — AC 61-27C, page 22

If the airspeed is increased in a turn, the angle of attack must be decreased and/or the angle of bank increased in order to maintain level flight. As airspeed is increased in a constant-rate level turn, both the radius of turn and centrifugal force increase. This increase in centrifugal force must be balanced by an increase in the horizontal lift component, which can be accomplished only by increasing the angle of bank. Thus, to maintain a turn at a constant rate, the angle of bank must be varied with changes in airspeed.

ALL

4879 [C] — (I05) — AC 61-27C, pages 32 and 33

The vertical speed indicator operates by indicating a change in pressure between the instantaneous static pressure in the diaphragm and the trapped static pressure inside the case. If the static ports are iced over prior to entering a climb or descent, the VSI pointer will remain at zero because the diaphragm static pressure and the trapped static pressure inside the case remain equal to each other.

ALL

4880 [C] — (I05) — AC 61-27C, page 31

With the altimeter set to the current reported altimeter setting, note any variation between the known field elevation and the altimeter indication. If the variation exceeds ±75 feet, the accuracy of the altimeter is questionable and the problem should be referred to an appropriately-rated repair station for evaluation and possible correction.

ALL

4881 [C] — (L59) — AC 91-46, 5(a)(4)

The following are items that should be checked before starting the engine(s):

1. If instruments are electrical, turn on and listen for any unusual or irregular mechanical noise.
2. Check instruments for poor condition, mounting, marking, broken or loose knobs. Also check the power-off indications of the instrument pointers and warning flags.

ALL

4882 [B] — (L59) — AC 91.46(5)(a)(2)

The following are items that should be checked before starting the engine(s):

1. Check turn and slip indicator and magnetic compass for fluid level (should be full).
2. If instruments are electrical, turn on and listen for any unusual or irregular mechanical noise.
3. Check instruments for poor condition, mounting, marking, broken or loose knobs. Also check the power-off indications of the instrument pointers and warning flags.

ALL

4883 [A] — (L59) — AC 91.46, page 89

You should observe the following on the turn-and-slip indicator during taxi:

1. Check and ensure the turn needle indicates proper direction of turn.
2. Check the ball for freedom of movement in the glass tube. Centrifugal force causes the ball to move to the outside of the turn.

ALL

4884 [C] — (I04) — AC 61-27C, page 44

The ball of the turn coordinator is a balance indicator, and is used as a visual aid to determine coordinated use of the aileron and rudder control. During a turn it indicates the relationship between the angle of bank and rate-of-turn. It indicates the "quality" of the turn, or whether the aircraft has the correct angle of bank for the rate-of-turn. When the ball is centered, the turn is coordinated, and neither slipping or skidding.

ALL

4885 [A] — (I05) — AC 61-27C, page 59

Allow 5 minutes after starting engines for the gyro rotor of the vacuum-operated heading indicator to attain normal operating speed. Before taxiing, or while taxiing straight, set the heading indicator to correspond with the magnetic compass heading. Be sure the instrument is fully uncaged if it has a caging feature. Before takeoff, recheck the heading indicator. If your magnetic compass and deviation card are accurate, the heading indicator should show the known taxiway or runway direction when the airplane is aligned with them (within 5°).

ALL

4886 [C] — (I05) — AC 61-27C, page 85

When on an east or west heading, no error is apparent while entering a turn to north or south.

ALL

4887 [A] — (I05) — AC 61-27C, page 85

If on a southerly heading and a turn is made toward the east or west, the initial indication of the compass needle will indicate a greater amount of turn than is actually made. This lead diminishes as the turn progresses toward east or west where there is no turn error.

ALL
4888 [B] — (I05) — AC 61-27C, page 85
A 15° bank is an approximate standard rate turn for most general aviation aircraft. Either lag or lead diminishes as the turn progresses toward east or west, where there is no turn error.

ALL
4889 [C] — (I05) — AC 61-27C, page 48
Northerly turning error is the most pronounced of the dip errors. Due to the mounting of the magnetic compass, its center of gravity is below the pivot point on the pedestal and the card is well balanced in the fluid. When the aircraft is banked, the card is also banked as a result of centrifugal force. While the card is in the banked attitude, the vertical component of the Earth's magnetic field causes the north-seeking ends of the compass to dip to the low side of the turn, giving an erroneous turn indication.

ALL
4890 [A] — (I05) — AC 61-27C, page 85
If on a southerly heading and a turn is made toward the east or west, the initial indication of the compass needle will indicate a greater amount of turn than is actually made. This lead diminishes as the turn progresses toward east or west where there is no turn error.

ALL
4891 [C] — (I05) — AC 61-27C, page 85
When on an east or west heading, no error is apparent while entering a turn to north or south.

ALL
4892 [B] — (I05) — AC 61-27C, page 85
If on a northerly heading and a turn is made toward east or west, the initial indication of the compass lags or indicates a turn in the opposite direction. This lag diminishes as the turn progresses toward east or west where there is no turn error.

ALL
4893 [C] — (I05) — AC 61-27C, page 85
When on an east or west heading, no error is apparent while entering a turn to north or south.

ALL
4894 [B] — (I05) — AC 61-27C, page 85
If on a northerly heading and a turn is made toward east or west, the initial indication of the compass lags or indicates a turn in the opposite direction. This lag diminishes as the turn progresses toward east or west where there is no turn error.

ALL
4895 [C] — (I05) — AC 61-27C, page 84
A standard rate turn (3°/second) takes 2 minutes to complete a 360° turn. A half-standard rate turn (1.5°/second) takes 4 minutes complete a 360° turn.

ALL
4896 [A] — (I05) — AC 61-27C, page 84
A standard rate turn is one during which the heading changes 3° per second. Therefore, at a rate of 3° of turn each second, 180° of turn would require 180/3 or 60 seconds (1 minute).

ALL
4897 [B] — (I05) — AC 61-27C, page 84
A half-standard rate turn is one in which the heading changes 1.5° per second. A heading change of 090° to 180° is a total turn of 90°. At a rate-of-turn of 1.5° each second would require 90/1.5 or 60 seconds (1 minute).

ALL
4898 [C] — (I04) — AC 61-27C, page 22
Both the rate of turn and the radius of turn vary with changes in airspeed. If airspeed is increased, then the rate of turn is decreased, and the radius of the turn is increased.

ALL
4899 [B] — (I04) — AC 61-27C, page 18
Three factors affect attitude in maintaining level flight: airspeed, air density, and aircraft weight. At a constant angle of attack, any change in airspeed will vary the lift. Lift varies directly with changes in air density. To support heavier loads (weight) at a given airspeed, the angle of attack must be relatively greater to provide the necessary lift.

ALL
4900 [A] — (I05) — AC 61-27C, page 40
Errors in both pitch and bank indications occur during normal coordinated turns. These errors are caused by the movement of the pendulous vanes by centrifugal force, resulting in the precession of the gyro towards the inside of the turn. The error is greatest in a 180° steep turn.

ALL
4901 [A] — (I05) — AC 61-27C, page 40
Errors in both pitch and bank indications occur during normal coordinated turns. The error is greatest in a 180° steep turn. If, for example, a 180° steep turn is made to the right and the aircraft is rolled out in straight-and-level flight by visual references, the miniature aircraft will show a slight climb and turn to the left.

ALL
4902 [B] — (I04) — AC 61-27C, page 35
Any rotating body exhibits gyroscopic properties according to Newton's laws of motion. The first law states: a body at rest will remain at rest; or if in motion in a straight line, it will continue in motion in a straight line unless acted upon by an outside force. The second law states: the deflection of a moving body is proportional to the deflective force applied and is inversely proportional to its weight and speed. A gyro is a wheel or disc designed to utilize these principles.

ALL
4903 [A] — (I05) — AC 61-27C, page 84
A standard rate turn is one during which the heading changes 3° per second. A heading change from 090° to 270° is a total turn of 180°. At a rate-of-turn of 3° each second, this turn would require 180/3 or 60 seconds (1 minute).

ALL
4904 [C] — (I04) — AC 61-27C, page 84
A standard rate turn is one during which the heading changes 3° per second. A turn to the left from 090° to 300° is a total turn of 150°. At a turn rate of 3° each second, this turn would require 150/3 or 50 seconds.

ALL
4905 [C] — (I04) — AC 61-27C, page 84
A standard rate turn is one during which the heading changes 3° per second. In this case, the aircraft is turning one-half that rate, or 1.5° per second. Therefore, a turn of 135° would require 135/1.5 or 90 seconds (1 minute 30 seconds).

ALL
4906 [A] — (I05) — AC 61-27C, page 78
The amount of lead varies with rate of climb and pilot technique. An effective practice is to lead the altitude by 10 percent of the vertical speed shown (500 fpm/50-foot lead—1,000 fpm/100-foot lead).

ALL
4907 [A] — (I05) — AC 61-27C, page 79
The level-off from a descent at descent airspeed must be started before you reach the desired altitude. The amount of lead depends upon the rate of descent and control technique. With too little lead, you will tend to overshoot the selected altitude, unless your technique is rapid. An effective practice is to lead the altitude by 10 percent of the vertical speed shown. Assuming a 500-fpm rate of descent, lead the desired altitude by approximately 50 feet.

ALL
4908 [C] — (I04) — AC 61-27C, page 26
If the alternate static source is vented inside the airplane, where static pressure is usually lower than outside static pressure, selection of the alternate source may result in the following instrument indications:

1. The altimeter reads higher than normal;
2. Indicated airspeed greater than normal; and
3. The vertical-velocity indicator momentarily shows a climb.

ALL
4909 [A] — (I04) — AC 61-27C, page 26
Clogging of the pitot opening by ice or dirt (or failure to remove the pitot cover) affects the airspeed indicator only.

ALL
4910 [B] — (I04) — AC 61-27C, page 30
The altimeter setting system provides the means that must be used to correct the altimeter for pressure variations. The system is necessary to ensure safe terrain clearance for instrument approaches and landings, and to maintain vertical separation between aircraft during instrument weather conditions.

ALL
4911 [B] — (I04) — AC 61-27C, page 30
Pressure altitude is the altitude indicated when the altimeter setting window is adjusted to 29.92" Hg. (the theoretical Standard Datum Plane). To find pressure altitude, follow these steps:

1. Find the difference between the given pressure and standard pressure:

   30.42 – 29.92 = 0.50" Hg.

2. Find the change in indicated altitude. The Standard Pressure Lapse Rate is about 1 inch per 1,000 feet:

   0.50 x 1000 = 500 feet difference

3. Find pressure altitude. The Standard Datum Plane pressure is lower than the current station setting and therefore, pressure altitude will also be lower:

   6,500 feet MSL – 500 feet = 6,000 feet Pressure Altitude

ALL
4912 [B] — (I04) — AC 61-27C, page 28
Pressure Altitude is the altitude read on the altimeter when the instrument is adjusted to indicate height above the Standard Datum Plane. The Standard Datum Plane is a theoretical level where the weight of the atmosphere is 29.92" Hg.

ALL
4913 [C] — (I04) — AC 61-27C, page 18
As temperature increases, air density decreases. A decreased air density causes true airspeed and true altitude to increase.

ALL
4914 [B] — (I03) — AC 61-27C, page 21
The horizontal component of lift varies directly in proportion to bank in a correctly executed turn. Thus, the rate of turn increases (and radius of turn decreases) as the angle of bank increases, and the airspeed decreases.

ALL
4915 [B] — (I03) — AC 61-27C, page 21
The division of lift into horizontal and vertical components during a turn reduces the amount of lift supporting the weight of the aircraft. Consequently, the reduced vertical component results in the loss of altitude unless the total lift is increased by:

1. Increasing the angle of attack of the wing,
2. Increasing the airspeed, or
3. Increasing the angle of attack and airspeed in combination.

ALL
4916 [B] — (H66) — AC 61-21A, page 294
If during flight, rough air or severe turbulence is encountered, the airspeed should be reduced to maneuvering speed or less, to minimize the stress on the airplane structure.

ALL
4917 [B] — (K04) — AC 00-54A
There are two potentially hazardous shear situations:

1. A tailwind may shear to either a calm or headwind component. In this instance, initially the airspeed increases, the aircraft pitches up and the altitude increases.
2. A headwind may shear to a calm or tailwind component. In this situation, initially the airspeed decreases, the aircraft pitches down, and the altitude decreases.

ALL
4918 [A] — (I04) — AC 61-27C, page 40
Acceleration and deceleration induce precession errors, depending upon the amount and extent of the force applied. During acceleration, the horizon bar moves down, indicating a climb.

ALL
4919 [C] — (I04) — AC 61-27C, page 40
Acceleration and deceleration induce precession errors, depending upon the amount and extent of the force applied. During deceleration, the horizon bar moves up, indicating a descent.

ALL
4920 [C] — (I04) — AC 61-27C, page 63
The attitude indicator would be the least appropriate instrument for determining the need for a pitch change in level flight at constant thrust. Until level flight, as indicated by the attitude indicator, is identified and established by reference to the altimeter and VSI, there is no way of knowing if it is truly level flight. With constant thrust, any change in the altimeter or VSI indications shows a need for a pitch change.

ALL
4921 [C] — (I05) — AC 61-27C, page 44
The miniature aircraft of the turn coordinator displays only rate-of-roll and rate-of-turn. It does not directly display the bank angle of the aircraft. The displacement of a turn coordinator increases as angle of bank increases.

ALL
4922 [B] — (I04) — AC 61-27C, page 31
The altimeter setting broadcast by each Flight Service Station is a computed correction for nonstandard surface pressure only, for a specific location and elevation. Consequently, altimeter indications based upon a local altimeter setting indicate true altitude at field elevation.

ALL
4923 [C] — (I04) — AC 61-27C, page 28
The pressure altimeter is a barometer that senses change in atmospheric pressure, and through a gearing mechanism, converts the pressure to an altitude indication in number of feet. The altimeter indicates the aircraft altitude in relation to the pressure level set in the barometric window.

ALL
4924 [C] — (I05) — AC 61-27C, page 79
The following method for entering descents is effective either with or without an attitude indicator:

1. Reduce airspeed to your selected descent airspeed while maintaining straight-and-level flight, then make a further reduction in power (to a predetermined setting); and
2. As the power is adjusted, simultaneously lower the nose to maintain constant airspeed, and trim off control pressures.

ALL
4925 [B] — (I05) — AC 61-27C, page 79
The level-off from a descent must be started before you reach the desired altitude. The amount of lead depends upon the rate of descent and control technique. Assuming a 500-fpm rate of descent, lead the altitude by 100-150 feet for level-off at an airspeed higher than descending speed. At the lead point, add power to the appropriate level flight cruise setting.

ALL
4926 [B] — (I05) — AC 61-27C, page 79
To level-off from a descent at descent airspeed, lead the desired altitude by approximately 50 feet, simultaneously adjusting the pitch attitude to level flight and adding power to a setting that will hold the airspeed constant.

ALL
4927 [C] — (I05) — AC 61-27C, page 90
A level-pitch attitude is indicated by the reversal and stabilization of the airspeed indicator and altimeter needles.

ALL
4928 [C] — (I05) — AC 61-27C, page 77
To enter a constant airspeed climb from cruising airspeed, raise the miniature aircraft on the attitude indicator to the approximate nose-high indication appropriate to the predetermined climb speed. The attitude will vary according to the type of airplane you are flying.

ALL
4929 [A] — (I05) — AC 61-27C, page 77
To enter a constant airspeed climb from cruising airspeed, raise the miniature aircraft on the attitude indicator to the approximate nose-high indication appropriate to the predetermined climb speed. The attitude will vary according to the type of airplane you are flying.

ALL
4930 [B] — (I05) — AC 61-27C, page 26
If the alternate static source is vented inside the airplane, where static pressure is usually lower than outside static pressure, selection of the alternate source may result in the following instrument indications:

1. The altimeter reads higher than normal;
2. Indicated airspeed greater than normal; and
3. The vertical-velocity indicator momentarily shows a climb.

ALL
4931 [A] — (I04) — AC 61-27C, page 44
Illustration 2 shows a slipping, half-standard rate turn. To perform a coordinated standard rate turn, we must increase our rate-of-turn and correct the slip by applying additional left rudder.

ALL
4932 [A] — (I04) — AC 61-27C, page 44
Illustration 3 indicates a coordinated turn, since the ball is centered.

ALL
4933 [B] — (I04) — AC 61-27C, page 44
Illustration 1 indicates a skidding turn, since the ball is to the outside of the direction of the turn. In a skid, the rate of turn is too great for the angle of bank, and excessive centrifugal force causes the ball to move to the outside of the turn.

ALL
4934 [B] — (I04) — AC 61-27C, page 44
Illustration 1 shows a skidding, half-standard rate turn. To perform a coordinated standard rate turn, we must increase our rate of turn and correct the skid by applying additional right rudder.

ALL
4935 [C] — (I04) — AC 61-27C, page 44
Illustration 2 indicates a slipping turn, since the ball is to the inside of the direction of the turn. In a slip, the rate of turn is too slow for the angle of bank, and the lack of centrifugal force causes the ball to move to the inside of the turn.

ALL
4936 [B] — (I05) — AC 61-27C, page 90
The conditions associated with this unusual attitude are: airspeed is high, nose is below horizon, vertical speed shows excessive rate of descent, compass indicates right turn, and turn coordinator shows uncoordinated greater-than-standard rate turn to the right. To correct for this nose-low condition the correct sequence is: reduce power to prevent excessive airspeed and loss of altitude, correct the bank attitude with coordinated aileron and rudder pressure to straight flight by referring to the turn coordinator, and raise the nose to level flight attitude by smooth back elevator pressure.

ALL
4937 [A] — (I05) — AC 61-27C, pages 33 and 90
In this situation, the vacuum instruments both show a climbing right turn. They are supported by the turn coordinator which shows a right turn, and the altimeter and VSI which indicate a climb. These are contradicted by the airspeed indicator which is rising (airspeed would not be increasing in this situation). The airspeed indicator then, is the one instrument which conflicts with the remaining unrelated systems and indicates that the pitot tube and drain holes are blocked, and the airspeed indicator is acting as an altimeter. The correct recovery procedure for a nose-high attitude is to lower the nose and level the wings to level flight attitude by use of the attitude indicator.

ALL
4938 [B] — (I05) — AC 61-27C, page 90
The conditions associated with this unusual attitude are: airspeed is decreasing, altitude is increasing, nose is above the horizon, vertical speed indicator shows an excessive rate of climb, compass indicates a right turn, and turn coordinator shows uncoordinated greater-than-standard rate turn to the right. To correct for this nose-high condition the correct sequence is: increase power, apply forward elevator pressure to lower the nose and prevent a stall, and correct the bank by applying coordinated aileron and rudder pressure to level the miniature aircraft and center the ball of the turn coordinator.

ALL
4939 [B] — (I05) — AC 61-27C, page 90
The attitude indicator and directional gyro show a right turn. The airspeed indicator, altimeter, and VSI show a climb. The turn coordinator (showing a lack of turn) is not in agreement with the other instruments.

ALL
4940 [C] — (I05) — AC 61-27C, page 90
The attitude indicator and the heading indicator (both vacuum driven) are presenting conflicting indications. The attitude indicator shows a bank to the right, while the heading indicator shows a left turn; therefore, there must be a vacuum system malfunction. All other instruments are in agreement with straight-and-level flight.

ALL

4941 [A] — (I05) — AC 61-27C, page 90

The question states that one instrument (not system) has malfunctioned. The attitude indicator disagrees with both the turn coordinator and the DG in its bank indication. It also disagrees with the altimeter, VSI, and airspeed indicator in its pitch indication. Elimination of the attitude indicator shows the airplane to be in a climbing turn to the right.

ALL

4942 [B] — (I05) — AC 61-27C, page 81

The question states that one instrument (not system) has malfunctioned. Although the airspeed indicator shows a decrease, the other pitch instruments (attitude indicator, altimeter, and VSI) all agree and show the aircraft in level flight. The airspeed indicator has malfunctioned. The three bank instruments are in agreement, showing a level turn to the right.

ALL

4943 [C] — (I05) — AC 61-27C, page 88

Notice the question states that there has been an entire system failure. The turn coordinator, DG, and attitude indicator show the aircraft to be turning right (based on the vacuum and electrical systems backing each other up). The altimeter is showing a descent. The airspeed indicator and vertical speed indicators are the only instrument in disagreement with a descending turn to the right (the airspeed should not be lower than stall speed, and the VSI indication appears erroneous).

Cross-Reference:

Answer, Subject Matter Knowledge Code & Category

The FAA does not publish a category list for any exam; therefore, the category assigned below is based on historical data and the best judgement of our researchers.

ALL All aircraft
AIR Airplane
RTC Rotorcraft (applies to both helicopter and gyroplane)
LTA Lighter-Than-Air (applies to hot air balloon, gas balloon and airship)

4001	[C]	(A20)	LTA
4002	[C]	(A24)	AIR
4003	[C]	(B08)	ALL
4004	[A]	(B07)	LTA
4005	[B]	(B10)	AIR
4006	[C]	(B10)	ALL
4007	[B]	(B11)	ALL
4008	[A]	(A20)	ALL
4009	[B]	(A20)	ALL
4010	[B]	(A20)	ALL
4011	[A]	(B08)	ALL
4012	[A]	(A20)	ALL
4013	[A]	(A20)	ALL
4014	[B]	(A20)	ALL
4015	[A]	(A20)	ALL
4016	[B]	(A20)	RTC
4017	[A]	(A20)	ALL
4018	[C]	(A20)	RTC
4019	[B]	(A20)	RTC
4020	[A]	(A20)	ALL
4021	[B]	(A20)	ALL
4022	[A]	(A20)	RTC
4023	[A]	(A20)	AIR
4024	[C]	(A20)	ALL
4025	[B]	(A20)	ALL
4026	[A]	(A20)	AIR
4027	[A]	(A20)	ALL
4028	[C]	(A20)	AIR
4029	[B]	(A20)	AIR
4030	[B]	(A20)	RTC
4031	[B]	(A20)	ALL
4032	[B]	(B10)	AIR
4033	[C]	(B08)	ALL
4034	[C]	(A20)	AIR
4035	[C]	(A20)	AIR
4036	[C]	(B10)	AIR
4037	[B]	(B11)	ALL
4038	[A]	(B11)	ALL
4039	[C]	(B07)	ALL
4040	[A]	(B10)	RTC
4041	[A]	(B10)	RTC
4042	[C]	(B11)	ALL
4043	[C]	(B11)	RTC
4044	[B]	(B10)	ALL
4045	[C]	(B11)	ALL
4046	[A]	(B10)	ALL
4047	[C]	(B13)	ALL
4048	[A]	(B10)	ALL
4049	[C]	(B13)	ALL
4050	[A]	(B11)	ALL
4051	[C]	(B11)	ALL
4052	[C]	(B11)	ALL
4053	[C]	(B11)	ALL
4054	[A]	(B10)	ALL
4055	[C]	(B11)	ALL
4056	[B]	(I05)	ALL
4057	[A]	(B11)	RTC
4058	[C]	(J15)	ALL
4059	[B]	(J15)	ALL
4060	[A]	(J15)	ALL
4061	[A]	(J15)	ALL
4062	[C]	(B10)	ALL
4063	[B]	(B10)	ALL
4064	[A]	(B10)	ALL
4065	[C]	(B10)	ALL
4066	[B]	(B10)	ALL
4067	[B]	(B10)	ALL
4068	[C]	(B10)	ALL
4069	[A]	(I07)	ALL
4070	[B]	(J34)	ALL
4071	[B]	(J15)	ALL
4072	[C]	(J15)	ALL
4073	[B]	(J15)	ALL
4074	[A]	(J15)	ALL
4075	[C]	(J15)	ALL
4076	[C]	(J15)	ALL
4077	[A]	(J15)	ALL
4078	[A]	(J15)	ALL
4079	[C]	(J34)	ALL
4080	[C]	(J06)	ALL
4081	[B]	(B10)	ALL
4082	[C]	(B10)	ALL
4083	[C]	(B10)	ALL
4084	[B]	(B10)	RTC
4085	[A]	(B10)	ALL
4086	[C]	(B10)	ALL
4087	[A]	(B10)	ALL
4088	[C]	(G10)	ALL
4089	[A]	(I22)	ALL
4090	[B]	(I22)	ALL
4091	[C]	(I22)	ALL
4092	[A]	(I31)	ALL
4093	[B]	(I22)	ALL
4094	[C]	(I21)	ALL
4095	[C]	(I21)	ALL
4096	[A]	(I21)	ALL

4097	[C]	(I20)	ALL
4098	[C]	(I27)	ALL
4099	[C]	(I24)	ALL
4100	[C]	(I29)	ALL
4101	[A]	(I24)	ALL
4102	[A]	(I24)	ALL
4103	[B]	(I24)	ALL
4104	[A]	(I24)	ALL
4105	[B]	(I23)	ALL
4106	[B]	(I23)	ALL
4107	[B]	(I23)	ALL
4108	[C]	(I23)	ALL
4109	[B]	(I22)	ALL
4110	[B]	(I22)	ALL
4111	[A]	(I22)	ALL
4112	[A]	(I21)	ALL
4113	[B]	(I21)	ALL
4114	[A]	(I21)	ALL
4115	[C]	(I25)	ALL
4116	[A]	(I27)	ALL
4117	[B]	(I27)	ALL
4118	[C]	(I25)	ALL
4119	[C]	(I27)	ALL
4120	[C]	(I27)	ALL
4121	[B]	(I25)	ALL
4122	[B]	(I25)	ALL
4123	[C]	(I25)	ALL
4124	[A]	(I25)	ALL
4125	[A]	(I21)	ALL
4126	[A]	(I30)	ALL
4127	[A]	(I27)	ALL
4128	[B]	(I27)	ALL
4129	[B]	(I28)	ALL
4130	[C]	(I28)	ALL
4131	[B]	(I26)	ALL
4132	[B]	(I28)	ALL
4133	[B]	(I29)	ALL
4134	[C]	(I26)	ALL
4135	[C]	(I30)	ALL
4136	[A]	(I27)	ALL
4137	[C]	(I30)	ALL
4138	[C]	(I28)	ALL
4139	[C]	(I28)	ALL
4140	[A]	(I28)	ALL
4141	[C]	(I27)	ALL
4142	[A]	(I43)	ALL
4143	[B]	(I30)	ALL
4144	[A]	(I30)	ALL
4145	[B]	(I30)	ALL
4146	[C]	(I30)	ALL
4147	[B]	(I30)	ALL
4148	[C]	(I30)	ALL
4149	[A]	(I28)	ALL
4150	[C]	(I28)	ALL
4151	[C]	(I29)	ALL
4152	[C]	(I29)	ALL
4153	[B]	(I29)	ALL
4154	[C]	(I20)	ALL
4155	[C]	(I32)	ALL
4156	[A]	(I31)	ALL
4157	[C]	(I26)	ALL
4158	[C]	(I27)	ALL
4159	[B]	(I24)	ALL
4160	[B]	(I30)	ALL
4161	[C]	(I24)	ALL
4162	[C]	(I32)	ALL
4163	[B]	(I31)	ALL
4164	[A]	(I31)	ALL
4165	[A]	(I31)	ALL
4166	[C]	(I31)	ALL
4167	[A]	(I31)	ALL
4168	[B]	(I32)	ALL
4169	[C]	(I31)	ALL
4170	[A]	(I43)	ALL
4171	[C]	(I29)	ALL
4172	[A]	(I48)	ALL
4173	[B]	(I53)	ALL
4174	[A]	(I46)	ALL
4175	[A]	(I47)	ALL
4176	[C]	(I43)	ALL
4177	[B]	(I43)	ALL
4178	[B]	(I43)	ALL
4179	[C]	(I43)	ALL
4180	[A]	(I43)	ALL
4181	[B]	(I43)	ALL
4182	[B]	(I42)	ALL
4183	[A]	(I43)	ALL
4184	[C]	(I43)	ALL
4185	[C]	(I43)	ALL
4186	[A]	(I43)	ALL
4187	[C]	(I43)	ALL
4188	[B]	(I43)	ALL
4189	[C]	(I43)	ALL
4190	[B]	(I43)	ALL
4191	[C]	(I43)	ALL
4192	[C]	(I48)	ALL
4193	[C]	(I48)	ALL
4194	[A]	(I48)	ALL
4195	[C]	(I51)	ALL
4196	[A]	(I41)	ALL
4197	[A]	(I50)	ALL
4198	[A]	(I41)	ALL
4199	[B]	(I48)	ALL
4200	[A]	(I21)	ALL
4201	[C]	(I43)	ALL
4202	[C]	(I41)	ALL
4203	[A]	(I43)	ALL
4204	[C]	(I41)	ALL
4205	[C]	(I41)	ALL
4206	[C]	(I45)	ALL
4207	[B]	(I45)	ALL
4208	[C]	(I45)	ALL
4209	[C]	(I44)	ALL
4210	[A]	(I53)	ALL
4211	[A]	(I47)	ALL
4212	[C]	(I47)	ALL
4213	[A]	(I47)	ALL
4214	[C]	(I47)	ALL
4215	[B]	(I43)	ALL
4216	[B]	(I47)	ALL
4217	[A]	(I47)	ALL
4218	[A]	(I47)	ALL
4219	[A]	(I47)	ALL
4220	[C]	(I42)	ALL
4221	[C]	(I47)	ALL
4222	[B]	(I47)	ALL

4223	[A]	(I47)	ALL
4224	[B]	(I47)	ALL
4225	[C]	(I47)	ALL
4226	[B]	(I43)	ALL
4227	[C]	(I20)	ALL
4228	[C]	(I43)	ALL
4229	[B]	(I47)	ALL
4230	[C]	(I46)	ALL
4231	[B]	(I46)	ALL
4232	[C]	(I46)	ALL
4233	[B]	(I46)	ALL
4234	[A]	(I46)	ALL
4235	[B]	(I46)	ALL
4236	[A]	(I46)	ALL
4237	[A]	(I46)	ALL
4238	[C]	(I23)	ALL
4239	[C]	(I50)	ALL
4240	[C]	(I50)	ALL
4241	[C]	(I43)	ALL
4242	[B]	(I52)	ALL
4243	[C]	(I52)	ALL
4244	[C]	(I52)	ALL
4245	[A]	(I52)	ALL
4246	[A]	(I48)	ALL
4247	[C]	(I48)	ALL
4248	[B]	(I50)	ALL
4249	[C]	(I48)	ALL
4250	[B]	(I48)	ALL
4251	[C]	(J25)	ALL
4252	[C]	(J25)	ALL
4253	[C]	(J25)	ALL
4254	[C]	(J25)	ALL
4255	[C]	(J25)	ALL
4256	[C]	(J25)	ALL
4257	[A]	(J25)	ALL
4258	[B]	(J25)	ALL
4259	[A]	(B07)	ALL
4260	[A]	(H06)	ALL
4261	[A]	(J17)	ALL
4262	[B]	(H06)	ALL
4263	[B]	(J35)	ALL
4264	[C]	(J35)	ALL
4265	[A]	(H06)	ALL
4266	[C]	(J15)	ALL
4267	[B]	(H06)	ALL
4268	[C]	(H06)	ALL
4269	[A]	(I08)	ALL
4270	[C]	(H06)	ALL
4271	[A]	(J16)	ALL
4272	[B]	(I05)	ALL
4273	[C]	(J01)	ALL
4274	[A]	(I10)	ALL
4275	[B]	(I10)	ALL
4276	[B]	(I10)	ALL
4277	[C]	(J15)	ALL
4278	[B]	(H06)	ALL
4279	[A]	(H06)	ALL
4280	[A]	(J35)	ALL
4281	[A]	(I08)	ALL
4282	[B]	(I07)	ALL
4283	[A]	(I07)	ALL
4284	[A]	(K24)	ALL
4285	[B]	(I10)	ALL
4286	[C]	(J18)	ALL
4287	[A]	(J35)	ALL
4288	[B]	(J15)	ALL
4289	[A]	(H06)	ALL
4290	[C]	(H06)	ALL
4291	[A]	(J35)	ALL
4292	[B]	(J41)	ALL
4293	[C]	(I12)	ALL
4294	[C]	(J41)	ALL
4295	[C]	(I12)	ALL
4296	[C]	(J42)	ALL
4297	[A]	(J42)	ALL
4298	[C]	(J42)	ALL
4299	[A]	(I08)	ALL
4300	[C]	(J15)	ALL
4301	[A]	(H06)	ALL
4302	[B]	(H06)	ALL
4303	[C]	(I10)	ALL
4304	[A]	(I08)	ALL
4305	[B]	(J34)	ALL
4306	[A]	(I12)	ALL
4307	[B]	(I10)	ALL
4308	[C]	(I10)	ALL
4309	[C]	(I10)	ALL
4310	[C]	(I10)	ALL
4311	[C]	(I10)	ALL
4312	[A]	(J15)	ALL
4313	[B]	(H06)	ALL
4314	[C]	(H06)	ALL
4315	[B]	(I08)	ALL
4316	[C]	(J40)	ALL
4317	[A]	(J35)	ALL
4318	[B]	(J35)	ALL
4319	[A]	(I10)	ALL
4320	[C]	(I08)	ALL
4321	[C]	(J35)	ALL
4322	[C]	(J15)	RTC
4323	[A]	(H06)	RTC
4324	[C]	(H06)	RTC
4325	[B]	(J35)	ALL
4326	[A]	(J01)	ALL
4327	[A]	(J34)	ALL
4328	[C]	(I10)	RTC
4329	[A]	(J08)	RTC
4330	[B]	(I10)	RTC
4331	[C]	(I08)	ALL
4332	[C]	(I10)	ALL
4333	[C]	(J15)	RTC
4334	[B]	(H06)	RTC
4335	[B]	(H06)	RTC
4336	[C]	(J35)	ALL
4337	[A]	(J01)	ALL
4338	[A]	(I08)	ALL
4339	[B]	(I10)	ALL
4340	[B]	(I10)	RTC
4341	[B]	(I10)	RTC
4342	[A]	(I10)	RTC
4343	[C]	(I10)	RTC
4344	[A]	(J15)	ALL
4345	[A]	(H06)	ALL
4346	[B]	(H06)	ALL
4347	[A]	(I08)	ALL
4348	[C]	(J35)	ALL

4349	[B]	(J41)	ALL
4350	[B]	(J42)	ALL
4351	[B]	(J42)	ALL
4352	[A]	(J42)	ALL
4353	[C]	(J42)	ALL
4354	[B]	(J42)	ALL
4355	[A]	(J42)	ALL
4356	[B]	(J42)	ALL
4357	[C]	(J42)	ALL
4358	[C]	(J15)	ALL
4359	[B]	(H06)	ALL
4360	[C]	(H06)	ALL
4361	[B]	(J40)	ALL
4362	[C]	(J01)	ALL
4363	[C]	(J10)	ALL
4364	[C]	(J40)	ALL
4365	[C]	(J35)	ALL
4366	[B]	(J35)	ALL
4367	[C]	(I08)	ALL
4368	[B]	(J42)	ALL
4369	[B]	(J42)	ALL
4370	[B]	(J35)	ALL
4371	[B]	(J42)	ALL
4372	[A]	(J01)	ALL
4373	[B]	(J01)	ALL
4374	[A]	(J24)	LTA
4375	[B]	(B13)	ALL
4376	[C]	(J01)	ALL
4377	[B]	(J01)	ALL
4378	[B]	(J01)	ALL
4379	[C]	(J19)	ALL
4380	[B]	(J14)	ALL
4381	[A]	(J14)	ALL
4382	[A]	(J01)	ALL
4383	[B]	(J01)	ALL
4384	[B]	(J01)	ALL
4385	[C]	(J01)	ALL
4386	[C]	(J01)	ALL
4387	[C]	(J01)	ALL
4388	[C]	(J01)	ALL
4389	[A]	(J01)	ALL
4390	[B]	(J11)	ALL
4391	[A]	(J01)	ALL
4392	[B]	(J14)	ALL
4393	[C]	(J14)	ALL
4394	[B]	(J16)	ALL
4395	[B]	(J14)	ALL
4396	[B]	(J14)	ALL
4397	[A]	(I08)	ALL
4398	[C]	(J14)	ALL
4399	[B]	(I08)	ALL
4400	[C]	(J01)	ALL
4401	[B]	(J25)	ALL
4402	[C]	(J28)	ALL
4403	[B]	(J11)	ALL
4404	[A]	(J11)	ALL
4405	[B]	(J15)	ALL
4406	[B]	(J15)	ALL
4407	[A]	(B07)	ALL
4408	[B]	(J03)	ALL
4409	[C]	(J11)	ALL
4410	[A]	(J10)	ALL
4411	[C]	(J01)	ALL
4412	[C]	(J01)	ALL
4413	[B]	(I08)	ALL
4414	[B]	(J16)	ALL
4415	[C]	(J11)	ALL
4416	[B]	(J11)	ALL
4417	[C]	(J40)	ALL
4418	[C]	(J40)	ALL
4419	[C]	(J40)	ALL
4420	[B]	(J16)	ALL
4421	[A]	(J11)	ALL
4422	[A]	(J33)	ALL
4423	[B]	(J33)	ALL
4424	[C]	(J33)	ALL
4425	[A]	(B07)	RTC
4426	[C]	(B08)	ALL
4427	[C]	(B08)	ALL
4428	[C]	(B08)	RTC
4429	[A]	(J33)	ALL
4430	[B]	(B11)	ALL
4431	[C]	(B11)	ALL
4432	[C]	(J33)	ALL
4433	[C]	(B11)	ALL
4434	[B]	(J09)	ALL
4435	[A]	(J33)	ALL
4436	[C]	(B06)	ALL
4437	[A]	(J33)	ALL
4438	[A]	(B08)	ALL
4439	[C]	(B08)	ALL
4440	[A]	(B08)	ALL
4441	[A]	(B08)	ALL
4442	[C]	(J14)	ALL
4443	[B]	(J14)	ALL
4444	[C]	(J14)	ALL
4445	[C]	(J26)	ALL
4446	[B]	(J26)	ALL
4447	[A]	(J08)	ALL
4448	[C]	(B08)	ALL
4449	[B]	(J14)	ALL
4450	[A]	(J14)	ALL
4451	[A]	(J14)	ALL
4452	[B]	(J14)	ALL
4453	[C]	(J14)	ALL
4454	[B]	(J14)	ALL
4455	[A]	(J14)	ALL
4456	[B]	(B07)	ALL
4457	[C]	(J14)	ALL
4458	[A]	(J14)	ALL
4459	[C]	(B08)	ALL
4460	[A]	(B08)	ALL
4461	[A]	(B08)	ALL
4462	[B]	(B08)	ALL
4463	[B]	(B08)	ALL
4464	[A]	(J24)	ALL
4465	[A]	(J24)	ALL
4466	[A]	(J24)	ALL
4467	[C]	(J25)	ALL
4468	[A]	(J25)	ALL
4469	[C]	(J10)	ALL
4470	[C]	(J16)	ALL
4471	[C]	(J19)	ALL
4472	[B]	(I07)	ALL
4473	[A]	(J08)	ALL
4474	[B]	(J08)	ALL

4475 .... [A] .... (J07) .... ALL
4476 .... [C] .... (J08) .... ALL
4477 .... [A] .... (I04) .... ALL
4478 .... [A] .... (I04) .... ALL
4479 .... [B] .... (I04) .... ALL
4480 .... [C] .... (J26) .... ALL
4481 .... [C] .... (J26) .... ALL
4482 .... [C] .... (J26) .... ALL
4483 .... [C] .... (I04) .... ALL
4484 .... [B] .... (I04) .... ALL
4485 .... [A] .... (J17) .... ALL
4486 .... [B] .... (J16) .... ALL
4487 .... [B] .... (I07) .... ALL
4488 .... [C] .... (J07) .... ALL
4489 .... [A] .... (J16) .... ALL
4490 .... [B] .... (J16) .... ALL
4491 .... [A] .... (J16) .... ALL
4492 .... [C] .... (J16) .... ALL
4493 .... [C] .... (J17) .... ALL
4494 .... [A] .... (J17) .... ALL
4495 .... [A] .... (I08) .... ALL
4496 .... [C] .... (J35) .... ALL
4497 .... [C] .... (J35) .... ALL
4498 .... [C] .... (J35) .... ALL
4499 .... [A] .... (J35) .... ALL
4500 .... [C] .... (J24) .... ALL
4501 .... [B] .... (J35) .... ALL
4502 .... [B] .... (J35) .... ALL
4503 .... [C] .... (B11) .... ALL
4504 .... [C] .... (J35) .... ALL
4505 .... [B] .... (B10) .... ALL
4506 .... [C] .... (J35) .... ALL
4507 .... [A] .... (I08) .... ALL
4508 .... [C] .... (J35) .... ALL
4509 .... [B] .... (J35) .... ALL
4510 .... [C] .... (J35) .... ALL
4511 .... [C] .... (I13) .... ALL
4512 .... [C] .... (J03) .... ALL
4513 .... [C] .... (B11) .... ALL
4514 .... [B] .... (I13) .... ALL
4515 .... [B] .... (S32) .... ALL
4516 .... [B] .... (J35) .... ALL
4517 .... [B] .... (J35) .... ALL
4518 .... [B] .... (B09) .... ALL
4519 .... [A] .... (B09) .... ALL
4520 .... [C] .... (B09) .... ALL
4521 .... [B] .... (B09) .... ALL
4522 .... [C] .... (B09) .... ALL
4523 .... [A] .... (B09) .... ALL
4524 .... [B] .... (B09) .... ALL
4525 .... [C] .... (B09) .... ALL
4526 .... [B] .... (J06) .... ALL
4527 .... [B] .... (J06) .... ALL
4528 .... [B] .... (J06) .... ALL
4529 .... [C] .... (J06) .... ALL
4530 .... [B] .... (J06) .... ALL
4531 .... [B] .... (J06) .... ALL
4532 .... [C] .... (J06) .... ALL
4533 .... [C] .... (J06) .... ALL
4534 .... [B] .... (J05) .... ALL
4535 .... [B] .... (J05) .... ALL
4536 .... [A] .... (J05) .... ALL
4537 .... [B] .... (J05) .... ALL
4538 .... [C] .... (J13) .... ALL
4539 .... [A] .... (J06) .... ALL
4540 .... [B] .... (J18) .... ALL
4541 .... [C] .... (B08) .... ALL
4542 .... [C] .... (B08) .... ALL
4543 .... [C] .... (B08) .... ALL
4544 .... [B] .... (J19) .... ALL
4545 .... [A] .... (J19) .... ALL
4546 .... [B] .... (J10) .... ALL
4547 .... [B] .... (J19) .... ALL
4548 .... [C] .... (I07) .... ALL
4549 .... [C] .... (I08) .... ALL
4550 .... [C] .... (I08) .... ALL
4551 .... [C] .... (I08) .... ALL
4552 .... [B] .... (I08) .... ALL
4553 .... [B] .... (I08) .... ALL
4554 .... [C] .... (I07) .... ALL
4555 .... [C] .... (J14) .... ALL
4556 .... [C] .... (I08) .... ALL
4557 .... [A] .... (I08) .... ALL
4558 .... [C] .... (I08) .... ALL
4559 .... [B] .... (I08) .... ALL
4560 .... [C] .... (I08) .... ALL
4561 .... [A] .... (I08) .... ALL
4562 .... [C] .... (I08) .... ALL
4563 .... [A] .... (I08) .... ALL
4564 .... [B] .... (I08) .... ALL
4565 .... [C] .... (I08) .... ALL
4566 .... [C] .... (I08) .... ALL
4567 .... [C] .... (I08) .... ALL
4568 .... [A] .... (I08) .... ALL
4569 .... [B] .... (I08) .... ALL
4570 .... [B] .... (I08) .... ALL
4571 .... [C] .... (I08) .... ALL
4572 .... [C] .... (I08) .... ALL
4573 .... [B] .... (I08) .... ALL
4574 .... [C] .... (I08) .... ALL
4575 .... [A] .... (I08) .... ALL
4576 .... [C] .... (I08) .... ALL
4577 .... [C] .... (I08) .... ALL
4578 .... [B] .... (I08) .... ALL
4579 .... [B] .... (I08) .... ALL
4580 .... [B] .... (I08) .... ALL
4581 .... [A] .... (I08) .... ALL
4582 .... [B] .... (I08) .... ALL
4583 .... [C] .... (I08) .... ALL
4584 .... [A] .... (I08) .... ALL
4585 .... [B] .... (I08) .... ALL
4586 .... [B] .... (I08) .... ALL
4587 .... [A] .... (I08) .... ALL
4588 .... [C] .... (I08) .... ALL
4589 .... [A] .... (I08) .... ALL
4590 .... [C] .... (I08) .... ALL
4591 .... [C] .... (I08) .... ALL
4592 .... [C] .... (I08) .... ALL
4593 .... [B] .... (I08) .... ALL
4594 .... [A] .... (I08) .... ALL
4595 .... [B] .... (I08) .... ALL
4596 .... [A] .... (I08) .... ALL
4597 .... [B] .... (I08) .... ALL
4598 .... [C] .... (I08) .... ALL
4599 .... [A] .... (I08) .... ALL
4600 .... [C] .... (I08) .... ALL

4601	[B]	(I08)	ALL
4602	[B]	(I08)	ALL
4603	[A]	(I08)	ALL
4604	[B]	(I08)	ALL
4605	[B]	(I12)	ALL
4606	[A]	(I08)	ALL
4607	[C]	(I08)	ALL
4608	[C]	(I08)	ALL
4609	[B]	(J17)	ALL
4610	[A]	(J17)	ALL
4611	[B]	(J17)	ALL
4612	[C]	(J17)	ALL
4613	[B]	(J17)	ALL
4614	[C]	(J17)	ALL
4615	[C]	(J17)	AIR
4616	[A]	(J17)	ALL
4617	[B]	(J17)	ALL
4618	[C]	(J17)	ALL
4619	[B]	(J17)	ALL
4620	[C]	(J17)	ALL
4621	[C]	(J17)	ALL
4622	[B]	(J17)	ALL
4623	[A]	(J17)	ALL
4624	[A]	(J17)	ALL
4625	[C]	(I12)	ALL
4626	[B]	(J17)	AIR
4627	[B]	(J18)	ALL
4628	[B]	(J18)	ALL
4629	[C]	(J18)	ALL
4630	[A]	(B08)	ALL
4631	[C]	(J18)	ALL
4632	[C]	(J18)	ALL
4633	[C]	(J19)	ALL
4634	[C]	(B08)	ALL
4635	[B]	(I10)	ALL
4636	[A]	(I10)	ALL
4637	[C]	(I10)	ALL
4638	[B]	(J16)	ALL
4639	[C]	(J15)	LTA
4640	[B]	(J17)	ALL
4641	[C]	(J18)	ALL
4642	[B]	(J33)	ALL
4643	[C]	(J33)	LTA
4644	[C]	(J14)	LTA
4645	[C]	(J35)	ALL
4646	[A]	(J35)	ALL
4647	[C]	(J35)	ALL
4648	[B]	(J42)	ALL
4649	[A]	(J42)	ALL
4650	[B]	(J42)	ALL
4651	[B]	(J42)	ALL
4652	[C]	(J42)	ALL
4653	[B]	(B11)	ALL
4654	[B]	(J42)	ALL
4655	[A]	(J18)	ALL
4656	[A]	(B97)	ALL
4657	[B]	(B97)	ALL
4658	[C]	(B97)	ALL
4659	[B]	(B97)	ALL
4660	[C]	(B97)	ALL
4661	[A]	(B97)	ALL
4662	[A]	(B97)	ALL
4663	[B]	(J01)	ALL
4664	[A]	(J01)	ALL
4665	[C]	(J01)	ALL
4666	[B]	(I07)	ALL
4667	[A]	(J18)	ALL
4668	[C]	(J17)	ALL
4669	[A]	(J01)	ALL
4670	[A]	(J18)	ALL
4671	[A]	(J18)	ALL
4672	[A]	(J18)	ALL
4673	[C]	(I10)	RTC
4674	[C]	(J01)	ALL
4675	[A]	(I10)	ALL
4676	[C]	(I10)	RTC
4677	[B]	(I10)	ALL
4678	[A]	(I10)	ALL
4679	[A]	(I10)	RTC
4680	[A]	(I10)	ALL
4681	[C]	(J17)	ALL
4682	[A]	(I10)	ALL
4683	[C]	(I10)	ALL
4684	[C]	(I10)	ALL
4685	[B]	(I07)	ALL
4686	[A]	(B97)	ALL
4687	[B]	(B97)	ALL
4688	[A]	(B97)	ALL
4689	[A]	(B97)	ALL
4690	[A]	(B97)	ALL
4691	[C]	(B97)	ALL
4692	[B]	(J18)	ALL
4693	[B]	(B97)	ALL
4694	[A]	(B97)	ALL
4695	[A]	(B97)	ALL
4696	[B]	(B97)	ALL
4697	[A]	(B97)	RTC
4698	[A]	(B97)	ALL
4699	[C]	(B10)	ALL
4700	[C]	(B97)	ALL
4701	[C]	(B97)	ALL
4702	[A]	(I07)	ALL
4703	[A]	(I07)	ALL
4704	[B]	(I07)	ALL
4705	[B]	(I07)	ALL
4706	[B]	(B10)	ALL
4707	[C]	(J27)	ALL
4708	[C]	(J27)	ALL
4709	[B]	(J27)	ALL
4710	[C]	(J27)	ALL
4711	[C]	(B10)	ALL
4712	[A]	(B09)	RTC
4713	[B]	(B10)	RTC
4714	[A]	(J33)	ALL
4715	[A]	(B97)	ALL
4716	[A]	(B10)	ALL
4717	[B]	(J33)	ALL
4718	[A]	(J18)	ALL
4719	[C]	(B10)	ALL
4720	[B]	(K04)	ALL
4721	[B]	(I10)	ALL
4722	[B]	(J42)	RTC
4723	[A]	(J42)	RTC
4724	[C]	(J42)	RTC
4725	[C]	(J19)	ALL
4726	[B]	(J18)	ALL

4727	[C]	(K04)	ALL
4728	[C]	(J18)	ALL
4729	[C]	(J10)	ALL
4730	[C]	(J01)	ALL
4731	[A]	(I12)	ALL
4732	[A]	(I12)	ALL
4733	[B]	(I12)	ALL
4734	[A]	(J19)	ALL
4735	[C]	(J19)	ALL
4736	[A]	(J19)	ALL
4737	[C]	(J19)	ALL
4738	[B]	(V14)	ALL
4739	[B]	(K04)	ALL
4740	[B]	(J18)	ALL
4741	[C]	(J18)	ALL
4742	[B]	(J01)	ALL
4743	[C]	(J18)	ALL
4744	[A]	(J18)	ALL
4745	[C]	(K04)	ALL
4746	[C]	(B97)	ALL
4747	[B]	(I07)	ALL
4748	[C]	(K04)	ALL
4749	[C]	(J18)	ALL
4750	[A]	(J18)	ALL
4751	[C]	(J18)	ALL
4752	[C]	(J01)	ALL
4753	[B]	(J01)	ALL
4754	[C]	(B10)	ALL
4755	[A]	(K04)	ALL
4756	[B]	(I05)	ALL
4757	[C]	(J18)	ALL
4758	[B]	(J13)	ALL
4759	[A]	(B10)	ALL
4760	[B]	(B10)	ALL
4761	[C]	(B10)	ALL
4762	[C]	(B10)	ALL
4763	[A]	(B10)	ALL
4764	[B]	(B10)	ALL
4765	[C]	(B10)	ALL
4766	[B]	(J17)	ALL
4767	[A]	(J17)	ALL
4768	[B]	(J17)	ALL
4769	[C]	(B10)	ALL
4770	[B]	(J01)	ALL
4771	[A]	(J18)	ALL
4772	[B]	(K04)	ALL
4773	[A]	(I08)	ALL
4774	[B]	(J03)	ALL
4775	[B]	(J03)	ALL
4776	[C]	(J03)	ALL
4777	[A]	(J03)	ALL
4778	[A]	(J03)	ALL
4779	[C]	(J03)	ALL
4780	[A]	(J03)	ALL
4781	[B]	(J03)	ALL
4782	[C]	(J03)	ALL
4783	[B]	(J03)	ALL
4784	[C]	(J03)	ALL
4785	[A]	(J03)	ALL
4786	[B]	(J03)	ALL
4787	[C]	(J03)	ALL
4788	[A]	(J03)	ALL
4789	[B]	(J03)	ALL
4790	[C]	(J03)	ALL
4791	[B]	(J05)	ALL
4792	[B]	(J05)	ALL
4793	[B]	(J05)	ALL
4794	[A]	(J05)	ALL
4795	[C]	(J03)	ALL
4796	[A]	(J03)	ALL
4797	[C]	(J05)	ALL
4798	[B]	(J01)	ALL
4799	[A]	(J01)	ALL
4800	[C]	(J01)	ALL
4801	[C]	(J01)	ALL
4802	[A]	(I02)	ALL
4803	[A]	(J31)	ALL
4804	[A]	(J31)	ALL
4805	[A]	(J31)	ALL
4806	[C]	(J31)	ALL
4807	[A]	(J31)	ALL
4808	[B]	(J31)	ALL
4809	[B]	(J31)	ALL
4810	[C]	(J31)	ALL
4811	[C]	(J31)	ALL
4812	[B]	(J31)	ALL
4813	[B]	(J31)	ALL
4814	[C]	(J31)	ALL
4815	[C]	(J31)	ALL
4816	[B]	(J31)	ALL
4817	[A]	(J31)	ALL
4818	[A]	(J31)	ALL
4819	[C]	(J31)	ALL
4820	[B]	(I05)	ALL
4821	[A]	(L57)	ALL
4822	[C]	(J18)	ALL
4823	[A]	(J18)	ALL
4824	[B]	(I07)	ALL
4825	[C]	(I07)	ALL
4826	[A]	(I07)	ALL
4827	[C]	(I04)	ALL
4828	[A]	(I04)	ALL
4829	[C]	(I04)	ALL
4830	[C]	(L57)	ALL
4831	[B]	(I04)	ALL
4832	[A]	(I05)	ALL
4833	[A]	(I05)	ALL
4834	[C]	(I05)	ALL
4835	[A]	(I05)	ALL
4836	[A]	(I05)	ALL
4837	[A]	(I05)	ALL
4838	[B]	(I05)	ALL
4839	[A]	(I05)	ALL
4840	[C]	(I05)	ALL
4841	[C]	(I06)	RTC
4842	[C]	(I05)	ALL
4843	[A]	(I03)	ALL
4844	[B]	(I03)	ALL
4845	[A]	(I05)	ALL
4846	[B]	(I06)	RTC
4847	[A]	(I03)	ALL
4848	[C]	(I05)	ALL
4849	[B]	(I06)	RTC
4850	[A]	(I05)	ALL
4851	[C]	(I05)	ALL
4852	[C]	(I06)	RTC

4853	[B]	(I06)	ALL
4854	[A]	(L57)	ALL
4855	[B]	(I05)	ALL
4856	[A]	(I05)	ALL
4857	[C]	(I04)	ALL
4858	[A]	(I05)	ALL
4859	[C]	(I05)	ALL
4860	[B]	(I04)	ALL
4861	[C]	(I04)	ALL
4862	[B]	(I05)	ALL
4863	[C]	(I05)	ALL
4864	[A]	(I04)	ALL
4865	[C]	(I05)	ALL
4866	[B]	(I05)	ALL
4867	[A]	(I05)	ALL
4868	[B]	(I04)	ALL
4869	[C]	(I05)	ALL
4870	[C]	(I05)	ALL
4871	[C]	(I05)	ALL
4872	[C]	(I05)	ALL
4873	[B]	(I05)	ALL
4874	[A]	(I05)	ALL
4875	[B]	(I05)	ALL
4876	[C]	(I05)	ALL
4877	[C]	(I05)	ALL
4878	[B]	(I05)	ALL
4879	[C]	(I05)	ALL
4880	[C]	(I05)	ALL
4881	[C]	(L59)	ALL
4882	[B]	(L59)	ALL
4883	[A]	(L59)	ALL
4884	[C]	(I04)	ALL
4885	[A]	(I05)	ALL
4886	[C]	(I05)	ALL
4887	[A]	(I05)	ALL
4888	[B]	(I05)	ALL
4889	[C]	(I05)	ALL
4890	[A]	(I05)	ALL
4891	[C]	(I05)	ALL
4892	[B]	(I05)	ALL
4893	[C]	(I05)	ALL
4894	[B]	(I05)	ALL
4895	[C]	(I05)	ALL
4896	[A]	(I05)	ALL
4897	[B]	(I05)	ALL
4898	[C]	(I04)	ALL
4899	[B]	(I04)	ALL
4900	[A]	(I05)	ALL
4901	[A]	(I05)	ALL
4902	[B]	(I04)	ALL
4903	[A]	(I05)	ALL
4904	[C]	(I04)	ALL
4905	[C]	(I04)	ALL
4906	[A]	(I05)	ALL
4907	[A]	(I05)	ALL
4908	[C]	(I04)	ALL
4909	[A]	(I04)	ALL
4910	[B]	(I04)	ALL
4911	[B]	(I04)	ALL
4912	[B]	(I04)	ALL
4913	[C]	(I04)	ALL
4914	[B]	(I03)	ALL
4915	[B]	(I03)	ALL
4916	[B]	(H66)	ALL
4917	[B]	(K04)	ALL
4918	[A]	(I04)	ALL
4919	[C]	(I04)	ALL
4920	[C]	(I04)	ALL
4921	[C]	(I05)	ALL
4922	[B]	(I04)	ALL
4923	[C]	(I04)	ALL
4924	[C]	(I05)	ALL
4925	[B]	(I05)	ALL
4926	[B]	(I05)	ALL
4927	[C]	(I05)	ALL
4928	[C]	(I05)	ALL
4929	[A]	(I05)	ALL
4930	[B]	(I05)	ALL
4931	[A]	(I04)	ALL
4932	[A]	(I04)	ALL
4933	[B]	(I04)	ALL
4934	[B]	(I04)	ALL
4935	[C]	(I04)	ALL
4936	[B]	(I05)	ALL
4937	[A]	(I05)	ALL
4938	[B]	(I05)	ALL
4939	[B]	(I05)	ALL
4940	[C]	(I05)	ALL
4941	[A]	(I05)	ALL
4942	[B]	(I05)	ALL
4943	[C]	(I05)	ALL

Cross-Reference:
Subject Matter Knowledge Code & Question Number

The subject matter knowledge codes establish the specific reference for the knowledge standard. When reviewing results of your knowledge test, you should compare the subject matter knowledge code(s) on your test report to the ones found below. All the questions on the Instrument test have been broken down into their subject matter knowledge codes and listed under the appropriate reference. This will be helpful for both review and preparation for the practical test.

FAR 61: Certification: Pilots and Flight Instructors

A20..... General
4001, 4008, 4009, 4010, 4012, 4013, 4014, 4015, 4016, 4017, 4018, 4019, 4020, 4021, 4022, 4023, 4024, 4025, 4026, 4027, 4028, 4029, 4030, 4031, 4034, 4035

A21..... Aircraft Ratings and Special Certificates

A23..... Private Pilots

A24..... Commercial Pilots

A26..... Flight Instructors

FAR 91: General Operating and Flight Rules

B06..... (FAR §91)
4436

B07..... General
4004, 4039, 4259, 4407, 4425, 4456

B08..... Flight Rules: General
4003, 4011, 4033, 4426, 4427, 4428, 4438, 4439, 4440, 4441, 4448, 4459, 4460, 4461, 4462, 4463, 4541, 4542, 4543, 4630, 4634

B09..... Visual Flight Rules
4518, 4519, 4520, 4521, 4522, 4523, 4524, 4525, 4712

B10..... Instrument Flight Rules
4005, 4006, 4032, 4036, 4040, 4041, 4044, 4046, 4048, 4054, 4062, 4063, 4064, 4065, 4066, 4067, 4068, 4081, 4082, 4083, 4084, 4085, 4086, 4087, 4505, 4699, 4706, 4711, 4713, 4716, 4719, 4754, 4759, 4760, 4761, 4762, 4763, 4764, 4765, 4769

B11..... Equipment, Instrument, and Certification Requirements
4007, 4037, 4038, 4042, 4043, 4045, 4050, 4051, 4052, 4053, 4055, 4057, 4430, 4431, 4433, 4503, 4513, 4653

B12..... Special Flight Operations

B13..... Maintenance, Preventive Maintenance, and Alterations
4047, 4049, 4375

FAR 97: Standard Instrument Approach Procedures

B97..... General
4656, 4657, 4658, 4659, 4660, 4661, 4662, 4686, 4687, 4688, 4689, 4690, 4691, 4693, 4694, 4695, 4696, 4697, 4698, 4700, 4701, 4715, 4746

NTSB 830: Rules Pertaining to the Notification and Reporting of Aircraft Accidents or Incidents and Overdue Aircraft, and Preservation of Aircraft Wreckage, Mail, Cargo, and Records

G10 .... General
4088

G11 .... Initial Notification of Aircraft Accidents, Incidents, and Overdue Aircraft

G12 .... Preservation of Aircraft Wreckage, Mail, Cargo, and Records

G13 .... Reporting of Aircraft Accidents, Incidents, and Overdue Aircraft

AC 61-23: Pilot's Handbook Of Aeronautical Knowledge

H03..... Flight Instruments

H04..... Airplane Performance

H05..... Weather

H06..... Basic Calculations Using Navigational Computers or Electronic Calculators
4260, 4262, 4265, 4267, 4268, 4270, 4278, 4279, 4289, 4290, 4301, 4302, 4313, 4314, 4323, 4324, 4334, 4335, 4345, 4346, 4359, 4360

H07..... Navigation

H09..... Appendix 1: Obtaining FAA Publications

AC 61-21: Flight Training Handbook

H62..... Emergency Flight by Reference to Instruments

H66..... (AC 61-21)
4916

AC 61-27: Instrument Flying Handbook

I01 ...... Training Considerations

I02 ...... Instrument Flying: Coping with Illusions in Flight
4802

I03 ...... Aerodynamic Factors Related to Instrument Flying
4843, 4844, 4847, 4914, 4915

I04 ...... Basic Flight Instruments
4477, 4478, 4479, 4483, 4484, 4827, 4828, 4829, 4831, 4857, 4860, 4861, 4864, 4868, 4884, 4898, 4899, 4902, 4904, 4905, 4908, 4909, 4910, 4911, 4912, 4913, 4918, 4919, 4920, 4922, 4923, 4931, 4932, 4933, 4934, 4935

I05 ...... Attitude Instrument Flying—Airplanes
4056, 4272, 4756, 4820, 4832, 4833, 4834, 4835, 4836, 4837, 4838, 4839, 4840, 4842, 4845, 4848, 4850, 4851, 4855, 4856, 4858, 4859, 4862, 4863, 4865, 4866, 4867, 4869, 4870, 4871, 4872, 4873, 4874, 4875, 4876, 4877, 4878, 4879, 4880, 4885, 4886, 4887, 4888, 4889, 4890, 4891, 4892, 4893, 4894, 4895, 4896, 4897, 4900, 4901, 4903, 4906, 4907, 4921, 4924, 4925, 4926, 4927, 4928, 4929, 4930, 4936, 4937, 4938, 4939, 4940, 4941, 4942, 4943

I06 ...... Attitude Instrument Flying—Helicopters
4841, 4846, 4849, 4852, 4853

I07 ...... Electronic Aids to Instrument Flying
4069, 4282, 4283, 4472, 4487, 4548, 4554, 4666, 4685, 4702, 4703, 4704, 4705, 4747, 4824, 4825, 4826

I08 ...... Using the Navigation Instruments
4269, 4281, 4299, 4304, 4315, 4320, 4331, 4338, 4347, 4367, 4397, 4399, 4413, 4495, 4507, 4549, 4550, 4551, 4552, 4553, 4556, 4557, 4558, 4559, 4560, 4561, 4562, 4563, 4564, 4565, 4566, 4567, 4568, 4569, 4570, 4571, 4572, 4573, 4574, 4575, 4576, 4577, 4578, 4579, 4580, 4581, 4582, 4583, 4584, 4585, 4586, 4587, 4588, 4589, 4590, 4591, 4592, 4593, 4594, 4595, 4596, 4597, 4598, 4599, 4600, 4601, 4602, 4603, 4604, 4606, 4607, 4608, 4773

I09 ...... Radio Communications Facilities and Equipment

I10 ...... The Federal Airways System and Controlled Airspace
4274, 4275, 4276, 4285, 4303, 4307, 4308, 4309, 4310, 4311, 4319, 4328, 4330, 4332, 4339, 4340, 4341, 4342, 4343, 4635, 4636, 4637, 4673, 4675, 4676, 4677, 4678, 4679, 4680, 4682, 4683, 4684, 4721

I11 ...... Air Traffic Control

I12 ...... ATC Operations and Procedures
4293, 4295, 4306, 4605, 4625, 4731, 4732, 4733

I13 ...... Flight Planning
4511, 4514

I14 ...... Appendix: Instrument Instructor Lesson Guide—Airplanes

I15 ...... Segment of En Route Low Altitude Chart

AC 00-6: Aviation Weather

I20 ...... The Earth's Atmosphere
4097, 4154, 4227

I21 ...... Temperature
4094, 4095, 4096, 4112, 4113, 4114, 4125, 4200

I22 ...... Atmospheric Pressure and Altimetry
4089, 4090, 4091, 4093, 4109, 4110, 4111

I23 ...... Wind
4105, 4106, 4107, 4108, 4238

I24 ...... Moisture, Cloud Formation, and Precipitation
4099, 4101, 4102, 4103, 4104, 4159, 4161

I25 ...... Stable and Unstable Air
4115, 4118, 4121, 4122, 4123, 4124

I26 ...... Clouds
4131, 4134, 4157

I27 ...... Air Masses and Fronts
4098, 4116, 4117, 4119, 4120, 4127, 4128, 4136, 4141, 4158

I28 ...... Turbulence
4129, 4130, 4132, 4138, 4139, 4140, 4149, 4150

I29 ...... Icing
4100, 4133, 4151, 4152, 4153, 4171

I30 ...... Thunderstorms
4126, 4135, 4137, 4143, 4144, 4145, 4146, 4147, 4148, 4160

I31 ...... Common IFR Producers
4092, 4156, 4163, 4164, 4165, 4166, 4167, 4169

I32 ...... High Altitude Weather
4155, 4162, 4168

I36 ...... Glossary of Weather Terms

AC 00-45: Aviation Weather Services

I40 ...... The Aviation Weather Service Program

I41 ...... Surface Aviation Weather Reports
4196, 4198, 4202, 4204, 4205

I42 ...... Pilot and Radar Reports and Satellite Pictures
4182, 4220

I43 ...... Aviation Weather Forecasts
4142, 4170, 4176, 4177, 4178, 4179, 4180, 4181, 4183, 4184, 4185, 4186, 4187, 4188, 4189, 4190, 4191, 4201, 4203, 4215, 4226, 4228, 4241

I44 ...... Surface Analysis Chart
4209

I45 ...... Weather Depiction Chart
4206, 4207, 4208

I46 ...... Radar Summary Chart
4174, 4230, 4231, 4232, 4233, 4234, 4235, 4236, 4237

I47 ...... Significant Weather Prognostics
4175, 4211, 4212, 4213, 4214, 4216, 4217, 4218, 4219, 4221, 4222, 4223, 4224, 4225, 4229

I48 ...... Winds and Temperatures Aloft
4172, 4192, 4193, 4194, 4199, 4246, 4247, 4249, 4250

I49 ...... Composite Moisture Stability Chart

I50 ...... Severe Weather Outlook Chart
4197, 4239, 4240, 4248

I51 ...... Constant Pressure Charts
4195

I52 ...... Tropopause Data Chart
4242, 4243, 4244, 4245

I53 ...... Tables and Conversion Graphs
4173, 4210

AIM: Aeronautical Information Manual

J01 ..... Air Navigation Radio Aids
4273, 4326, 4337, 4362, 4372, 4373, 4376, 4377, 4378, 4382, 4383, 4384, 4385, 4386, 4387, 4388, 4389, 4391, 4400, 4411, 4412, 4663, 4664, 4665, 4669, 4674, 4730, 4742, 4752, 4753, 4770, 4798, 4799, 4800, 4801

J02 ..... Radar Services and Procedures

J03 ..... Airport Lighting Aids
4408, 4512, 4774, 4775, 4776, 4777, 4778, 4779, 4780, 4781, 4782, 4783, 4784, 4785, 4786, 4787, 4788, 4789, 4790, 4795, 4796

J04 ..... Air Navigation and Obstruction Lighting

J05 ..... Airport Marking Aids and Signs
4534, 4535, 4536, 4537, 4791, 4792, 4793, 4794, 4797

J06 ..... Airspace: General
4080, 4526, 4527, 4528, 4529, 4530, 4531, 4532, 4533, 4539

J07 ..... Class G Airspace
4475, 4488

J08 ..... Controlled Airspace
4329, 4447, 4473, 4474, 4476

J09 ..... Special Use Airspace
4434

J10 ..... Other Airspace Areas
4363, 4410, 4469, 4546, 4729

J11 ..... Service Available to Pilots
4390, 4403, 4404, 4409, 4415, 4416, 4421

J12 ..... Radio Communications Phraseology and Techniques

J13 ..... Airport Operations
4538, 4758

J14 ..... ATC Clearance/Separations
4380, 4381, 4392, 4393, 4395, 4396, 4398, 4442, 4443, 4444, 4449, 4450, 4451, 4452, 4453, 4454, 4455, 4457, 4458, 4555, 4644

J15 ..... Preflight
4058, 4059, 4060, 4061, 4071, 4072, 4073, 4074, 4075, 4076, 4077, 4078, 4266, 4277, 4288, 4300, 4312, 4322, 4333, 4344, 4358, 4405, 4406, 4639

J16 ..... Departure Procedures
4271, 4394, 4414, 4420, 4470, 4486, 4489, 4490, 4491, 4492, 4638

J17 ..... En Route Procedures
4261, 4485, 4493, 4494, 4609, 4610, 4611, 4612, 4613, 4614, 4615, 4616, 4617, 4618, 4619, 4620, 4621, 4622, 4623, 4624, 4626, 4640, 4668, 4681, 4766, 4767, 4768

J18 ..... Arrival Procedures
4286, 4540, 4627, 4628, 4629, 4631, 4632, 4641, 4655, 4667, 4670, 4671, 4672, 4692, 4718, 4726, 4728, 4740, 4741, 4743, 4744, 4749, 4750, 4751, 4757, 4771, 4822, 4823

J19 ..... Pilot/Controller Roles and Responsibilities
4379, 4471, 4544, 4545, 4547, 4633, 4725, 4734, 4735, 4736, 4737

J21 ..... Emergency Procedures: General

J22 ..... Emergency Services Available to Pilots

J23 ..... Distress and Urgency Procedures

J24 ..... Two-Way Radio Communications Failure
4374, 4464, 4465, 4466, 4500

J25 ..... Meteorology
4251, 4252, 4253, 4254, 4255, 4256, 4257, 4258, 4401, 4467, 4468

J26 ..... Altimeter Setting Procedures
4445, 4446, 4480, 4481, 4482

J27 ..... Wake Turbulence
4707, 4708, 4709, 4710

J28 ..... (AIM)
4402

J29 ..... Potential Flight Hazards

J30 ..... Safety, Accident, and Hazard Reports

J31 ..... Fitness for Flight
4803, 4804, 4805, 4806, 4807, 4808, 4809, 4810, 4811, 4812, 4813, 4814, 4815, 4816, 4817, 4818, 4819

J32 ..... Type of Charts Available

J33 ..... Pilot Controller Glossary
4422, 4423, 4424, 4429, 4432, 4435, 4437, 4642, 4643, 4714, 4717

J34 ..... Airport/Facility Directory
4070, 4079, 4305, 4327

J35 ..... En Route Low Altitude Chart
4263, 4264, 4280, 4287, 4291, 4317, 4318, 4321, 4325, 4336, 4348, 4365, 4366, 4370, 4496, 4497, 4498, 4499, 4501, 4502, 4504, 4506, 4508, 4509, 4510, 4516, 4517, 4645, 4646, 4647

J36 ..... En Route High Altitude Chart

J39 ..... Terminal Area Chart

J40 ..... Standard Instrument Departure (SID) Chart
4316, 4361, 4364, 4417, 4418, 4419

J41 ..... Standard Terminal Arrival (STAR) Chart
4292, 4294, 4349

J42 ..... Instrument Approach Procedures (IAP)
4296, 4297, 4298, 4350, 4351, 4352, 4353, 4354, 4355, 4356, 4357, 4368, 4369, 4371, 4648, 4649, 4650, 4651, 4652, 4654, 4722, 4723, 4724

AC 67-2: Medical Handbook For Pilots

J52 ..... Hypoxia

J56 ..... Alcohol

J57 ..... Drugs and Flying

J58 ..... Carbon Monoxide

J59 ..... Vision

J60 ..... Night Flying

J61 ..... Cockpit Lighting

J62 ..... Disorientation (Vertigo)

J63 ..... Motion Sickness

J64 ..... Fatigue

J65 ..... Noise

J66 ..... Age

J67 ..... Some Psychological Aspects of Flying

Additional Advisory Circulars

K01 ..... AC 00-24, Thunderstorms

K02 ..... AC 00-30, Rules of Thumb for Avoiding or Minimizing Encounters with Clear Air Turbulence

K04 ..... AC 00-54, Pilot Wind Shear Guide
4720, 4727, 4739, 4745, 4748, 4755, 4772, 4917

K23 ..... AC 20-121, Airworthiness Approval of Airborne Loran C Systems for Use in the U.S. National Airspace System

K24 ..... AC 20-138, Airworthiness Approval of Global Positioning System (GPS)
4284

K40 ..... AC 25-4, Inertial Navigation System (INS)

K80 ..... AC 60-4, Pilot's Spatial Disorientation

L50 ..... AC 91-6, Water, Slush, and Snow on the Runway

L53 ..... AC 91-14, Altimeter Setting Sources

L57 ..... AC 91-43, Unreliable Airspeed Indications
4821, 4830, 4854

L59 ..... AC 91-46, Gyroscopic Instruments: Good Operating Practices
4881, 4882, 4883

L61 ..... AC 91-50, Importance of Transponder Operation and Altitude Reporting

L62 ..... AC 91-51, Airplane Deice and Anti-Ice Systems

L70 ..... AC 91-67, Minimum Equipment Requirements for General Aviation Operations Under FAR Part 91

M51 .... AC 20-117, Hazards Following Ground Deicing and Ground Operations in Conditions Conductive to Aircraft Icing

S32 ..... (Advisory Circular)
4515

FAA Accident Prevention Program Bulletins

V01 ..... FAA-P-8740-2, Density Altitude

V02 ..... FAA-P-8740-5, Weight and Balance

V03 ..... FAA-P-8740-12, Thunderstorms

V04 ..... FAA-P-8740-19, Flying Light Twins Safely

V05 ..... FAA-P-8740-23, Planning Your Takeoff

V06 ..... FAA-P-8740-24, Tips on Winter Flying

V07 ..... FAA-P-8740-25, Always Leave Yourself an Out

V08 ..... FAA-P-8740-30, How to Obtain a Good Weather Briefing

V09 ..... FAA-P-8740-40, Wind Shear

V10 ..... FAA-P-8740-41, Medical Facts for Pilots

V11 ..... FAA-P-8740-44, Impossible Turns

V12 ..... FAA-P-8740-48, On Landings, Part I

V13 ..... FAA-P-8740-49, On Landings, Part II

V14 ..... FAA-P-8740-50, On Landings, Part III
4738

V15 ..... FAA-P-8740-51, How to Avoid a Midair Collision

V16 ..... FAA-P-8740-52, The Silent Emergency

Note: AC 00-2, Advisory Circular Checklist, transmits the status of all FAA advisory circulars (ACs), as well as FAA internal publications and miscellaneous flight information such as Aeronautical Information Manual (AIM), Airport/Facility Directory, practical test standards, knowledge test guides, and other material directly related to airman certificates and ratings. To obtain a free copy of AC 00-2, send your request to:

U.S. Department of Transportation
General Services Section, M-45.3
Washington, DC 20590

New Products from ASA

Improved, expanded and updated: ASA's product line continues to evolve. The products listed here are brand new. For a copy of our latest ASA Catalog, which includes many other products, call 1-800-ASA-2-FLY.

New Pilot Supplies...

Lexan® Plotters

Durable Lexan® resin won't melt, warp or break in extreme temperatures—guaranteed! Each plotter features 8 scales to meet the needs of all VFR aeronautical charts—nautical on one side, statute on the flip. Nautical and Statute conversion scale also included.

Ultimate Fixed Plotter. 13¼" long.
#ASA-CP-1LX *$7.95*

Ultimate Rotating Plotter. 13¼" long. Features rotating azimuth which eases flight planning with direct readout of course.
#ASA-CP-RLX *$9.95*

Ultimate Micro Plotter. 8" long, convenient for flightbag, shirt-pocket, or airplane storage.
#ASA-CP-MLX *$4.95*

Lexan is a registered trademark of the General Electric Company.

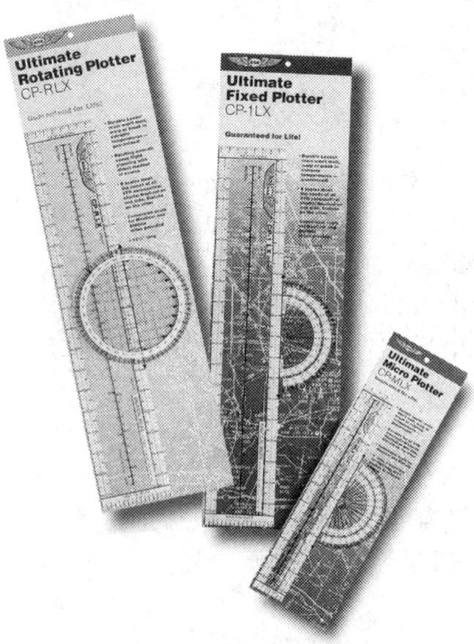

New Books...

Say Again, Please

Guide to Radio Communications
by Bob Gardner

Mr. Gardner's manual provides a clear, conversational approach to aviation communications. The text features examples of typical radio transmissions that explain how the air traffic control system works, and presents a simulated flight to demonstrate the correct procedures for communicating in each class of airspace. This hands-on book covers: the ABCs of communicating, understanding radio equipment, communication etiquette and rules, VFR, IFR and emergency communication procedures, air traffic control facilities and their functions, and a review of airspace definitions.
#ASA-SAP *$14.95*

New Practical Test Standards

These are fundamentals to a successful checkride and include the skills and knowledge required for each license or rating. Instructors, students, and examiners use these books to prepare, review, take, or issue the checkride.

Private, Single-Engine, Land	*#ASA-8081-14S*	*$4.95*
Baker's Dozen	*#ASA-8081-14S-13*	*$59.40*
Private, Multi-Engine, Land	*#ASA-8081-14M*	*$4.95*
Baker's Dozen	*#ASA-8081-14M-13*	*$59.40*
Private & Commercial, Single-Engine, Sea	*#ASA-8081-SEA*	*$4.95*
Baker's Dozen	*#ASA-8081-SEA-13*	*$59.40*

Flight Maneuvers Handbook and Manual

ASA now provides the standard in maneuvers manuals, previously published by Haldon Books, Inc. Both books provide excellent learning and teaching aids for instructors and students, and cover all maneuvers necessary for FAA certification. Each maneuver is described in step-by-step detail with many supporting graphics, and all are described according to the type of aircraft (high or low wing) the lesson will take place in. Through the use of these handbooks, the student will have a more thorough understanding of the maneuvers, creating a safer and more competent pilot—thereby shortening each flight instruction period.

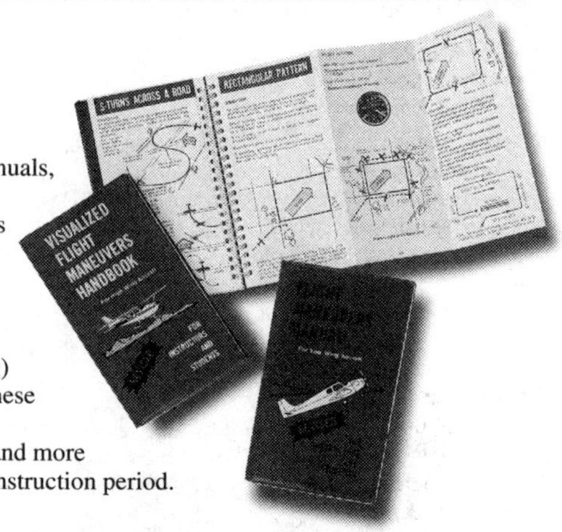

Visualized Flight Maneuvers Handbook
Based on high-wing aircraft. Known as the "Blue Maneuvers Book."
#ASA-FM-HI *$19.95*

Flight Maneuvers Manual
Based on low-wing aircraft. Known as the "Red Maneuvers Book."
#ASA-FM-LO *$19.95*

Pilot Examiner's Handbook
Reprint of the official FAA Order 8710.3B of September 1994. Required for Designated Pilot Examiners, and an excellent reference book for flight instructors.
#ASA-8710.3B *$14.95*

More New Books...

A Pilot's Guide Series

by Jeremy Pratt

Each Pilot's Guide is a comprehensive book on the fundamentals of flying the airplane. Subjects covered in detail include an overall description of the aircraft, limitations, handling characteristics, and loading/performance data. All the information is gleaned from flying experiences by experts in the industry, and is presented in an easy-to-read format. Pilots will find each guide in the series an invaluable companion to the aircraft's Flight Manual and an excellent sourcebook for the aircraft's principal characteristics.

Cessna 172: A Pilot's Guide	*#ASA-PG-C-172*	*$14.95*
Cessna 152: A Pilot's Guide	*#ASA-PG-C-152*	*$14.95*
Cessna 150: A Pilot's Guide	*#ASA-PG-C-150*	*$14.95*
Piper Warrior: A Pilot's Guide	*#ASA-PG-PA-28W*	*$14.95*
Piper Cherokee: A Pilot's Guide	*#ASA-PG-PA-28C*	*$14.95*
Piper Tomahawk: A Pilot's Guide	*#ASA-PG-PA-38*	*$14.95*

Aviation Weather Services Workbook

by John Holley

Designed to be used in conjunction with *Aviation Weather Services*. Contains text and exercises to increase reader's knowledge of weather and weather services. Previously published by IAP, Inc.

#ASA-WX-WK *$7.95*

Aviation Weather Services

Reprint of the newly updated FAA AC 00-45D we've all been waiting for! Includes information in all areas of weather reports, forecasts and analysis with all new charts.

#ASA-AC00-45D *$14.00*

Flying Ultralights

by Doug Chipman

Ultralights are becoming one of the fastest growing sectors in the general aviation market. These aircraft provide the experience of flying in a manner which is much less expensive, and much less regulated, than traditional airplanes. *Flying Ultralights* provides a framework for learning in a safe, efficient, and inexpensive environment. The book includes a series of lessons and exercises, making it an excellent syllabus for students and instructors alike.

#ASA-ULF *$14.95*

For FAA testing centers and flight schools:

Private Pilot Test Prep Course

Organized into a 2-day exam review seminar, this course can be used as final preparation for the FAA Knowledge Test. The Instructor's guide follows ASA's *Private Pilot Test Prep* chapters in subject order for classroom presentation, and offers airplane-only questions for faster review. Students are expected to follow along using the ASA *Private Pilot Test Prep*. Additional memory aids, hints, and many overhead transparencies are included.

#ASA-TPC-P *$100.00*

Global Navigation for Pilots

by Dale DeRemer and Donald McLean

Presents background needed by pilots to learn international and oceanic operations. Glossary and index included. Previously published by IAP, Inc. *421 pages.*

#ASA-GNP *$28.95*

Air Traffic Control Test Prep Study Guide

by Patrick Mattson

Prepares reader for the Air Traffic Control test, and includes questions, answers, application procedures, employment and training information, and test-taking strategies.

#ASA-ATC *$19.95*

Where Am I?

Previously published by Haldon Books, Inc., this is an instructional manual on the use of today's standard instruments in navigation. The book explains, with practical applications, the use and understanding of ADF and OMNI (VOR). The text is supported by many graphics throughout.

#ASA-WHERMI *$15.95*

The Pilot's Manual Series Syllabi

by Jackie Spanitz

These syllabi meet all Part 141 requirements and supplement Trevor Thom's textbooks: *Flight Training*, *Private & Commercial*, and *Instrument Flying*. The syllabi have been reviewed by the FAA, present a fresh approach, and offer the most integrated, comprehensive, and simple flight and ground training programs available. Flight lessons are presented side-by-side with their coordinating ground lessons. Each syllabus comes with appropriate Stage Exams.

Private Pilot Syllabus	*#ASA-PM-S-P*	*$10.95*
Instrument Rating Syllabus	*#ASA-PM-S-I*	*$10.95*
Commercial Pilot Syllabus	*#ASA-PM-S-C*	*$10.95*

All prices based on U.S. currency.
Prices subject to change without notice.

Call **1-800-ASA-2-FLY** for the retailer nearest you.

ASA's CX-1a Pathfinder™: State-of-the-Art in Flight Computers.

Pilots choose the CX-1a Pathfinder™ from ASA's AirClassics™ collection because it is easy to use and it has many outstanding features:

- Large LCD screen shows entire problem as it is being worked, including:
  a six-line display, prompts, units reference and menu
- Prompts guide you through each problem
- 18 formulas provide an incredible 31 aviation calculations, quickly and easily
- 14 conversion functions at your fingertips
- Calculator can be used alone or in any aviation mode
- Chain calculation results are easily stored and instantly recalled
- Easy battery access
- Authorized for use on FAA tests

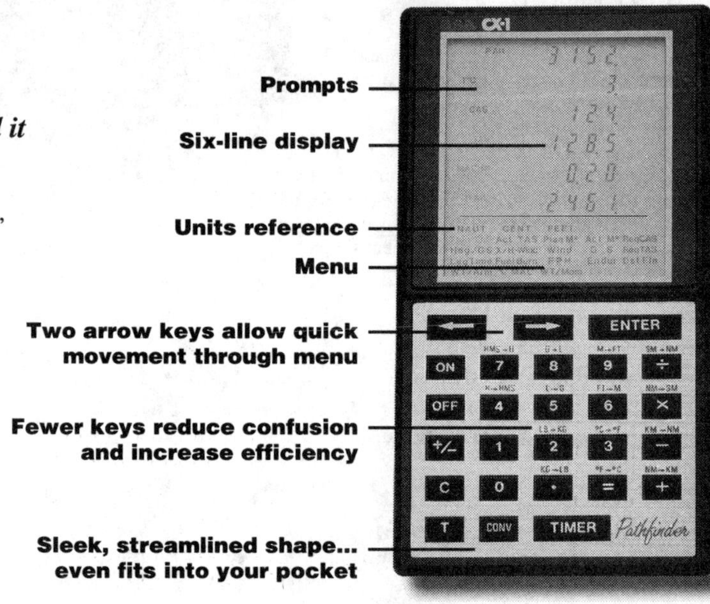

AirClassics and CX-1a Pathfinder are trademarks of Aviation Supplies & Academics, Inc.

The CX-1a Pathfinder™ is a favorite of pilots at all levels—from student pilots to 777 captains. This computer's time-proven success speaks for itself.

#ASA-CX-1a *$69.95*

ASA's AirClassics™ HS-1 Headset

The most advanced technology, high-quality components, sleek look, reasonable price, maximum comfort, and a lifetime warranty—your wish list is complete!

Features

- ***Earcups of high-density acoustic foam provide best passive noise attenuation***
- ***High fidelity speakers for clear, natural sound***
- ***Electret, noise-canceling microphone reduces background noise and allows clearest voice transmission***
- ***Gold-plated microphone and headphone plugs ensure best connection and resist corrosion***
- ***High-grade multi-strand wire improves cable life***
- ***Stereo/mono capability***
- ***Large, dual controls for quick, easy volume adjustment for each ear***
- ***Microphone muff***
- ***Adjustable headband with easy thumb screws for maximum comfort, eliminates "hot spots," and accommodates eyeglasses and any head size***
- ***Quality foam ear seals and light weight add to overall comfort***
- ***Sleek, all-black design with ASA wings attractively silk-screened on each earcup***
- ***Sturdy, high-quality and reliable— yet reasonably priced for maximum value***

LIFETIME WARRANTY

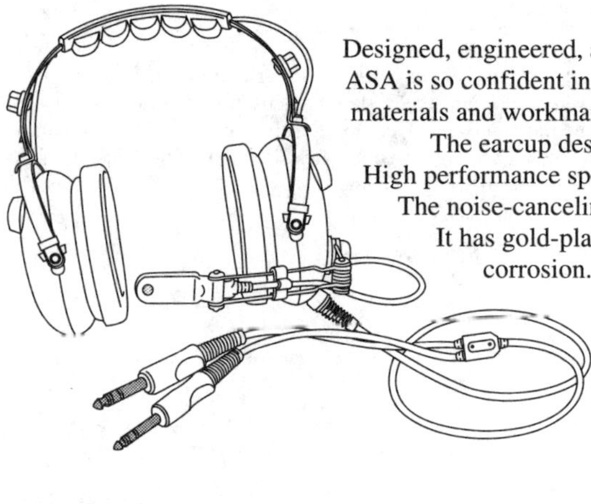

Designed, engineered, and manufactured by experts in the headset business, ASA is so confident in the quality of the HS-1 we are guaranteeing it against defects in materials and workmanship for the life of the original owner. It's really that good.

The earcup design and high density acoustic foam provides superb noise attenuation. High performance speakers are standard and provide excellent fidelity for clearer, more natural sounds. The noise-canceling electret microphone reduces cabin noise and allows clearer voice transmission. It has gold-plated microphone and headphone plugs to insure good connections and resist corrosion. Microphone and speaker cables incorporate multi-stranded wire that assure long cable life. The adjustable headband provides maximum comfort by eliminating "hot spots," and can be easily adjusted to accommodate glasses and any head size. Quality foam ear seals and light weight unit add to the overall comfort. The HS-1 has the additional benefits of mono/stereo capability, dual volume controls, and a microphone muff. *#ASA-HS-1* *$149.95*